Springer Series in
Surface Sciences

7

Editor: Robert Gomer

V. F. Kiselev   O. V. Krylov

Electronic Phenomena
in Adsorption and Catalysis
on Semiconductors and Dielectrics

With 76 Figures

Springer-Verlag Berlin
Heidelberg GmbH

Professor Dr. Vsevolod F. Kiselev

Department of Physics, Moscow State University,
SU-117234 Moscow, USSR

Professor Dr. Oleg V. Krylov

Institute of Chemical Physics, Academy of Sciences of the USSR,
SU-142092 Moscow, USSR

Series Editors

Professor Dr. Gerhard Ertl

Fritz-Haber-Institut der Max-Planck-Gesellschaft, Faradayweg 4–6
D-1000 Berlin 33

Professor Robert Gomer

The James Franck Institute, The University of Chicago, 5640 Ellis Avenue,
Chicago, IL 60637, USA

ISBN 978-3-642-83022-8   ISBN 978-3-642-83020-4 (eBook)
DOI 10.1007/978-3-642-83020-4

Library of Congress Cataloging-in-Publication Data. Kiselev, V.F. (Vsevolod Fedorovich). Electronic phenomena in adsorption and catalysis on semiconductors and dielectrics. (Springer series in surface sciences ; 7) Bibliography: p. 1. Semiconductors–Surfaces. 2. Dielectrics–Surfaces. 3. Adsorption. 4. Catalysis. I. Krylov, O. V. (Oleg Valentinovich) II. Title. III. Series. QC611.6.S9K58 1987 537.6'22 87-4768

Offsetprinting: Druckhaus Beltz, 6944 Hemsbach/Bergstr.
Bookbinding: J. Schäffer GmbH & Co. KG., 6718 Grünstadt
2153/3150-543210

Preface

It is now firmly established that various adsorptive and catalytic processes taking place on the surface of semiconductors and in MIS structures strongly influence their electronic properties and hence modify the parameters of semiconductor devices. The inverse problem of how the semiconductor's electronic subsystem influences adsorption and dissociation of molecules at the surface has been recognized but much less explored. The main purpose of the present book is to generalize the experimental data and explain the relationship between these two classes of phenomena. We also discuss tentative models of surface electronic states and their interaction with adsorbed molecules.

The subject of this book should attract the attention of researchers working in the overlapping areas of physics and chemistry, and of physics and biology. The research done in this field will help to widen the scope of semiconductor applications by finding novel ways of employing surface effects in the construction of microelectronic devices, semiconductor gas analysers, solar cells, etc.

The authors hope that this book will be useful to a wide circle of chemists and physicists concerned with the study of interphase phenomena and questions of adsorption and catalysis. Certain parts of the book will be helpful to physicists and technicians working in rapidly developing branches of semiconductor physics and technology. The book can also serve as a textbook for both under- and postgraduates specializing in this field.

For interesting and fruitful discussions and for help with the preparation of the book the authors are grateful to Yu.A. Zarif'jants, S.N. Kozlov, Yu.N. Rufov, K.N. Spiridonov, Mmes Z.L. Krylova, O.V. Nikitina and L.Ya. Margolis. The authors are very grateful to Mr. A.S. Dobrovolski for the excellent translation of this book from our native language.

Our thanks are due to Dr. H. Lotsch of Springer-Verlag for his friendly concern and help.

December, 1986   V.Kiselev

Moscow   O.Krylov

Contents

1. Introduction

The theoretical and experimental data, accumulated by now, indicate that molecular and electronic processes on a solid surface are closely interconnected. The adsorbed molecules create new surface states and/or change the parameters of the existing electron surface states of the solid. In some cases the electrons and holes, localized on these states, influence the nature of bonds between molecules and surface, the structure of adsorbed molecules and their reactivity. The energy spectrum of the surface predetermines the nature of chemisorptive and catalytic processes which take place on the surface. On the other hand, chemisorptive and catalytic phenomena largely influence many electronic processes which proceed in the surface region of a semiconductor and in heterostructures of an MIS-type (metal-insulator) semiconductor. Such phenomena play a crucial role in modern microelectronics.

The investigation of how chemisorptive and electronic processes are interrelated, constitutes one of the main tasks of the modern physical chemistry of solid surfaces. Unfortunately, in existing monographs this aspect has received less attention than it deserves. The chemists, concerned with chemisorption and catalysis, are not always aware of progress made by physicists, working in the field of surface electronics. At the same time the physicists, who are dealing with semiconductors, are often unaware of the existence of a large body of experimental material, accumulated by chemists in the study of chemisorption and catalysis. There still exists a certain gap between physical and chemical aspects of the general approach to surface phenomena, which hinders the development of this very important branch of the physical chemistry of solid surfaces.

The first part of the book is concerned with the discussion of the specific features of the electronic processes on real, as well as on atomically clean, semiconductor surfaces. The phenomenological theory of these processes is based on the presentation of the energy spectrum of the surface electron states as a set of energy levels, whose parameters can be obtained experimentally and are related to the macroscopic experimental data. Such questions are presented in a rather condensed form (Chap.2), since they have been discussed in detail elsewhere [1.1,2]. Our main concern is the nature of the energy spectrum of a real surface (Chap.3).

Every real surface is characterized by different surface states having different parameters. Adsorption changes the energy spectrum of these states and creates new states in the forbidden band of the crystal. Considering the high density of such states, the disturbed surface structure of the crystal, the disordered distribution of defects and adsorbed molecules on the surface, the most promising approach to an analysis of the energy spectrum is furnished by the modern electronic theory of disordered systems.

In adsorption and catalysis, as well as in MIS electronics, one usually deals with semiconductors having wide forbidden band, amorphous semiconductors and dielectrics. The peculiar features of electronic processes in these systems and the problem of investigation of the energy spectrum of these substances are discussed in Chap.4. In addition, information obtained from electrophysical measurements performed on polycrystalline semiconductors has been analyzed.

Chapters 5 to 9 are addressed to the main problems of catalysis and semiconductor surface physics: the investigation of the nature of the active centers of a real surface being responsible for adsorption and capture of charge carriers; and the understanding of the mechanism of adsorption and charging of the surface. This aspect is not considered explicitly in the phenomenological description of electronic processes. This is because the electronic theory of chemisorption deals with highly idealized models of a surface and of the elementary act of adsorption (Chap.5). A quantitative theory of chemisorption and surface charging seems not yet available.

In order to make progress in this direction, first of all one has to establish direct relationships between concrete types of adsorptive bonds on a real surface (with the necessary information being derived from spectroscopic data) and the corresponding changes of the energy spectrum of the surface, as indicated by electrophysical and optical measurements. Such an attempt has already been undertaken by KISELEV [1.3]. Recent years saw major innovations in surface-analysis techniques. Rapid development of spectroscopic techniques (UV and X-ray photoelectron spectroscopy, tunnelling spectroscopy, etc.), as well as the appearance of novel electrophysical techniques, enabled considerable progress in the study of the nature of surface states. Increased sensitivity and higher resolving power of those techniques allowed the study of molecular and electronic processes on the surface of monocrystals (Chap.6).

In Chap.7 we deal with the reactivity of adsorbed molecules and the connection between catalytic activity and electronic processes on the surface.

Based upon experimental data, we demonstrate that the appearance of a radical (or ion-radical) form of chemisorption due to the interaction of valence-saturated and neutral molecules with the surface is not always equivalent to the formation of charged complexes. In many cases an important role in charging of a surface is played by donor-acceptor (coordinated) adsorptive complexes. So far these nonvalence bonds were overlooked by the electronic theory of adsorption. Here we discuss the influence of surface charging on the activity of electron-acceptor, electron-donor and proton-donor catalytic centers.

In many cases the experimental data on adsorption and catalysis cannot be fully explained with the aid of a purely electronic mechanism of the elementary act of adsorption. In Chap.8 we discuss the question of dissipation of energy, released at capture and recombination of free charge carriers at a semiconductor surface, and the question of the role played by the phonon excitations created in processes of adsorption and catalysis. We supply experimental data which directly prove the participation of vibrational modes of adsorbed molecules in capture of electrons and holes on adsorptive electron states. The probable involvement of Auger processes in adsorption and catalysis will be analyzed.

The last chapter deals with the investigation of the surface proton processes. According to the donor-acceptor mechanism, in some cases the capture of holes by surface states can stimulate dissociation of adsorbed molecules, including water molecules. The originating protons contribute to the surface conductivity and alter many electrophysical parameters of the surface. We discuss the possible role of these processes in the performance of solar cells and in models of semiconductor photosynthesis systems.

In discussing the mechanism of electron transitions and phonon excitations on the surface, as well as the reactivity of the adsorbed molecules, the authors will freely employ information regarding the nature and the adsorptive properties of concrete solid surface, collected in [1.4].

2. The Phenomenological Description of Electronic Processes on Semiconductor Surfaces

This chapter gives a brief account of the main phenomenologic concepts of semi-conductor surfaces, the knowledge of which is essential to understand subsequent discussions. We consider the general properties of surface electron states and their location in the crystal band model. Here we also state the basic relationships, which link the net charge of the surface and its potential with space-charge conduction, and list the major techniques for investigating the parameters of surface states.

2.1 Surface-Electron States. Basic Concepts

Surface-Electron states, being the centers of localization of free charge carriers on a surface, play a desicive role in the electronic processes that take place on semiconductor surfaces. Half a century ago, in 1932, TAMM [2.1] published his fundamental work in which a basic difference was shown to exist between the electron energy spectra in infinite and finite (limited) crystals. He demonstrated that the creation of a surface leads to the appearance, in the forbidden band, of new allowed electron energy states, which later came to be known as Tamm states. Seven years later, SHOCKLEY [2.2] considered another possible way in which surface states might be created. He proposed that when strong localized bonds on the surface are caused to rupture due to a decrease in interatomic distances and a crossing of the limits of allowed bands, two local levels in the forbidden band (Shockley states) split off. The conditions under which Tamm and Shockley states originate will be discussed later. What is important here is that even with an ideal impurity-free crystal, the creation of a surface leads to the appearance of a set of intrinsic surface electron states.

The real pattern of the energy spectrum of a crystal surface, even an atomically clean one, is much more complex. As demonstrated in [2.3], the formation of an atomically clean surface, owing to rehybridization of surface atoms, leads to reconstruction of the surface. The surface becomes covered with domains having superstructures which differ in symmetry and in the size of their elementary crystalline cells. Owing to the disorderly configuration of some of the surface atoms, and to the existence of domains, steps, dislocation exits, vacancies, and other defects, the forbidden band of the crystal will exhibit an additional set of electron surface states; the spectrum of existing Tamm and Shockley states will also be changed. All these states, whose parameters are largely determined by the "history" of the creation and treatment of the surface, we shall call biographic states.

When the surface comes in contact with the environment, the energy spectrum of the biographic states is supplemented by the electron states which owe their existence to adsorbed atoms and molecules. In 1947 VOLKENSTEIN [2.4] proved, within the framework of the method of molecular oribtals (MO), that local levels are created in the forbidden band when univalent atoms are adsorbed on the surface of an ionic crystal. The adsorbed atoms and molecules not only create their own system of adsorptive surface states, but also interact with surface flaws to change the parameters of electron traps of biographic origin. Later we shall see that collective effects may take place in this mechanism.

Of especial interest in practice are investigations of the electron states on real semiconductor surfaces. Under ordinary conditions, many important monatomic semiconductors (Ge, Si), as well as binary semiconductors of types $A^{II}B^{VI}$ and $A^{III}B^{V}$, become covered with a thin oxide film after etching and treatment. Such a surface is conventionally called a real surface. What we are actually dealing with here is a dielectric-semiconductor heterojunction (DS system), where one must speak not of a two-dimensional distribution of surface states over an idealized flat surface, but of a three-dimensional distribution in a layer of finite thickness, including the interface. The term "surface states" is thus a purely conventional one.

On the one hand, the formation of an oxide film leads to saturation of some of the dangling bonds on the surface and to rehybridization of surface atoms. As a result, a number of intrinsic electron surface states are erased from the energy spectrum of the crystal. On the other hand, owing to the dissimilarities in the structures of the semiconductor and the oxide film, "mismatching" defects appear in the boundary region, giving rise to a new set of levels in the forbidden band. In addition, new states appear due to the presence of foreign atoms near the boundary which have been captured in the course of etching or have come up to the surface from the bulk, as well as those due to vacancy-type defects, which are formed parallel to the creation of the oxide film. It has been shown experimentally that with an appropriate choice of conditions for etching and oxidation of the surface, the total concentration of biographic surface states on a real surface can be several orders of magnitude lower than that of an atomically clean surface.

The charge Q_s arrested on the surface states exerts a strong influence on the electron processes taking place in the subsurface region of the semiconductor. On the other hand, the capture of charge carriers on the surface states alters the activity of those states as potential centers of adsorption and catalysis. For this reason the study of surface states is of equal interest to researchers working in semiconductor physics and to those concerned with the electron theory of adsorption and catalysis.

The ultimate objective of this research is to elucidate the nature of defects and chemisorptive bonds on the surface and to establish quantitative relationships between their physicochemical characteristics and the electrophysical parameters of the surface. Because of the nonuniformity of the surface, the peculiarities of hybridization of surface atoms, and the great varity of chemisorptive bonds, this problem is far from being solved. The quantum-chemical theory of surface states can at best give only a very general notion of the nature of the energy spectrum of different kinds of surface defects. We are not yet able to calculate the basic parameters of the surface states, their concentration N_t, the position in the zone ε_t, and the cross sections of capture of an electron c_n and a hole c_p on the basis of our knowledge about the structure of adsorbed molecules and the solid. This explains the tendency in surface science to give phenomenological and statistical descriptions of electron processes on the surface. In such an approach the parameters of the energy spectrum are either postulated in advance, or (if possible) found experimentally, and then the quantitative relationship is derived between the charge of the surface and its main electrophysical properties: surface conductivity, work function, photoelectric characteristics, etc.

We shall not enter here into a lengthy discussion of phenomenological theories in the physics of electron processes in semiconductors, since they have already been exhaustively treated [2.5-10]. In this chapter we shall restrict ourselves to a brief summary of the basic concepts necessary for an understanding of our subsequent discussion. In doing this we shall adhere to the sequence and nomenclature adopted in the monograph by RZHANOV [2.8]. We shall also analyze critically the experimental data on the energy spectrum of atomically clean and real semiconductor surfaces.

Under ordinary (real) conditions the surface interacts with the environment. As we have already pointed out, the adsorbed molecules not only create their own system of surface states, but also change the spectrum of already existing biographic states. In the unified system of solid and environment, the processes which occur in the electronic and molecular subsystems are closely inter-connected. Since these relationships are of great importance in adsorption and catalysis, the interaction between surface states and adsorbed atoms and mole-cules will be discussed separately in Chaps. from 5-9.

2.2 The Space-Charge Region

At thermodynamic equilibrium, by virtue of the electroneutrality of the crystal as a whole, the electric charge Q_s captured by the surface states is neutralized by an equal charge of opposite sign, Q_{sc}, in the subsurface region. In metals the space-charge region (SCR), owing to the high concentration of free charge carriers, penetrates the bulk to a depth of only several lattice constants. In semiconductors the length of Debye shielding, L_D, amounts to 10^{-4} or 10^{-5} cm. For an intrinsic semiconductor,

$$L_D = \left(\frac{\varepsilon kT}{2\pi q^2 n_i}\right)^{1/2} \quad ,$$

where ε is the dielectric permittivity of the crystal, q is the electron charge and n_i is the concentration of charge carriers in an intrinsic semiconductor.

The charge Q_{sc} in the SCR is composed of the charge of ionized acceptor and donor traps and the charge of free carriers, whose equilibrium concentration differs from that in the bulk. In the band scheme this corresponds to bending of the allowed energy bands with respect to the Fermi level (Fig.2.1). The

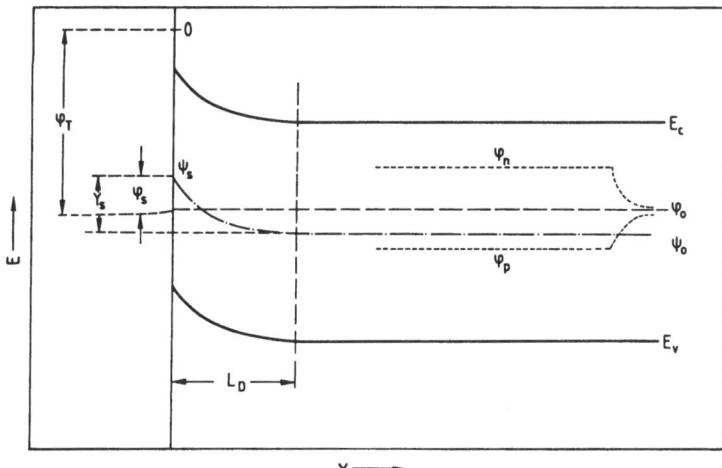

Fig.2.1. Band scheme of the space-charge region of a semiconductor (negative charging of the surface). E_c is the bottom of the conduction band, E_v is the top of the valence band

existence of the SCR strongly influences many electrophysical properties of the surface, such as the surface conductivity, work function, and photoelectromotive force. In order to describe these quantitatively, one must know the distribution of the electrostatic potential Ψ (Fig.2.1) in the SCR, which is described by the Poisson equation

$$\nabla^2 \Psi = (4\pi / \varepsilon)\rho \quad , \tag{2.1}$$

where ρ is the charge density in the SCR.

This problem was solved most consistently by GARRET and BRATTAIN [2.11]. They considered the case of a nondegenerate semiconductor having a nondegenerate subsurface region, which allowed them to employ Boltzmann distribution. In addition it was assumed that all the impurities in the bulk were completely ionized. Such is the case, for example, with germanium and silicon at room temperature.

With these assumptions, and assuming also that Ψ and ρ change only in the direction normal to the surface (x), we get

$$\rho(x) = q(N_D - N_A + p - n) \quad ,$$

where N_D and N_A are the concentrations of ionized donors and acceptors respectively, p and n are the concentrations of holes and electrons in the SCR, and q is the electron charge.

Since in the bulk of the semiconductor the condition of electroneutrality is satisfied, $N_D - N_A + p_0 - n_0 = 0$, where n_0 and p_0 are the equilibrium bulk concentrations of electrons and holes. Hence

$$\rho(x) = q[(p - p_0) - (n - n_0)] \quad . \tag{2.2}$$

At the thermodynamic equilibrium the distribution of free charge carriers in a semiconductor is determined by a single Fermi level (the electrochemical potential φ_0 in Fig.2.1). The Boltzmann distribution being valid, the concentration of electrons and holes in the bands is known to be

$$n = n_i \exp[\beta(\Psi - \varphi_0)] \quad ; \quad p = p_i \exp[\beta(\varphi_0 - \Psi)] \quad , \tag{2.3}$$

where n_i and p_i are the concentrations of charge carriers in the intrinsic semiconductor, $\beta = q / kT$, and Ψ is the electrostatic potential (in the bulk $\Psi = \Psi_0$).

In order to determine the concentration of the charge carriers in the SCR, the value of $q\Psi_0$ inside the crystal is usually assumed to fall into the middle of the forbidden band (Fig.2.1). Then the potential energy of an electron on the surface will be described as $q(\Psi_s - \Psi_0)$, and the position of the Fermi level on the surface with respect to the middle of the band is $-q\varphi_s = q(\Psi_s - \varphi_0)$. Henceforth it will be more convenient to use the dimensionless electrostatic potentials $Y = (\Psi - \Psi_0)\beta$ or $\beta\varphi_s$.

In the general case of thermodynamic nonequilibrium, the concentrations of electrons and holes n and p exceed the equilibrium values by Δn and Δp, respectively. If the nonequilibrium carriers are not captured in the bulk, then $\Delta n = \Delta p$, and the extent to which the concentration of carriers diverges from the equilibrium value is usually expressed as a dimensionless injection level [2.8] which is the same for both electrons and holes

$$\delta = \Delta n / n_i = \Delta p / p_i \quad . \tag{2.4}$$

The Boltzmann distribution being satisfied, the concentrations of free charge carriers will now be determined via the Fermi quasi levels $q\varphi_n$ and $q\varphi_p$ for electrons and holes respectively (Fig.2.1)

$$n = n_i \exp[\beta(\Psi - \varphi_n)] \quad ; \quad p = p_i \exp[\beta(\varphi_p - \Psi)] \quad . \tag{2.5}$$

At equilibrium $\varphi_n = \varphi_p = \varphi_0$, and (2.5) is identical with (2.3).

Substituting (2.5) into the expression for charge density, see (2.2), one can integrate the Poisson equation (2.1) over the x coordinate[1]. Using the boundary conditions $Y = Y_s$ at $x = 0$ (surface) and $\Psi = \Psi_0$, $dY / dx = 0$ at $x \to \infty$ (Fig.2.1), we get the first integral of this equation expressed via the dimensionless quantities Y, λ, and δ, which determine the distribution of the electric field strength in SCR:

$$\frac{dY}{dx} = \frac{2}{L_D} F(Y, \lambda, \delta) \quad , \tag{2.6}$$

where the dimensionless quantity

$$\lambda = p_0 / n_i = n_i / n_0 = \exp \beta(\varphi_0 - \Psi_0) \tag{2.7}$$

characterizes the extent to which the semiconductor is not an intrinsic one. For an intrinsic semiconductor $\lambda = 1$; for a p-type semiconductor $\lambda > 1$ and the Fermi level lies below the middle of the forbidden band Ψ_0; and for an n-type

---

[1] Fermi quasi levels are assumed to be independent of x in the SCR.

semiconductor $\lambda < 1$ and the Fermi level lies above ψ_0 (Fig.2.1). The function F in (2.6) has the form

$$F(Y, \lambda, \delta) = \pm[(\delta + \lambda)(e^{-Y} - 1) + (\delta + \lambda^{-1})(e^{-Y} - 1) + (\lambda - \lambda^{-1})Y]^{1/2} \quad .$$

(2.8)

At thermodynamic equilibrium

$$F(Y, \lambda) = \pm[\lambda(e^{-Y} - 1) + \lambda^{-1}(e^{-Y} - 1) + (\lambda - \lambda^{-1})Y]^{1/2} \quad .$$

(2.9)

The plus sign corresponds to upward bending of bands (negative charging of the surface, $Y < 0$), while the minus sign corresponds to downward bending (positive charging, $Y > 0$); see Fig.2.1.

According to the Gauss theorem, the charge in the SCR is linked to the field strength by the expression

$$Q_{sc} = \frac{\varepsilon}{4\pi} \left| \frac{d\psi}{dx} \right|_{x=x_0} = qn_i L_D F(Y_s, \lambda, \delta) \quad ,$$

(2.10)

while at equilibrium

$$Q_{sc} = qn_i L_D F(Y_s, \lambda) \quad .$$

(2.11)

Since the surface charge $Q_{sc} = -Q_s$, these equations establish a direct connection between the charge localized on surface states and the surface potential Y.

The second integral of the Poisson equation (2.1), $Y(x)$, defines the shape of the potential barrier on the semiconductor surface. Fig.2.2 shows the most typical shapes of these barriers. Part (a) corresponds to accumulation layers: the concentration of the majority carriers in the SCR is greater than that in the bulk. Part (b) corresponds to depletion layers; here the concentration of majority charge carriers in the SCR is lower than that in the bulk. Finally, Part (c) corresponds to inversion layers; in this case the conductivity in the SCR is opposite in type to that in the bulk, and the Fermi level crosses the curve depicting the middle of the crystal's forbidden band.

In the general case the integral of (2.6) can be calculated only approximately. The dependences $Y(x)$ are considered in detail in [2.7,8] for some special cases of accumulation and depletion layers. Here we shall discuss only one of the parameters determining the potential barrier: the thermionic work function φ_t, which can be easily obtained from experiment.

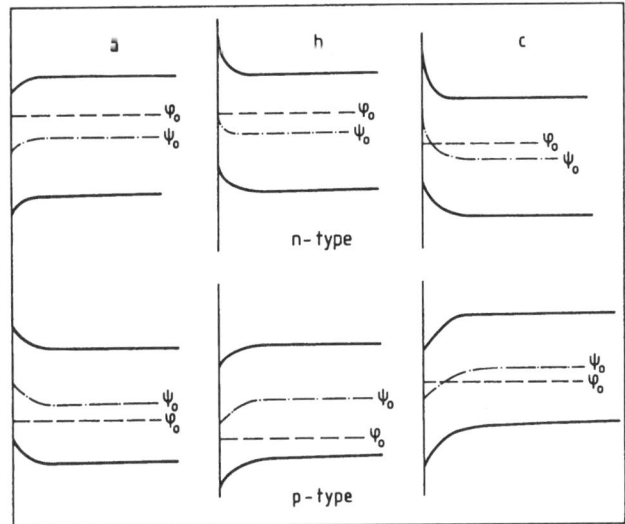

Fig.2.2a-c. Typical shapes of potential barriers on the surface of n-type and p-type semiconductors

By definition, the thermionic work function equals the difference between the energy of a free electron (at vacuum level) and the energy of an electron at the Fermi level (Fig.2.1). In the presence of potential barrier Y_s,

$$\varphi_t = x + q(\varphi_0 - \Psi_0) - Y_s kT \quad . \tag{2.12}$$

The quantity x is determined by the properties of the semiconductor lattice: $x = x_0 + E_g / 2$, where x_0 is the electron affinity and E_g is the width of the forbidden band; $\varphi_0 - \Psi_0$ is the position of the Fermi level in the bulk with respect to the middle of the forbidden band. The value of φ_t decreases with positive Y_s (downward bending of the band) and increases with negative Y_s.

Under real conditions, (2.12) becomes more complicated. In the presence of an oxide film on the surface of the semiconductor the right-hand side must include the term $q\varphi_{ox}$, which describes the drop in potential in the oxide layer. In adsorption one must also consider the dipole component of work function φ_D, which describes the jump in potential on the potential barrier created by the dipoles of adsorbed molecules. From simple electrostatic relationships one can obtain the value of φ_D for below-monolayer coverage:

$$\varphi_D = 4\pi\mu_\perp n^0 \quad , \tag{2.13}$$

where $\mu_\perp$ is the projection of the dipole moment onto the normal to the surface, and n^0 is the concentration of adsorbed molecules.

The appearance of excess charge carriers in the SCR due to a change in the surface potential will affect the value of the electroconductivity of the surface layer. By "excess carrier concentration" in the SCR, we mean, by analogy with the Gibbs surface excess in the theory of adsorption [2.3,Chap.2], the difference in the surface concentrations of electrons and holes of the semiconductor at a given value of surface potential Y_s and at zero potential (flat bands):

$$\Gamma_n = \int_0^\infty (n - n^*)dx \quad , \qquad \Gamma_p = \int_0^\infty (p - p^*)dx \quad , \tag{2.14}$$

where n^* and p^* are the stationary values of carrier concentrations outside SCR.

If the injection levels for electrons and holes are the same, then

$$n^* = n_0 + \Delta n = n_i(\lambda^{-1} + \delta) \quad ,$$

$$p^* = p_0 + \Delta p = p_i(\lambda + \delta) \quad . \tag{2.15}$$

Turning now to integration over Y, and taking into account (2.5,6,8,15), we get the following expressions for Γ_n and Γ_p

$$\Gamma_n = \int_{Y_s}^0 \frac{(n - n^*)dY}{dY / dx} = 0.5n_i(\lambda^{-1} + \delta)L_D \int_{Y_s}^0 \frac{(e^Y - 1)}{F(Y, \lambda, \delta)} dY \quad , \tag{2.16}$$

$$\Gamma_p = \int_{Y_s}^0 \frac{(p - p^*)dY}{dY / dx} = 0.5p_i(\lambda + \delta)L_D \int_{Y_s}^0 \frac{(e^{-Y} - 1)}{F(Y, \lambda, \delta)}dY \quad .$$

In accordance with the definition of excess charge carriers, the specific surface conductivity of the specimen σ_s (measured in Ω^{-1} per unit area) is expressed as

$$\sigma_s = q(\mu_{ns}\Gamma_n + \mu_{ps}\Gamma_p) \quad , \tag{2.17}$$

where μ_{ns} and μ_{ps} are the mobilities of electrons and holes in SCR. By virtue of (2.16),

$$\sigma_s(Y_s, \lambda, \delta) = 0.5qn_i\mu_{ps}L_D \int_{Y_s}^0 \frac{(\lambda + \delta)(e^{-Y} - 1) + b_s(\lambda^{-1} + \delta)(e^Y - 1)}{F(Y, \lambda, \delta)}dY \quad , \tag{2.18}$$

where $b_s = \mu_{ns} / \mu_{ps}$.

At thermodynamic equilibrium ($\delta = 0$), the expression for σ_s becomes considerably simpler, see (2.9)

$$\sigma_s(Y_s, \lambda) = 0.5qn_i\mu_{ps}L_D \int_{Y_s}^{0} \frac{\lambda(e^{-Y}- 1) + b_s\lambda^{-1}(e^{Y} - 1)}{F(Y, \lambda)}dY \quad . \qquad (2.19)$$

In the general case the integrals entering (2.18 and 19) cannot be taken analytically. Therefore one usually employs precalculated values of these integrals within certain limits of variation of Y, λ, presented in the form of tables [2.12] or monograms [2.7].

The mobilities of electrons μ_{ns} and holes μ_{ps} can differ from the corresponding values for the bulk (μ_n, μ_p) owing to additional scattering on the surface. The two extreme cases are (i) purely elastic scattering, which changes only the normal component of the momentum of the charge carrier, while the tangential component remains the same, and (ii) entirely diffuse scattering, when the vector of the momentum of the charge carrier, after collision with the surface, takes any direction within a hemisphere whose flat side coincides with the surface. In the first case the values of μ_{ns} and μ_{ps} are usually taken to equal the corresponding bulk values.

The macroscopic theory of entirely diffuse scattering was developed by SCHRIEFFER [2.13]. Having introduced the effective values of the mobilities of carriers in SCR, Schrieffer calculated the dependence of the ratio of mobilities μ_{ns}^{eff} / μ_n, μ_{ps}^{eff} / μ_p on the surface potential (Fig.2.3). This is now

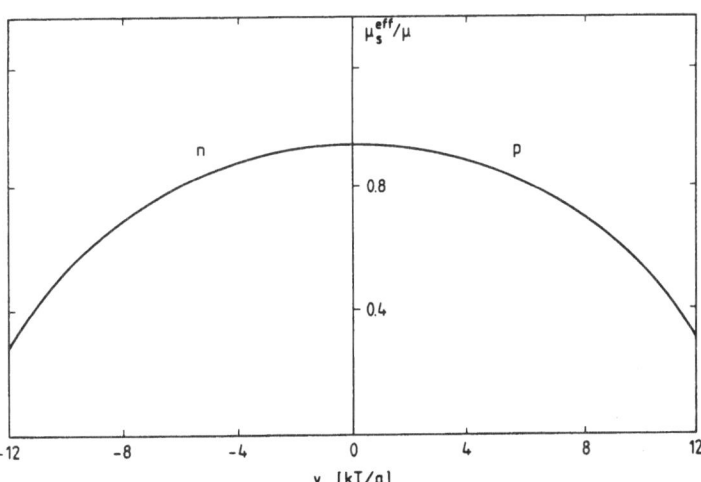

Fig.2.3. The ratio of effective surface mobility μ_{eff} to bulk mobility μ_v as a function of the surface potential [2.13]

widely employed in introducing the corresponding corrections to the value of
the surface conductivity calculated according to the Garret-Brattain theory.
The experimental results, however, indicate that in some cases the diffuse-
ness of the scattering can vary substantially. The question of how to draw a
rigorous theoretical distinction between the contributions made by diffuse
scattering and elastic scattering, under varying surface conditions and with
widely varying surface potential and temperature, is far from solved [2.5,6,
8]. The peculiar features of the surface structure can be linked to the mechanism of scattering only within a microscopic theory which considers the
specific features of the scattering of charge carriers on surface phonons,
macroscopic defects, and localized surface charges [2.14,15].

The Garret-Brattain theory outlined above deals with the case when all bulk
impurities in the semiconductor are ionized. This naturally restricts the
number of pertinent semiconductors and limits the temperature range in which
the electrophysical properties of the surface can be measured. More recently
SEIWATZ and GREEN [2.16] and GORKUN [2.17] elaborated this theory, extending
it to apply to cases in which the extent of ionization of impurities is arbitrary, and in which there possible degeneration of one of the bands in the
SCR. The expressions for dependences $\sigma_s(Y_s)$ take on a rather complicated form
in such cases. Numerical integration of the obtained expressions was carried
out for a number of intermetallic compounds.

During the past decade, special attention has been directed to the study
of quantum effects in SCR. If the bendings of bands are great enough, the potential pit for charge carriers on the surface can become quite narrow. Such
a situation is often encountered in the case of high inversion (Fig.2.2c).
Then the motion of charge carriers in the subsurface crystal region becomes
quantized in the direction normal to the surface, and their wave function becomes zero at the interface and in the bulk. The latter circumstance leads
to appearance of two-dimensional discrete surface subbands. As a result, the
distribution of charge $\rho(x)$ in the subsurface region deviates from that predicted by the classical theory. The two-dimensional movement of charge carriers
within the confines of surface subbands has a number of specific features
which result from the scattering of carriers on surface phonons, Coulomb centers,
and fluctuations in potential. The transitions between the surface subbands
are manifested as extrema in the spectra of photoconductivity, as oscillations
of the SCR capacitance and as certain other phenomena. Enormous oscillations
in conductivity (of the Shubnikov-de Gaaz type) are observed in a magnetic
field. A discussion of quantum effects in SCR and in thin films would go
beyond the scope of the present book; for more information the reader is referred to an excellent review by STERN [2.18].

2.3 Field Effect as a Method of Investigating the Energy Spectrum of the Surface

The dependences of Q_{sc} and σ_s on surface potential Y_s, stated in the theory of space charge, form the basis of the field-effect technique, which is most widely used in the investigation of the energy spectrum of semiconductor surfaces. The field effect, first observed by SHOCKLEY and PEARSON [2.19] on cuprous oxide, consists essentially in the following. If a capacitor having one plate of metal and the other of the semiconductor studied is furnished with a difference in potentials, then the charge in the SCR of the semiconductor will change. The total charge Q_{ind}, induced at a given instant by the transverse field, will be redistributed between the charge arrested on the surface states, Q_s, and the charge Q_{sc} in the SCR, i.e., $Q_{ind} = Q_s + Q_{sc}$. The charge Q_{ind} can be found from the well-known formula $Q_{ind} = C_c V$, where C_c is the capacitance, and V is the difference in potential applied to the plates. Thus, by changing the external field, one can vary the surface potential of semiconductor Y_s and its surface conductivity σ_s within certain limits.

As follows from the Garret-Brattain theory, Q_{sc} and σ_s both depend on the potential, see (2.10,11,18,19). An analysis of dependences $\sigma_s(Y_s)$, see (2.18, 19), reveals that they have a minimum. The value of surface potential which corresponds to the minimum of surface conductivity is

$$Y_{s0} = \ln(\lambda^2 / b_s) \quad . \tag{2.20}$$

If Schrieffer's correction for mobilities μ_{ns}^{eff} / μ_n, μ_p^{eff} / μ_p is neglected (Fig.2.3), the minimum on the plot $\sigma_s(Y_s)$ corresponds to the value of surface potential Y_{s0} such that the concentration of minority carriers in SCR equals the concentration of majority carriers in the bulk. Since the surface scattering only slightly affects σ_s in the neighborhood of the minimum, the value of Y_{s0}, according to (2.20), depends only on the bulk properties of the substance which are determined by λ, see (2.7); the point Y_{s0} on the potential axis is a good reference point for measuring the surface conductivity, as it does not depend upon the condition of the semiconductor surface.

Experiments with the field effect have revealed that with real surfaces of Ge, Si, PbS, CdS, InS, and GaAs, the experimental curves of σ_s plotted against potential V, applied to the field-effect cell, display certain minima. The existence of a minimum makes it possible to compare the experimental and the theoretical curves, and thus calculate the dependence of charge Q_s, captured on surface states, on the surface potential Y_s. This dependence is one of the most important characteristics of the surface. To do this, one plots on the same diagram the experimental dependence $\sigma_s(Q_{ind})$ and the theoretical values of σ_s and Q_{sc} [calculated from (2.11,19)] for the same values of the surface potential (Fig.2.4a) [2.8].

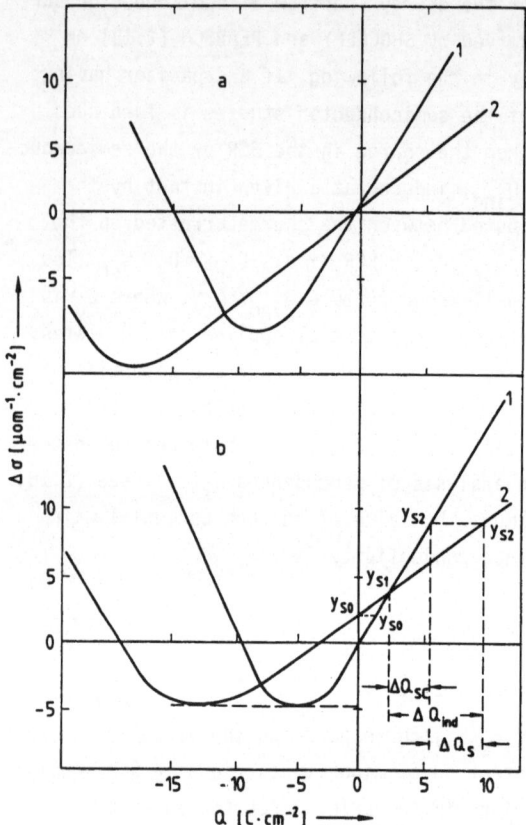

Fig.2.4. (a) Theoretical curve $\Delta\sigma_s(Q_{sc})$ (1) and experimental curve $\Delta\sigma_s(Q_{ind})$ (2). (b) The same curves after alignment of their minima [2.8]

Since the differences in the ordinates of the two curves are due only to the different choice of reference point, it is sufficient to shift the experimental curve along the y axis to bring together the minima of both curves (Fig.2.4b). After such alignment, equal ordinates of the right-hand and left-hand branches of both curves will correspond to equal values of Y_s, which can be determined by transposing the corresponding points from the theoretical curve (1) to the experimental curve (2) parallel to the x axis. The potential Y_{s0} in Fig.2.4b corresponds to zero external field.

The same plots can be employed for easy determination of the values of charge captured on surface states. As seen from Fig.2.4b, the change in potential from crossing-point Y_{s1} to Y_{s2} on the theoretical curve corresponds to the change in charge in SCR, Q_{sc}, which is smaller than the change in induced charge ΔQ_{ind}, which is necessary to bring about a similar change in potential on the experimental curve. The difference in the abscissas represents the captured charge ΔQ_s. Taking the potential at zero field Y_{s0} as the zero point in the reference frame, one can plot the dependence of the captured charge upon the potential.

The curve $\Delta Q_s(Y)$ depicts the dependence of the total charge of surface states upon the potential. Some researchers prefer differential characteristics. For this purpose they introduce the concept of mobility of field effect μ_{fe}, defined as the slope of the curve $\sigma_s(Q_{ind})$ in Fig.2.4, i.e., $\mu_{fe} = d\sigma_s / dQ_{ind}$; $\mu_{fe} > 0$ when hole-type conductivity prevails in SCR, and $\mu_{fe} < 0$ when electron-type conductivity is predominant. At the point of minimum $\mu_{fe} = 0$. The sign of μ_{fe} allows one to determine the type of surface conductivity instantly from the experimental curve of the field effect. On the basis of measurements of μ_{fe} at small amplitudes of external field, when the variation of surface potential $\Delta Y \ll kT / q$, the author of [2.7] proposed a technique for calculating the differential curves of capture $(dQ_s / dY_s)(Y_s)$. The differential dependences are naturally more sensitive to any kind of anomalous behavior on the part of the charge capture curve.

The first studies of the field effect in germanium demonstrated that the kinetics of variation of the surface conductivity is governed by two transition processes after the transverse field is switched on. In the first of these, during a time lapse comparable to the lifetime of nonequilibrium charge carriers in the specimen, the value of σ_s changes from a certain instantaneous magnitude attained after the establishment of the field to a certain new quasi-equilibrium value. This first process has a characteristic time of 10^{-5} to 10^{-7} s and is connected with the attainment of equilibrium between bands and the so-called fast surface states. After that in a process (taking from milliseconds to hours) σ_s slowly resumes the value it had before the field was switched on. The second, slow process is connected with the occupation of slow states whose charge shields the specimen from the external field.

The substantial difference in the relaxation times of fast (τ_f) and slow (τ_s) states makes it possible to study the capture of charge on these states separately. This is done by employing pulse fields or alternating fields, which vary according to the sine law. If the pulse duration (or period) $t \ll \tau_s$, the experimental curve $Q_s(Y_s)$ represents the curve of capture on fast states. In future we shall denote the charge captured on fast states as Q_{sf}. On the curve of the field effect (Fig.2.4) we are dealing with the occupation of fast states and $Q_s = Q_{sf}$.

The quasi-equilibrium capture curve for slow states $Q_{ss}(Y_s)$ is virtually impossible to study by means of the field-effect technique because the relaxation times are extremely long. Moreover, owing to the high density of slow surface states, even at high-amplitude external electric fields the bulk of the semiconductor is effectively shielded by these states. Slow states can be investigated only with the aid of kinetic techniques, which will be discussed later.

The existence of two types of surface states with strongly differing relaxation times has been conclusively established for germanium and silicon, as well as for certain semiconductors of types $A^{III}B^V$ and $A^{II}B^{VI}$. However, it would be premature to generalize this classification into one of fast and slow states, because depending on the experimental conditions and the nature of the semiconductor surface, states can be encountered which have intermediate characteristic times. Especially interesting in this connection are investigations of the frequency-dependent field effect, which enable one to evaluate the kinetic parameters of surface states: the cross section of capture of electron c_n and hole c_p.

The majority of theoretical and experimental works have studied the kinetics of capture on fast surface states. Experimental investigations [2.20,21] of the frequency dependence of field-effect mobility have revealed that for every shape of potential barrier (Fig.2.2) the region of high dispersion starts from frequencies of the order of 10^7 Hz. This is connected with the relaxation of the capture processes in fast surface states. The capture time in fast states for germanium at room temperature is about 10^{-7} to 10^{-8} s. This time increases with a decrease in temperature. RUPPRECHT [2.22] and RZHANOV [2.8], using an analysis of the temperature dependence of the kinetics of relaxation, made estimates of capture cross sections c_n and c_p, which for germanium, for example, depending upon the position of the energy level, vary from 10^{-15} to 10^{-20} cm^2 for c_n and 10^{-13} to 10^{-21} cm^2 for c_p. It should be noted that analyzing nonequilibrium electron porcesses in SCR of semiconductors is rather complicated, and the interpretation of the kinetic capture curves is still quite controversial. The field effect technique is the most informative instrument for investigating the energy spectrum of energy states in semiconductors. In some cases, however, serious obstacles are encountered in its use. So far the minimum on the experimental curve $\sigma_s(V)$ or $\sigma_s(Q_{ind})$, from which the quantitative dependence of the charge in surface states upon the potential can be derived, has been observed with only a limited number of semiconductors. For many heavily doped and wide-band semiconductors the minimum on the field-effect curve could not be obtained. Furthermore, this technique cannot be used with surfaces displaying a high density of surface states ($10^{12} - 10^{13}$ cm^{-2}). For this reason researchers are seeking to alternative surface-oriented electrophysical techniques which would supply information on surface states. Some of these techniques will be discussed in the next section.

2.4 Alternative Electrophysical Techniques for the Study of Surface States

2.4.1 Photoelectric Techniques

Irradiation of a semiconductor with light having a wavelength near the long-wavelength limit of the main absorption band generates electron-hole pairs. The appearance of excess charge carriers shifts the equilibrium between the charge in the surface states and the charge in SCR. If the irradiation is steady, a new quasi-equilibrium distribution of charge carriers in the SCR is established, as well as a new quasi-equilibrium surface potential Y_s. The quantity $\Delta Y_s = Y_{s0} - Y_s$ (Y_{s0} being the zero potential) is often called the surface photoelectromotive force. From a comparison of (2.10,11) it is evident that the equilibrium function $F(Y_s, \lambda)$ is now replaced by the quasi-equilibrium one, $F(Y_s, \lambda, \delta)$ whose argument includes the level of injection of carriers in the bulk, see (2.4). The excess nonequilibrium carriers are responsible for the change in conductivity (the so-called stationary excess photoconductivity $\Delta\sigma_{ph}$).

We shall not go into the theory of photoelectric phenomena on semiconductor surfaces since it has already been discussed in detail [2.5-8,23]. We shall discuss only the essential features of the physical phenomena that govern the redistribution of charge carriers in the SCR and the change in the surface potential when the surface is irradiated. First, there is the diffusion of light-produced electron-hole pairs and their recombination both in the bulk and on the surface. With thin specimens the characteristic time for the establishment of quasi-equilibrium is the longer time of the recombination process τ_{rec}. Second, some of the nonequilibrium carriers in SCR will be captured by surface states, which will result in a change of surface charge Q_s. Since the processes of surface recombination and capture involve the surface states, the data on stationary photoconductivity and photoelectromotive force (with an adequate choice of experimental conditions) can supply information on the parameters of surface states. Let us first consider photoconductivity.

Consider the simplest case, in which the relaxation of excess conductivity in a thin semiconductor specimen is wholly determined by the recombination of electron-hole pairs, while the capture of nonequilibrium carriers by surface states can be neglected. This can be accomplished for a number of semiconductors by performing the experiment at a low injection level and limiting the duration of irradiation in such a way that the relaxation of excess conductivity is determined only by the recombination process. If the absorption coefficient for the exciting light is sufficiently high, then for a thin specimen (whose thickness d is much smaller than the length of diffusional displacement $L_d = \sqrt{D\tau_0}$, where D is the diffusion coefficient and τ_0 is the lifetime of bulk charge carriers) the value of stationary photoconductivity $\Delta\sigma_{ph}$ is linked to the rate of surface recombination S via the relation

$$\Delta\sigma_{ph} = q(\mu_n + \mu_p) \frac{b}{\ell} J_0 \left(\frac{1}{\tau_0} + \frac{2S}{d}\right)^{-1} \quad , \tag{2.21}$$

where b and ℓ are the width and length of the wafer (both much greater than d), and J_0 is the rate of production of excess charge carriers.

The phenomenological value of the effective rate of surface recombination S, first introduced by Shockley, is defined as the ratio of the overall recombination speed U, expressed in the number of electron-hole pairs per unit surface and unit time, to the excess number of minority carriers Δn (or Δp) in the plane separating the SCR from the bulk, where $\Psi = \Psi_0$ (Fig.2.1), i.e.,

$S = U / \Delta n$, or $S = U / \Delta p$.

It is noteworthy that the phenomenological value of S introduced in this manner is by no means always a universal characteristic of the recombination processes on the surface. The concept of the rate of surface recombination loses its meaning at high values of S for highly enriched or highly exhausted layers, as well as for inversion layers. An important assumption that is made in order to obtain the expression for S is that the Fermi quasi levels in the SCR are constant. Often this assumption is not realized, for example in semiconductors with a small value of diffusional displacement L_d [2.23]. The limits of applicability of the concept of the rate of surface recombination were discussed in detail in [2.8,23].

On the basis of statistics on Shockley-Read recombination processes [2.5-8] the relationships were found which connect the rate of surface recombination to the parameters of recombination centers and to the surface potential. For instance, for a low level of injection of nonequilibrium carriers, the expression for the rate of surface recombination in the case of a set of similar discrete recombination levels has the following form:

$$S = \frac{N_t^r (\alpha_n \alpha_p)^{1/2} \cdot 1/2(\lambda + \lambda^{-1})}{ch[q(\varphi_s - \xi)/kT] + ch[(\varepsilon_t - q\xi)/kT]} \quad , \tag{2.22}$$

where N_t^r is the surface concentration of recombination centers; ε_t is their position in the forbidden band with respect to the middle of the band; α_n and α_p are the probabilities of capture of an electron and a hole, respectively by the recombination center, which are equal to the product of the effective cross section of capture (c_n, c_p) and the mean thermal velocity of the relevant carriers; φ_s and λ have the same meaning, as in Sect.2.2; and $q\xi/kT = 0.5 \ln(\alpha_p/\alpha_n)$.

As follows from (2.22), the dependence $S(\varphi_s)$ or $S(Y_s)$ is represented by a bell-shaped curve. In the general case the experimental dependence $S(Y_s)$, obtained with the aid of the combined techniques of field effect and photoconductivity [2.8,24], can be considered a sum of two components, one of which is monotonic and the other is bell-shaped (Fig.3.5).

The rate of surface recombination is very sensitive to surface conditions. We shall encounter this important characteristic of the surface again in our discussion of the energy spectrum of the surface (Chap.3) and its dependence on various adsorption processes (Chaps.6,9).

The surface photoelectromotive force, which arises when a thin semiconductor wafer is irradiated, can, like the photoconductivity, be split into two components. The first component results from the appearance of nonequilibrium carriers in the semiconductor, while the second is connected with the change in surface potential due to the additional capture of electrons and holes on surface states.[2] Theoretical calculations performed by ZUEV et al. [2.23] made it possible to relate the quasi-stationary value of photoelectromotive force to the capture cross sections of surface centers. Since the photoelectromotive force depends strongly upon the conditions of surface treatment and adsorption, the results suggest that the surface photoelectromotive force technique offers great possibilities for studying the parameters of surface states and how they

---

[2] When dealing with thick specimens, where thickness d is greater than the length of diffusional displacement, one has to account for a third component of the photo emf, the so-called Dember emf, which is connected with nonuniformity in the production of charge carriers and the fact that they have different diffusion coefficients.

vary when different processes take place on the surface. The technique is most promising at high levels of injection of nonequilibrium charge carriers [2.25]. It is widely used to determine the surface potential of semiconductors having a high concentration of surface states, when the field-effect technique is unworkable due to the absence of a minimum on the conductivity curve [2.26].

2.4.2 Capacitance Techniques

The change that occurs in the space charge Q_{sc} when the surface potential Y_s varies can be presented as differential SCR capacitance $C_{sc} = \beta \mid \partial Q_{sc} / \partial Y_s \mid$. Substituting Q_{sc} from (2.10) we get

$$C_{sc} = \beta q n_i L_D \left| \frac{\partial F(Y_s, \lambda, \delta)}{\partial Y_s} \right| . \tag{2.23}$$

At thermodynamic equilibrium ($\delta = 0$) for an intrinsic semiconductor ($\lambda = 1$) we have

$$C_{sc} = 1/2 q n_i \beta L_D \left| \frac{e^{Y_s} - e^{-Y_s}}{(e^{Y_s} - e^{-Y_s} + 2)^{1/2}} \right| . \tag{2.24}$$

The dependence $C_{sc}(Y_s)$ is represented by an upturned bell-shaped curve, whose minimum corresponds to $Y_s = 0$. If the semiconductor is not an intrinsic one, the minimum capacitance shifts along the potential axis, and the branches of the curve are no longer symmetrical, one corresponding to the positive value of Y_s and the other to the negative one.

The presence of charge Q_s in surface states can be formally characterized by differential capacitance $C_s = \beta \mid \partial Q_s / \partial Y_s \mid$, connected in parallel with C_{sc}. Thus the total surface capacitance is $C_0 = C_{sc} + C_s$.

The dependence $C_0(Y_s)$ can also be found experimentally. As pointed out in Sect.2.3, the charge induced in the specimen in field-effect investigations is $Q_{ind} = C_c V$. The capacitance of the condenser formed in the field-effect cell is composed of two capacitances connected in a series: the capacitance of the gap between the metallic electrode and the semiconductor C_g, and the surface capacitance C_0. The resulting capacitance is $C_c = C_g C_0 / (C_g + C_0)$. When experimenting with the field effect it is advisable that $C_g \ll C_0$; then $C_c \approx C_g$. If the opposite relation is realized ($C_g \gg C_0$), then the capacitance of the condenser formed in the field-effect cell will be about equal to the surface capacitance. This is achieved by depositing the metallic electrode directly on the dielectric film of the oxide or other dielectric compounds covering the semiconductor, thus creating a MIS structure.

22

We see that in MIS structures one can directly measure the dependence of the surface capacitance on the applied difference in potential. The vaue of differential capacitance is measured with the aid of a low-intensity ac signal, and the potential is changed by applying dc bias to the condenser. Since the dependence $C_{sc}(Y_s)$ can be calculated theoretically, the capture curves can, in principle, be calculated by comparing theoretical and experimental data.

The above discussion pertains to cases in which the rates of change in the surface potential are sufficiently low, allowing quasi equilibrium to be established in the SCR and on the surface states. In reality, calculations of CV characteristics are much more complicated, owing to the presence of states with different characteristic times on the surface. With highly pronounced inversion layers the generation-recombination processes will be inert. To separate the transition processes, the surface capacitance is usually measured at varying frequencies. Then Y_s and the capture curve can be efficiently calculated using the method of equivalent circuits. The capacitance characteristics of the surface were discussed in detail in [2.27-32].

3. The Energy Spectrum of Semiconductor Surfaces

The principal characteristic of semiconductor surface on a phenomenological basis is the energy spectrum of surface states. In this chapter we discuss the specific features of energy spectra for atomically clean and real germanium and silicon surfaces. These features are demonstrated to be most consistently accounted for by the electron theory of disordered systems.

3.1 The Fast States and Recombination Electron States on Real Semiconductor Surfaces

As we pointed out in Sect.2.3, the most abundant information on the energy spectrum of surface states is obtained from the study of the quasi-equilibrium field effect. In field-effect experiments using an ac signal, whose period is much smaller than the characteristic times of capture on slow surface states, the curve of capture on fast surface states $Q_{sf}(Y_s)$ is directly obtained from the curve of the field effect (Fig.2.4).

Under equilibrium conditions the population of levels is uniquely determined by the position of the Fermi level on the surface, $q\varphi_s / kT$ (Fig.2.1). Let us consider the simplest case, when just two types of surface states are present on the surface: acceptor (A) and donor (D) states. The acceptor centers, whose levels are shifted by ε_{tA} with respect to the middle of the forbidden band, are charged negatively after capturing an electron. The empty donor centers are charged positively; their energy position is ε_{tD}. Then the charge in these states is

$$Q_{sf}^A = - \frac{qN_{tA}}{1 + \exp[(\varepsilon_{tA} - q\varphi_s)/kT]} \quad , \tag{3.1}$$

$$Q_{sf}^D = \frac{qN_{tD}}{1 + \exp(q\varphi_s - \varepsilon_{tD})/kT} \quad ,$$

where N_{tA} and N_{tD} are the concentrations of the states.

If m acceptor and n donor discrete levels are present on the surface, then the total charge in the fast surface states is

$$Q_{sf} = q\left(\sum_{i=1}^{n} \frac{N_{itD}}{1 + \exp[(q\varphi_s - \varepsilon_{itD})/kT]} - \sum_{i=1}^{m} \frac{N_{itA}}{1 + \exp[(\varepsilon_{itA} - q\varphi_s)/kT]} \right) .$$

$$(3.2)$$

The derivative of Q_{sf}, which defines the slope of curve $Q_{sf}(Y_s)$, equals $qN_t/4$ at the points where the position of the ith level ε_{it} coincides with the Fermi level, $q\varphi_s = \varepsilon_{it}$. Hence, if the experimental curves $Q_{sf}(Y_s)$ have points of inflection ($q\varphi_s = \varepsilon_{it}$), then the latter's positions will directly determine the position of the ith energy level, while the slope of the curve describes the concentration N_{it}.

Detailed investigations of the field effect carried out on real surfaces of Ge [3.1-3], Si [3.4], InSb [3.5,6], GaAs [3.7], and PbS [3.8] convincingly indicate that the dependence of an integral charge captured in fast states Q_{sf} upon the surface potential is depicted by monotonic curves. Examples are given in Figs.3.1,2.

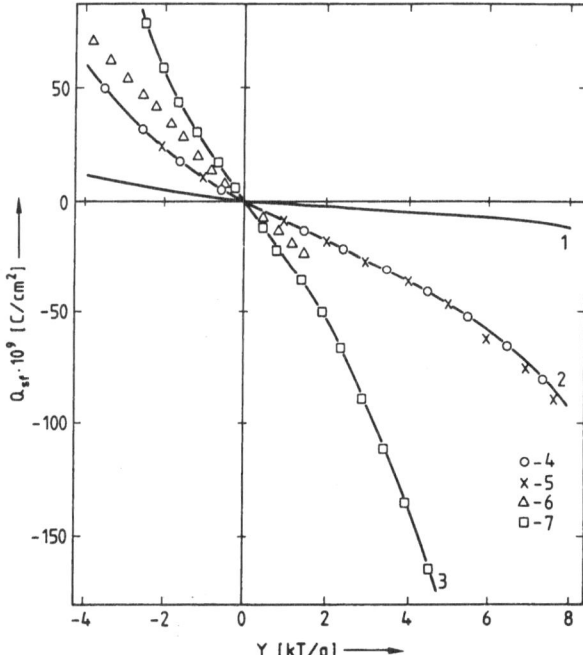

Fig.3.1. Curves of capture on fast states on germanium surfaces heat- treated in a vacuum at various temperatures T_{cal}. (1) Prior to treatment, (2) T_{cal} = 500 K, (3) T_{cal} =750 K, (4) after adsorption of ozone on Ge$_{500}$, (5) the same for ammonia, (6) the same for p-Bq, (7) after adsorption of water on Ge$_{750}$ [3.3]

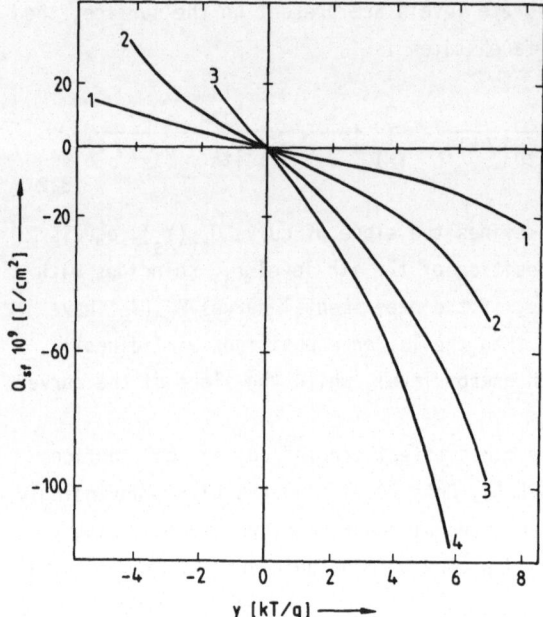

Fig.3.2. Curves of capture on fast states on silicon surfaces heat-treated in a vacuum at various temperatures. (1) Prior to treatment, (2) $T_{cal} = 650$ K, (3) $T_{cal} = 800$ K, (4) $T_{cal} = 900$ K [3.4]

The question of how the smooth capture curves should be interpreted is still controversial. Some researchers attribute them to a discrete spectrum of fast states [3.9,10]. In order to explain the absence of inflection points on the experimental curves $Q_{sf}(Y_s)$ they assume that at a high density of levels the inflection point of the ith level is masked by the tail of the Fermi curve of the occupation of an adjacent level. Under these assumptions the energy spectrum is characterized by a small number of predominant discrete levels, which are considered to be certain characteristics of the surface states. For instance, for real germanium and silicon surfaces, the energy spectrum was characterized by sets of four (Ge) or five (Si) discrete levels, whose density increased from 10^{10} (near the middle of the forbidden band) to 10^{11} cm^{-2} for Ge, and from 10^{11} to 10^{12} cm^{-2} for Si [3.9,10].

If a set of discrete levels actually exists in the energy spectrum, then it must be resolved at low temperatures; i.e., with a decrease in temperature the curves $Q_{sf}(Y_s)$ must show their structure. Analysis of the temperature dependences of $Q_{sf}(Y_s)$ for germanium clearly indicates [3.11] that the curves retain their monotonicity down to quite low temperatures. These results prove the prevalence of a continuous (or quasi-continuous) spectrum of fast surface

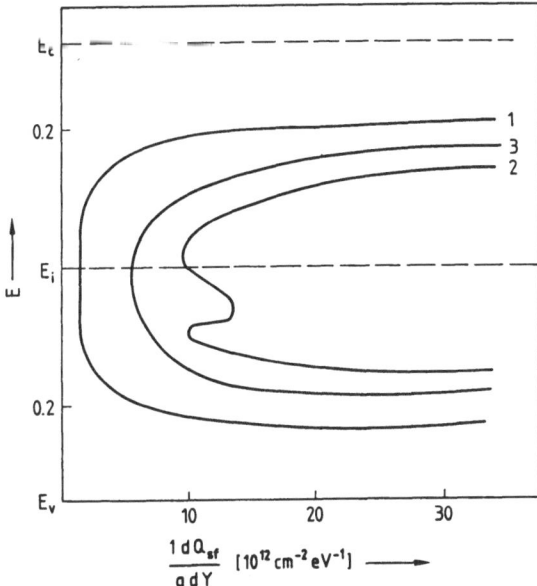

Fig.3.3. Energy spectrum of fast states on a germanium surface. (*1*) Freshly etched specimen, (*2*) after doping of the surface with gold [3.13], (*3*) after vacuum heat treatment at 500 K. E_C and E_V are the conduction band and the valence band respectively

states. RZHANOV [3.1] undertook to approximate these spectra by adopting an exponentiel law for variations in the concentrations of acceptor and donor traps in the energy spectrum:

$$N_{tA} = A \exp(a\ \varepsilon_{tA} / kT) \quad ; \quad N_{tD} = B \exp(-b\ \varepsilon_{tD} / kT) \tag{3.3}$$

The values of Q_{sfA} and Q_{sfD} obtained through integration turned out to be in good agreement with the experimental curves $Q_{sf}(Y_s)$.

An example is given in Fig.3.3 which shows the energy spectrum of fast states on a real germanium surface, calculated from experimental curves $Q_{sf}(Y_s)$ [3.3]. It is evident from the figure that the densities of fast states increase rapidly and continuously from the center towards the edges of the forbidden band. Let us consider the effects of various external factors on the pattern of the energy spectrum. A monotonic spectrum similar to that depicted in Fig.3.3 was observed in [3.1] for sufficiently perfect growth of germanium surfaces emerging from the gaseous phase, and for surfaces subjected to mechanical treatment with subsequent etching. The effect of various etchants on the shape of curve $Q_{sf}(Y_s)$ and on the parameters of fast states also turned out to be surprisingly small [3.1].

27

Detailed investigations of the influence which different processes of adsorption and desorption have upon the energy spectrum of fast states were performed in [3.2,12]. Heat treatment of Ge specimens causes an increase in capture that is monotonic in the whole range of Y_s values studied (Fig.3.1). Subsequent adsorption of water almost completely restores the initial spectrum of traps. In spite of the fact that the overall density of states increases sharply in the course of heat treatment the continuous nature of the spectrum is retained (Fig.3.3). As opposed to adsorption of water, adsorption of ammonia and ozone on Ge_{500} leaves the density of states almost unchanged, while adsorption of para-benzoquinone (p-Bq) elevates the density of states only in the upper half of the forbidden band (Fig.3.1). From Fig.3.1 it is also evident that heat treatment at $T_{cal} > 700$ K results in a sharp increase of Q_{sf}, and adsorption of water on such a surface does not alter the density of states.

A similar change in Q_{sf} was observed with silicon as well (Fig.3.2). The curves Q_{sf} retain their shape up to $T_{cal} = 650$ K, and the capture starts to increase noticeably at $T_{cal} > 700$ K.

Analysis of published data reveals that as a rule the dominant role of the continuous spectrum also persists when the surface is modified in other ways: doping with metals [3.13], ion bombardment [3.14], thermal oxidation [3.15], and after replacement of the intrinsic oxide on germanium and silicon with other inorganic compounds, e.g., $SiO_2(Ge)$, $Ge_3N_4(Ge)$, $Si_3N_4(Ge)$ [3.15-18], or $Si_3N_4(Si)$ and $Al_2O_3(Si)$ [3.19,20].

Note that the energy range accessible to investigation by the field-effect technique is considerably narrower than the width of the forbidden band E_g. The possible variations in the position of the Fermi level on the surface $(\beta\varphi_s)$ are restricted by pre-breakdown occurrences in the field-effect capacitor. With germanium one can study at best the energy spectrum the range of about $(2/3)E_g$; in other words, as yet we can only speak with certainty about the parameters of states near the middle of the band. This range can be widened by increasing the condenser capacitance with the help of MIS structures (e.g., using CV characteristics; see Sect.2.4.2). These experiments indicate that at the edges of the allowed bands, where the density of levels increases, the spectrum of fast states remains continuous. The energy spectrum of these states on the Si-SiO$_2$ interface in MIS structures, which was reproduced in [3.21,22], looks surprisingly like the energy spectrum of real surfaces (Curve 2 in Fig. 3.2).

The specific features of the spectrum of fast states discussed above are not restricted exclusively to real surfaces of germanium and silicon covered with an oxide phase. Similar variations in the spectrum were observed with

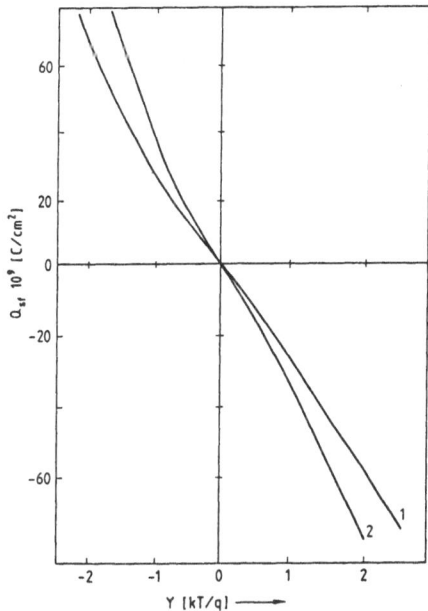

Fig.3.4. Curves of capture on fast states on the surface of $PbS_{(500)}$ [3.8]. (1) Prior to adsorption, (2) after adsorption of oxygen

monocrystalline PbS films after adsorption of oxygen [3.3,8] (Fig.3.4). In this case an uninterrupted oxide phase is practically absent.

The predominant rate of the continuous spectrum is evident from the fact that a continuous character is maintained in the curves of capture by fast states, whose density may vary widely under the action of diverse agents. Of course, this does not exclude the simultaneous existence of separate discrete levels; in the above cases, however, their density in any energy range is lower than the density of continuously distributed levels. The discrete levels appear to be resolved under ion bombardment [3.14] and when silicon and germanium are strongly doped with gold [3.13]. A certain structure was observed in capture curves obtained for silicon with the differential field-effect technique [3.10]. However, even in these cases the appearance of a discrete level was always accompanied by a simultaneous increase in the density of states in a continuous spectrum (see, e.g., Fig.3.3, Curve 2).

Similar conclusions regarding the existence of discrete capture levels on the surface of CdS monocrystals were reached by SHIRSHOV et al. [3.23], who also used the differential field-effect technique. The minimum observed of the curve of field-effect mobility μ_{fe} plotted against the potential of the field electrode was used to calculate the parameters of a single energy level. Similar minima of the curve μ_{fe} were observed for Si and CdSe, the latter being doped with indium [3.24]. However, the theoretical analysis carried out in [3.24] has demonstrated convincingly that the presence of extrema of the field-effect mobility function cannot serve as proof of the discrete nature of the energy spectrum of fast states.

Another circumstance often assumed to indicate the discrete nature of the energy spectrum is the exponential kinetics of the pulse field effect. This shape of the kinetic curves is frequently attributed to the monoenergetic level. However, the theoretical analysis performed by PANKIN et al. [3.25] shows that such kinetics can be consistently described under the assumption of a continuous energy spectrum, if one assumes the surface states to be continuously distributed over capture cross sections c_n and c_p.

In order to improve the resolution of the energy spectrum when taking the CV characteristics, PRIKHODKO and GULYAEV [3.26] used a series of mathematical methods of solving noncorrect problems that was developed by A.N. Tikhonov (the function regularization technique). Such treatment of experimental results led to the appearance of blurred maxima, corresponding to certain types of defects, against the background of a continuous spectrum. However, the density of states in these maxima was many orders of magnitude lower than the density of a quasi-continuous spectrum. Dominating on the disordered semiconductor-insulator interface was a quasi-continuous spectrum of surface states.

From our viewpoint, the absence of a specific response by the surface energy spectrum to various agents points to cooperative properties of the spectrum, which ought to be approached from the standpoint of the electron theory of disordered systems [3.27,28]. Since such an approach is very promising as regards an explanation of many electron phenomena on the surface, it will be discussed in detail in the next section; right now we shall briefly examine another important group of surface states, namely recombination centers.

As already mentioned in Sect.2.4, the dependence of the rate of surface recombination upon the surface potential $S(Y_s)$ can in some cases be presented as the sum of two components: a bell-shaped component S_b and a monotonic component S_m, the latter being but weakly dependent on Y_s. From Fig.3.5 it can be seen that both the absolute value of S and the contribution from S_b and S_m depend considerably on the conditions of thermal vacuum treatment of the samples and on adsorption. The first detailed investigations of the effects of adsorption and desorption upon the dependence $S(Y_s)$ were carried out by RZHANOV and NOVOTOTSKY-VLASOV [3.1,2]. Heat treatment of germanium specimens in a vacuum at 500 K resulted in a sharp increase of the maximum surface recombination rate S_{max}, while the bell-shaped curve remained generally the same (Fig.3.5). Subsequent adsorption of water restored the initial dependence $S(Y_s)$.

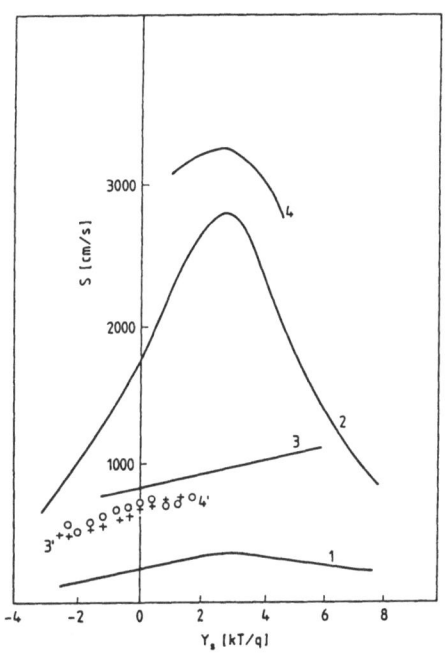

Fig.3.5. Effects of treatment on the rate of surface recombination as a function of the surface potential of germanium, $S(Y_s)$ [3.2]. (1) Before heat treatment, (2) $T_{cal} = 500$ K, (3) $T_{cal} = 750$ K, 5 min, (3') $T_{cal} = 750$ K, 1 h, (4) after adsorption of water on Ge_{750}, (4') after subsequent heat treatment at 750 K

The bell-shaped dependence $S_b(Y_s)$ is attributed in [3.1,2] to the recombination of charge carriers via a monoenergetic discrete center. For germanium the ratio c_p/c_n between the capture cross sections of a recombination center, calculated from the location of the maximum of the curve $S(Y_s)$, falls within the range 10 - 100, i.e., $c_p > c_n$ and the center is of the acceptor type. As indicated in [3.2], the quantity N_t^r (the concentration of recombination centers) is rather ambiguous. In experiments with the field effect on sulfurized germanium surfaces, NEIZVESTNY et al [3.28] managed to obtain estimates of $\sim 5 \cdot 10^8$ cm^{-2} for the upper limit of N_t^r for germanium, which is several orders of magnitude lower than the accepted values. If one accepts this value of N_t^r, then $c_p \sim 10^{-11} - 10^{-12}$ cm^2, and $c_n \sim 10^{-13} - 10^{-14}$ cm^2.

Investigations of the dependence $S(Y_s)$ for high-resistance silicon were reported in [3.4]. In order to exclude sticking effects and ensure the lowest possible injection level of minority carriers, the experiments were carried out at elevated temperatures (370 K). Under these conditions the dependence $S(Y_s)$ differs little from the case of germanium, and the parameters of the recombination center are also similar.

With a high-temperature treatment the results are different. At temperatures above 500 K one observes a decrease in S_{max}, accompanied by a distortion of the bell-shaped curve. Starting from 650 K the bell-shaped component almost totally disappears. An increase in T_{cal} increases the role of monotonic component S_m, which becomes most distinct after heat treatment at 750 K (Fig.3.5). In this temperature range the dominant recombination centers responsible for the bell-shaped component $S_b(Y_s)$ are destroyed. This is attributed in [3.1,2] to the removal of chemically bound water. It should be noted that with heat treatment of germanium in hydrogen sulfide, which results in complete substitution of the oxide film by sulfide coating, the bell-shaped component also completely disappears, and does not reappear after subsequent adsorption of water [3.28].

On the basis of all of the experimental results cited (Fig.3.5), RZHANOV and NOVOTOTSKY-VLASOV proposed the following model for the recombination centers:

1) The basis of the discrete recombination level responsible for the bell-shaped component S_b is a defect on the $Ge - GeO_2$ (or $Si - SiO_2$) interface [3.4]. A polar molecule of water, chemisorbed on this defect, changes one of the capture cross sections (c_n) by virtue of its dipole moment, thus making it an effective acceptor like surface recombination center. Desorption of such water molecules at 750 K causes destruction of the centers (Fig.3.5), while adsorption of water restores the centers responsible for S_b.

2) A reversible change in the concentration of discrete centers in the desorption-adsorption cycle of weakly bound water (up to $T_{cal} = 500$ K; see Fig.3.5) is due to screening, by water dipole molecules, of the electric field created by a dipole recombination center. A similar effect was observed by the authors with adsorption of polar molecules of alcohols, chlorobenzene, and nitrobenzene.

3) The monotonic component S_m is connected with recombination via a quasi-continuous set of fast surface states [3.29].

In the proposed model, all changes in the rate of surface recombination are attributed to direct interaction between polar molecules and the defects which form the basis of the recombination centers. A purely electrostatic treatment of this interaction, as we shall see later (Chap.6), does not agree well with the real adsorption mechanism of molecules such as H_2O, NH_3, alcohols, etc. Furthermore, it is hard to bring the concept of a monoenergetic adsorptive recombination level into agreement with the heterogeneity of the interface (the latter being supported by numerous spectroscopic and adsorption data [3.30]). Finally, if one considers the real interatomic bonds in adsorption complexes, it is hard to explain the high efficiency of a neutral dipole recombination center, which is close to the efficiency of an oppositely charged Coulomb center.

Experiments carried out at Moscow University and reported in [3.31-33] cast doubt on the purely electrostatic model proposed in [3.1,2]. In contrast to previous experiments[1], these studies were performed in a vacuum of about 10^{-8} torr, and the initial stage of adsorption was closely monitored. First we shall discuss the effect of neutralization. As seen from Fig.3.6, adsorption of water (dipole moment $\mu = 1.8$ D) on the surface of a germanium specimen vacuumized at 500 K, up to occupancies of $n_a \sim 10^{11}$ molecules $\cdot$ cm^{-2}, does not give rise to any changes in the bell-shaped component of surface recombination or in the charge arrested in fast states Q_{sf}. At higher n_a there is a decrease in S_b and Q_{sf}, which agrees well with Figs.3.1,5. As will be shown in Sect.9.3,

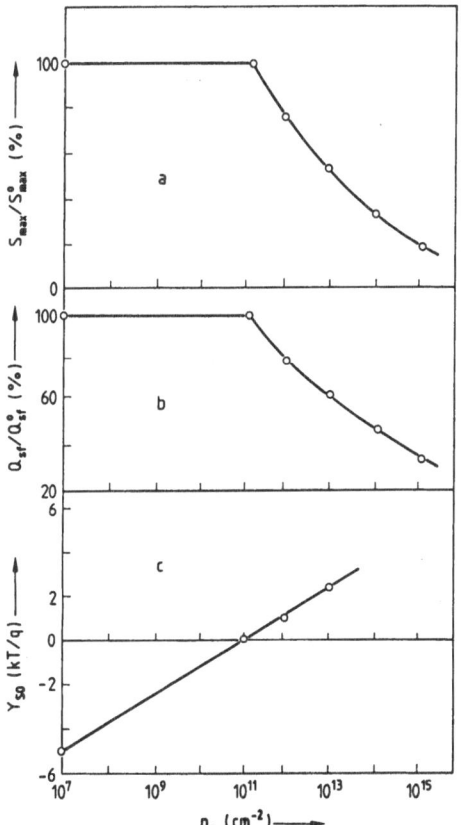

Fig.3.6a-c. Dependence of (a) S_{max}/S_{max}^0, (b) Q_{sf}/Q_{sf}^0, (c) Y_{s0} on the concentration of adsorbed H_2O molecules, n_a. Values S_{max}^0 and Q_{sf}^0 refer to the initial specimen ($T_{cal} = 500$ K); Q_{sf} corresponds to $Y_s = 4kT/q$

---

[1] All the experiments in [3.1,2] were carried out in a vacuum of about $10^{-5} - 10^{-6}$ torr.

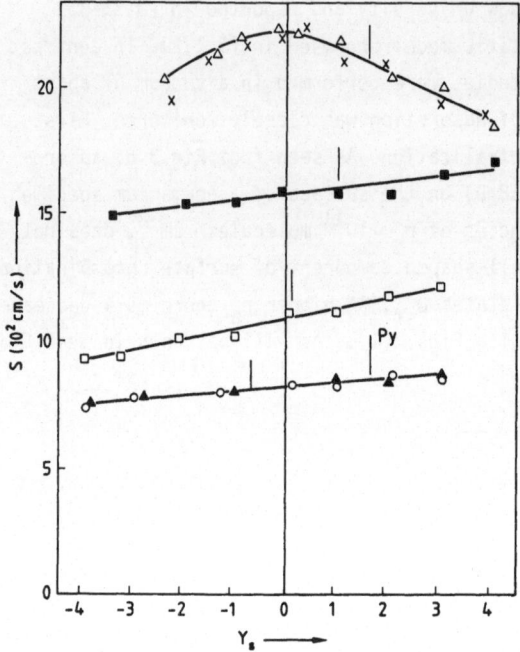

Fig.3.7. Changes in $S(Y_s)$ upon adsorption of H_2O and Py. $\circ$ - T_{cal} = 700 K, $\square$ - H_2O (10^{-3} torr, n_a = 10^{11} molecules cm^{-2}), $\blacksquare$ - H_2O (10^{-2} torr; n_a = 10^{12} molecules cm^{-2}), $\triangle$ - H_2O (10^{-1} torr, n_a = 10^{13} molecules cm^{-2}), X - H_2O (20 torr), $\blacktriangle$ - Py (10^{-1} torr). Marks indicate the values of Y_{s0} (zero electric field)

adsorption of other polar molecules having a dipole moment equal to or greater than that of water does not influence S_b and Q_{sf}. Hence follows the important conclusion that the centers responsible for the bell-shaped component of re-combination and fast capture are not the primary centers of adsorption of the above-mentioned molecules. The effect of neutralization is not directly con-nected with the dipole moment of adsorbed molecules. Further on we shall show that this effect can be explained more comprehensively through proton processes on the surface.

Interesting results were obtained in investigations of surface recombination on germanium specimens dehydrated at 750 K. As already indicated, under these conditions only the monotonic component is ovserved (Fig.3.5). From Fig.3.7 one sees that in the initial stage of adsorption, up to occupancies of $n_a \sim 10^{12}$ molecules $\cdot cm^{-2}$, only the monotonic component S_m increases. The bell-shaped component appears only at occupancies $n_a \geq 10^{13}$ molecules $\cdot cm^{-2}$, which does not contradict the results reported in [3.2] (Fig.3.5). All variations in surface recombination illustrated in Fig.3.7 were irreversible. In order to restore the state of the surface as it was before adsorption, the specimen had to be subjected to high-temperature (700 K) vacuum treatment. This suggests the formation of strong adsorptive bonds responsible for S_b. Adsorption of pyridine ($\mu = 2.2$ D), as seen from Fig.3.7, had no influence at all on recombination. In contrast to the variation in $S(Y_s)$, the slant of the capture curves for fast surface states decreased somewhat upon adsorption of H_2O and Py, which indicates a lack of correlation between $S(Y_s)$ and $Q_{sf}(Y_s)$. This is attributed in [3.1,2] to highly different concentrations of recombination centers and fast capture centers.

The pattern of variation in surface recombination as presented in Fig.3.7 cannot be brought into agreement with the model of a dipole recombination center [3.1,2]. Indeed, adsorption of pyridine, whose dipole moment is greater than that of water, did not cause changes in S, nor did adsorption of nitrobenzene ($\mu = 3.9$ D) [3.28]. The appearance of the bell-shaped component S_b, which takes place only when the surface is sufficiently well covered by water molecules ($n_a \geqslant 10^{13}$ molecules $\cdot$ cm^{-2}), cannot be explained by the formation of dipole complexes tightly bound to the surface, which are responsible for the discrete recombination level. The differential heat of adsorption, which describes the strength of the adsorptive bonds, drops abruptly with an increase in n_a [3.30]. The appearance of S_b is equally hard to attribute to the charge heterogeneity of the surface, which increases with an increase in adsorption. Adsorption of Py, which results in practically the same integral charge as adsorption of water, does not give rise to S_b. It is impossible to explain the pattern of variation of S illustrated in Fig.3.7 as being due to recombination on the centers in the SCR, although such a possibility is admitted by RZHANOV [3.1] as an explanation of the monotonic component S_m. SACHENKO [3.34] demonstrated that when the quasi levels in the SCR are not steady, as well as when the limiting stage of recombination is the appearance of nonequilibrium charge carriers in the SCR, the dependence of recombination in the SCR on the surface potential is also bell shaped. In our case (Fig.3.7) adsorption of H_2O and Py while the bending of bands was the same, the character of the variation in $S(Y_s)$ upon adsorption of these molecules was qualitatively different.

RZHANOV [3.1] and NOVOTOTSKY-VLASOV [3.2] drew a connection between surface recombination and point defects (vacancies, oxygen complexes, etc.) activated by adsorbed molecules and metal atoms. This standpoint hardly explains all variations in $S(Y_s)$ during adsorption, as presented in Figs.3.6,7. A more natural approach is provided by the theory of electron transitions in a disordered system, represented here by the interface.

Surface recombination via a quasi-continuous set of fast states is discussed in [3.29,34-37]. Recombination can proceed either by virtue of electron transition between the acceptor and the donor branches of a quasi-continuous spectrum and the relevant bands of the bulk, or owing to transitions between different groups of surface states [3.29]. In the first case a bell-shaped dependence is observed. RZHANOV [3.1] rejected the explanation of experimental curve $S(Y_s)$ via such transitions, pointing out that the linewidth exhibited by theoretical curve $S(Y_s)$ is much smaller. However, the parameters of $S(Y_s)$ may vary widely, depending on the choice of parameters for the spectrum [3.29,36].

Using the assumption that the probabilities of capture are independent of the location of surface states in the energy spectrum, NEIZVESTNY and OVSYUK [3.29] considered recombination due to transitions between the acceptor and donor branches of surface states. Two cases are possible here:

1) The quasi-continuous spectrum occupies the entire width of the forbidden band (Fig.3.8). In this case, depending upon the distribution coefficients A and B in (3.3), the monotonic component S_m will be the predominant one.

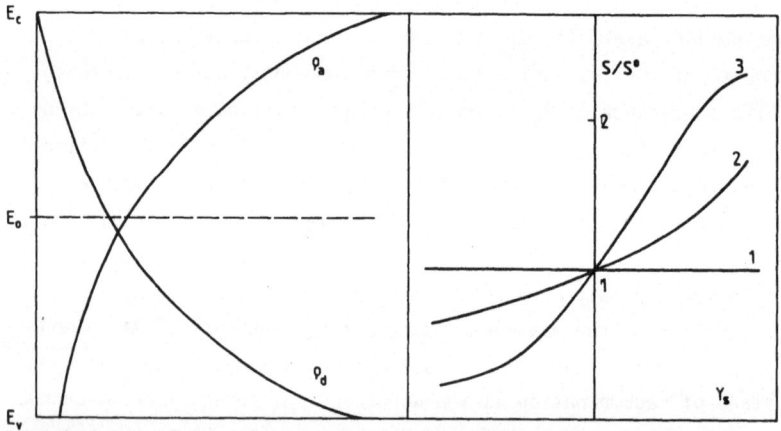

Fig.3.8. The rate of surface recombination S/S_0 as a function of surface potential Y_s (right) with quasi-continuous distribution of acceptor-type (ρ_a) and donor-type (ρ_d) fast states in the forbidden band (left). S^0 is a parameter; the digits on the S/S^0 curves denote the relations of the preexponential factors in (3.3) [3.29]

2) The quasi-continuous spectrum is truncated (Fig.3.9). In this case the bell-shaped component S_b is observable. The height and parameters of the "bell" strongly depend on the choice of capture coefficients.

We believe [3.32,33] that the proposed model correctly explains the experimental data displayed in Fig.3.7. Adsorption of water (with high occupancies)

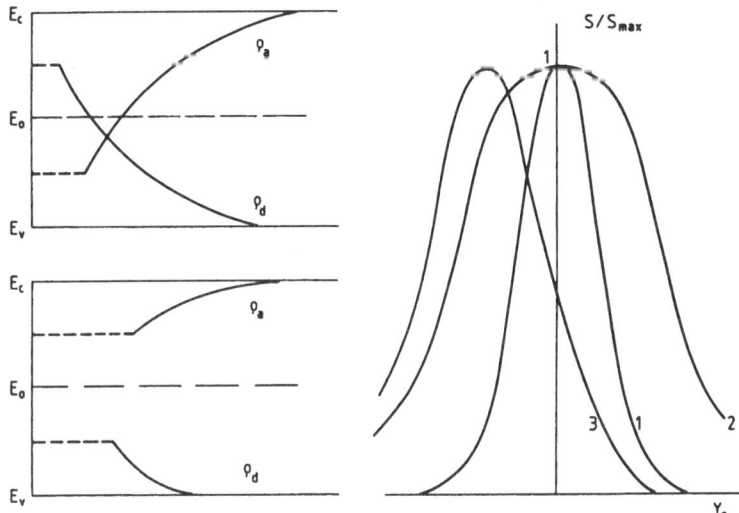

Fig.3.9. The rate of surface recombination S/S_{max} as a function of surface potential Y_s (*right*) with a truncated quasi-continuous spectrum of fast states (*left*). The digits on the S/S_{max} curves correspond to different parameters of the spectrum [3.29]

results in additional hydration [3.30] and in reconstruction of the surface. For this reason certain states may drop out of the quasi-continuous spectrum, and this, in turn, according to the model proposed in [3.29], will give rise to the bell-shaped component S_b. RZHANOV [3.1] considered the absence of cor-relations between $S(Y_s)$ and $Q_{sf}(Y_s)$ to be the main argument against recombination through the continuous spectrum of fast states. We have demonstrated that this objection can be easily discounted if one assumes the fast capture centers and recombination centers to belong to two independent groups of surface states which differ in concentration and in capture cross sections, and possibly in their location in the disordered surface phase.

3.2 The Energy Spectrum of the Surface from the Standpoint of the Theory of Disordered Systems

The development of the electron theory of disordered systems can be considered without exaggeration to be one of the most brilliant achievements in the physics of condensed phases in the past decade. The theory has succeeded not only in explaining numerous phenomena observed in experiments with amorphous and glass-like bodies and highly doped semiconductors but also in predicting new effects characteristic of these materials. Unfortunately, the implications of this theory, which was developed primarily with regard to three-dimensional phases, have not yet been fully appreciated in surface science. But there are good reasons to consider a surface a typical disordered system. With atomically clean surfaces the long-range order is violated in connection with reconstruction of the surface, and with real surfaces it is violated due to the disordered structure of the semiconductor-oxide interface. The presence of structural defects, chemical surface complexes, and high concentrations of adsorbed atoms and molecules justifies viewing the surface as a two-dimensional analogue of a highly doped semiconductor.

We shall not dwell at length on the theory of the behavior of electrons in disordered systems, since it has been exhaustively treated by researchers who themselves made important contributions to this essential branch of contemporary physics [3.38-43]. Here we shall just mention certain fundamental concepts and consequences of the theory[2] which will help us to comprehend better the peculiar features of the energy spectrum of surface states and its changes under the action of various agents.

The basic difference between the approach adopted for the construction of the energy spectrum in the theory of disordered systems and the approach taken in the conventional band theory is the following: in the conventional band theory of ideal crystals, the potential energy of a charge carrier is a periodic function of coordinates, as a consequence of the periodicity of the crystal structure. In a disordered system, owing to the violation of the long-range (and sometimes the short-range) order, and to the presence of randomly distributed impurity atoms and defects, the charge carriers are in random electric fields. The potential energy of an electron incorporates a component which is a random function of coordinates, to borrow a term from the theory of probability. To describe random fields the quantity $v(r)$ is replaced by the functional $P[v(r)]$, which describes the probability that the energy of a charge carrier with coordinate r will be equal to $v(r)$. The problem of the energy spectrum now takes on probability features. All the principal characteristics of the spectrum obtained from experiment are now represented by values averaged over random fields.

It has been demonstrated that the most general characteristic describing any system, including a disordered one, is the density of states $\rho(E)$ [3.42]. The density of states is known to be linked to the concentration of carriers via the expression

---

[2] Our treatment of theoretical questions is mainly based on the works of BONCH-BRUEVICH et al. [3.41,42].

$$n = \int_E \rho(E) n_F dE \quad , \tag{3.4}$$

where $n_F = \{\exp[(q\varphi_0 - E)/kT] + 1\}^{-1}$ is the Fermi function (on the surface $q\varphi_0 = q\varphi_s$). The concept of the Fermi energy as a purely thermodynamic characteristic of the system is, naturally, also valid for a disordered system. The most effective way to define explicitly the function $\rho(E)$ averaged over random fields is to employ Green's functions. The form of function $\rho(E)$ determines the character of the energy spectrum of the system in question: in the allowed bands $\rho(E)$ is nonzero and continuous; in the forbidden band $\rho(E) = 0$. The values of energies where $\rho(E)$ has isolated singular points correspond to discrete levels.

The nature of the energy spectrum depends on the extent of "disorderliness" of the system, i.e., on the relative contribution of the random component to the potential energy of an electron, $v(r)$. In a number of covalent glass-like and amorphous substances the short-range order is conserved. The potential energy of an electron in these systems will be primarily determined by the local configuration of atoms in the first coordination sphere. The distortion of the second, third, and higher-level coordination spheres is responsible for violation of the long-range order (topological disturbances). The contribution which these disturbances make to the potential energy of interaction between charge carriers and random fields can be neglected by virtue of their long-range nature. In such systems the random component of $v(r)$ is small, and the energy spectrum is wholly determined by the properties of the atomic functions and the atomic configuration in the first coordination sphere.

Consider, for example, amorphous germanium and silicon. Data obtained with the aid of electron or X-ray diffraction conclusively indicate that deviations of the interatomic distances and angles in the first coordination sphere from the corresponding values in the crystalline state are quite small [3.43]. According to data from measurements of electric and optical properties, amorphous germanium and silicon behave like semiconductors. When prepared in a certain way, they exhibit a forbidden band whose width is just a little narrower than in the respective crystals, and there are no allowed states in the forbidden band. We see that in substances with a nonperiodic function $v(r)$, provided the contribution of the random component to $v(r)$ is small, we can also speak of semiconductor bands, the term "band" being understood here in the statistical sense.

Let us now examine the change in the energy spectrum when the contribution of the random component to $v(r)$ can no longer be neglected. Consider the simplest example: the change in the energy spectrum of an ideal semiconductor crystal after the introduction of randomly distributed defects or impurity atoms ("doping" of the semiconductor). If the concentration of defects is very small, so that when analyzing the interaction of an electron with one of the defects we can neglect the existence of other defects, then the discrete energy levels of defects in the forbidden band will be determined solely by their nature, and there will be nothing random about them[3]. They will be represented on the curve of density of states $\rho(E)$ by narrow δ-like spikes.

The situation is quite different when the concentration of defects is so large that the Coulomb fields of charged defects overlap. With a uniform distribution of defects this occurs when the average separation between defects is lower than the Debye shielding length L_D for the given material. Now an electron interacting with a defect will fall under the influence of a number of other defects as well. In this way the random field created by randomly distributed defects acquires collective properties.

The random nature of the potential energy of an electron in such a field can give rise to deep potential pits, scattered at random, which function as centers of localization of free electrons. The corresponding discrete energy levels are usually called the fluctuation levels [3.38,40,41]. Note that by virtue of the cooperative properties of the random field, these levels can no longer be attributed to any particular defects or groups of defects [3.44]. The concentration of fluctuation levels having a given energy is proportional to the probability of their formation. This probability will exhibit a continuous dependence on their depth, that is, on the distance from the edges of the continuous spectrum (E_c for electrons and E_v for holes). In other words, greater energies of ionization of levels correspond to greater fluctuations of potential energy. However, the probability of great fluctuations in the random field is low, so that after averaging, the density of states decreases rapidly with increasing distance from the edges of the allowed bands.

Hence, only by virtue of the appearance of random fields is a dense spectrum of discrete acceptor-type fluctuation levels split off from the conduction band, and a dense spectrum of discrete donor-type fluctuation levels split off from the valence band [3.38]. Thanks to the high density of levels, in the case when the energy range under consideration is much greater than the mean distance between levels, one can speak of a continuous (envelope) function $\rho(E)$ in this interval. Depending on the statistical characteristics

---

[3] This gives us a random distribution of defects in space, but not a random electric field.

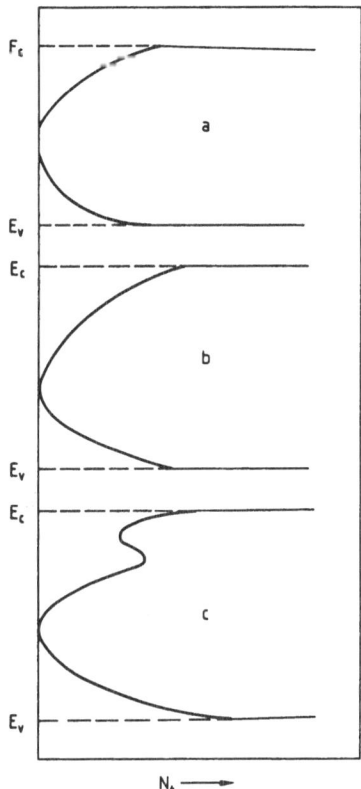

Fig.3.10a-c. Types of energy spectra of disordered systems

of the random field, which are determined by the form of the functional $P[v(r)]$, two extreme cases are possible [3.42]:

a) there are certain finite limits of the fluctuation levels (Fig.3.10a), and

b) the spectrum has no definite limits, and the whole energy range $E_c - E_v$ is filled by discrete levels (Fig.3.10b)[4]. In this case one can no longer speak of a forbidden band of electron energies, but must speak of limits of localized and delocalized states.

Closeness of levels in terms of energies by no means implies proximity of the corresponding centers of localization of electrons in space [3.44]. In a homogeneous system the appearance of equal-energy levels having nearly the

---

[4] At low fluctuations the differing tails of the two allowed bands are due to the different values of the effective mass of the charge carriers [3.38].

same coordinates is a rather rare occurrence. Transitions of electrons over discrete levels can take place only by means of jumps ("jump-type" conductivity). At $T \to 0$, jump-type conductivity tends toward zero, and discrete centers are nonconducting. This mechanism of electroconductivity differs from the mechanism of charge transfer in the range of the continuous spectrum, where the wave functions embrace the whole crystal, so that one can speak of "free charge carriers". The energy range occupied by discrete levels is often termed the mobility slit. When the boundary between the continuous and the discrete spectra is crossed, the mobility of carriers undergoes an abrupt change [3.43]. In the case of amorphous bodies the generally accepted concept of limits of the conduction and valence bands must be understood in a purely statistical sense as the limits of continuous and discrete spectra of states [3.41,42].

Theoretical estimates made by BONCH-BRUEVICH et al. [3.41,42] and MOTT and DAVIS [3.43] indicate that the concentration of fluctuation levels can be quite high. The question then arises: why does this not give rise to an overlapping of wave functions pertaining to different levels, to form a band of delocalized states as in the case of a crystal? The answer lies in the presence of random fields. For the levels to be smeared into a band, the centers of localization of electrons have to form a strictly periodic configuration.

Along with fluctuation levels, the disordered system may have certain idiosyncratic inherent defects, giving rise to deep potential pits, which also serve as centers of localization of free charge carriers. Such defects in a crystal would be represented by narrow δ-like spikes of the density of states. In the presence of random fields these spikes can either completely disappear or be greatly broadened. Figure 3.10c shows the form of the energy spectrum of a disordered system having a smeared-up level of an inherent defect.

Let us now discuss the major question of how the principal regularities in the spectrum of surface states are explained from the standpoint of the theory of disordered systems. For the time being we shall restrict ourselves to fast surface states, whose energy spectrum in a certain energy range can be obtained with the aid of the equilibrium field effect (Sect.2.3).

According to the capture curves presented in Figs.3.1,2,4 and to published data [3.1,8-10], the minimum densities of fast surface states, under the influence of diverse agents, vary from $2 \cdot 10^{11}$ to $6 \cdot 10^{12}$ $eV^{-1} \cdot cm^{-2}$ for Ge, $3 \cdot 10^{11}$ to $4 \cdot 10^{12}$ $eV^{-1} \cdot cm^{-2}$ for Si, and $2 \cdot 10^{12}$ to $3 \cdot 10^{12}$ $eV^{-1} \cdot cm^{-2}$ for PbS. The integral concentrations of fast states in the forbidden band in all cases are no less than 10^{12} to 10^{13} cm^{-2}. Considering the value of the Debye shielding length, it is easy to see that the Coulomb fields of fast capture centers do overlap [3.3]. This is a straightforward indication of the cooperative properties of random fields, and also explains the surprsingly low sensitivity of the spectrum of fast states to the particular type of defects introduced onto the surface (Sect.3.1). All disturbances lead to the formation of a dense spectrum of discrete levels, observed experimentally as a continuous spectrum. Note the striking similarity between the experimental (Fig.3.3) and the theoretical (Fig.3.10b) energy spectra of the surface.

Adsorption of the most varied molecules changes the capture curves smoothly and shifts the energy spectrum. No local levels characteristic of any type of adsorbed molecules have ever been observed in experiments on adsorption. All this agrees well with the assumption of the cooperative nature of the random field of the surface, in whose creation both biographic and adsorptive states participate.

In some cases the quasi-continuous spectrum may exhibit discrete levels as well. As we pointed out earlier, surface doping of germanium with gold [3.13] leads to the appearance of a blurred maximum in the energy spectrum of fast states (Fig.3.3, Curve 2). Broadened discrete levels have been observed on $Si - SiO_2$ interfaces and in amorphous silicon [3.21,22,45,46]. A broadening of levels, as indicated, is characteristic of systems with random fields (Fig.3.10c).

RZHANOV and OVSYUK [3.1,47] made the first attempt to consider the energy spectrum of fast surface states of a semiconductor on the basis of concepts from the theory of disordered systems. They attributed the appearance of a quasi-continuous spectrum to the violation of the symmetry of the atomic potential in the subsurface crystal region. The decrease in the density of fast states when polar molecules are adsorbed on germanium was explained in the following way: these molecules wander on the surface and settle at the points where fluctuations of the atomic potential are greatest, thus smoothing these fluctuations. This results in a decreased blurring of the allowed bands and a decrease in capture as observed in the field effect.

This purely electrostatic explanation does not, however, embrace all the experimental data. Indeed, adsorption of H_2O ($\mu = 1.8$ D) on Ge_{500} reduces the density of states (Fig.3.1), while adsorption of NH_3 ($\mu = 1.5$ D), as well as adsorption of ozone [3.2] ($\mu = 0.52$ D), has almost no influence on the spectrum, despite the fact that the dipole moments of these molecules have close values. Adsorption of oxygen ($\mu = 0$) on PbS increases capture (Fig.3.4). The electrostatic treatment of recombination processes was already critized in Sect.3.1.

In order to gain insight into the mechanism by which adsorption influences the energy spectrum of surface states within the framework of the theory of disordered systems, one must elucidate the possible nature of random field sources on the surface, as well as the causes of variations in term when adsorption occurs. We shall return to these questions in Chaps.6,7.

It is our opinion that in order for progress to be made in the phenomenological description of the energy spectrum of electron states on a disordered surface, it is first necessary to obtain detailed statistics on random fields on the surface. In the electron theory of disordered systems all the statistical characteristics are incorporated in the functional $P[v(r)]$, which determines the nature of the dependence of the density of states on energy, $\rho(E)$. The theory considers certain particular types of random fields (Gaussian, Lorentz, and Poisson distributions of random fields). Electrophysical experiments dealing with the field effect or with CV characteristics can provide information on the distribution of states over energies in the band scheme of a solid, but not on their distribution over space.

Data on the topography of surface defects and their interaction are hidden in any spectral characteristic. We illustrated this in [3.48] with EPR data from surface defects of a model semiconductor (rutile). From an analysis of the dependence of EPR signal linewidth on concentration, important conclusions were derived about the cluster configuration of defects, and an estimate was made of the local concentration of defects in a cluster [3.48]. In this case the concentration of defects was sufficient to cause overlapping of Coulomb fields and the formation of a system of random fields on the surface. The distribution of defects, which serve as sources of these fields, turned out to be satisfactorily described by Poisson distribution. The energy distribution of electron states on a rutile surface cannot be studied with the aid of the field effect owing to their high density. Germanium or silicon might be successfully employed for this purpose if a technique were developed of seeding their surface with controllable quantities of paramagnetic centers. An analysis of the correlation between the statistics of their distribution over energies would be of considerable help in the construction of a more reliable model of the surface.

Finally, let us discuss calculations of the concentration of carriers with respect to localized states in a disordered system. Despite the rigorous definition of function $\rho(E)$, it is quite cumbersome to use in calculations. For that reason, well-founded phenomenological parameters are introduced: effective levels ε_t^*. They determine the effective density of states on a chosen spectral energy interval via the local level [3.49]. Let us consider the case when the density of states in the forbidden band is not monotonic, but exhibits blurred extrema (of the kind illustrated in Fig.3.10c). Then the quantity ε_t^* is defined in such a way that the concentration of charge carriers in the region

Δ of the given extremum coincides with their concentration on the discrete level, i.e., according to (3.4),

$$\int_\Delta \rho(E)n_F(E)dE = N_t n_F(\varepsilon_t^*) \quad , \tag{3.5}$$

where N_t is the concentration of levels of this kind.

Introducing dimensionless moments of local density m (which, generally speaking, depend on the statistical characteristics of the random field), we get

$$\varepsilon_t^* = \varepsilon_t + \Delta\left[m + \frac{\Delta}{2kT}(m^2 - 1)\ \tanh\left(\frac{\varepsilon_t - q\varphi_0}{kT}\right)\right] \quad . \tag{3.6}$$

(For the surface, substitute $q\varphi_s$ for $q\varphi_0$).

By substituting ε_t^* into (3.1), we obtain the effective value of Q_{sf} for the surface.

Thus, by introducing effective levels we can use all the expressions describing the statistics of electrons and holes in a semiconductor. However, in contrast to the effective parameters employed in the conventional band theory (when the location of levels in the band is assumed to be independent of temperature), in disordered systems the value of ε_t^*, see (3.6), is a function of temperature and the concentration of states. It would be interesting to investigate the temperature and concentration dependence of ε_t^* to substantiate the model of the surface and a disordered system. A similar operation can be used to introduce the effective levels in the quasi-continuous regions of the spectrum. However, additional assumptions are necessary for the exact calculation of ε_t^* [3.49].

3.3 Slow Electron States on Real Semiconductor Surfaces

The relaxation time of slow states at room temperature, as was pointed out in Sect.2.1, is of the order of minutes or even hours in the case of germanium and silicon. This makes it practically impossible to employ the field-effect technique to obtain the equilibrium curve of capture, $Q_{ss}(Y_s)$, on these states. For this reason, slow processes are investigated by kinetic methods. The set of slow states is driven away from equilibrium by applying a constant transverse field (field effect at a constant potential difference), irradiating the surface with light, or changing the temperature. Then the relaxation of the surface charge, surface conductivity (σ_s) or surface potential (Y_s) is monitored.

45

The very first investigations of slow relaxation [3.50-52][5] indicated that its rate depends strongly upon the environmental humidity. KOC [3.54] and PILKUHN [3.55] observed a sharp decrease in relaxation time with an increase in the partial pressure of water vapor. KOZLOV et al. [3.56] demonstrated that adsorption not only of water, but also of a large number of other acceptor molecules (oxygen, iodine, p-Bq) and donor molecules (ammonia, diphenylamine) accelerates relaxation. In the presence of a medium the density of surface states reaches 10^{13} to 10^{14} cm^{-2}, which is several orders of magnitude greater than the density of fast states (Sect.3.1). The total charge of the surface, $Q_s = Q_{sf} + Q_{ss}$, and therefore surface potential Y_s, is in certain cases completely determined by the charge in slow states Q_{ss}. The charging of the semiconductor surface upon adsorption must then be understood as a change in the charge in slow states.

Such was the case in adsorption of various molecules on germanium specimens vacuumized at 500 K [3.31]. As seen from Fig.3.6, in contrast to the changes in S and Q_{sf}, initial injections of water lead to a rapid increase in surface potential Y_s, which is connected with the appearance of charged slow states on the surface. The defects which constitute the basis of these states are the most active adsorption centers in the initial state of surface occupation. Similar results were obtained during adsorption of other molecules (Chap.9). Desorption and vacuum heat treatment in contrast, lead to a sharp decrease in the concentration of slow states. This gives good reason to assume that the overwhelming majority of slow states (at least on Ge and Si surfaces) owe their existence to the formation of adsorption complexes on the surface. Let us discuss slow relaxation and the phenomenological parameters of slow states in greater detail.

Experimental data reveal that even at small deviations from equilibrium, the law of relaxation, for example of surface conductivity $\Delta\sigma_s(t)$, is not exponential. Attempts have been made to break down the experimental curves $\Delta\sigma_s(t)$ into separate exponential sections [3.53] and to approximate them with hyperbolic [3.57] and logarithmic [3.52] curves, but they have failed to provide a satisfactory explanation of all the experimentally observed regularities in slow relaxation. The most convenient empirical approximation of $\Delta\sigma_s(t)$ was proposed by KOC [3.58]

$$\Delta\sigma_s(t) = \Delta\sigma_0 \, \exp(-t/\tau_s)^a \quad , \qquad \text{or}$$

---

[5] A vast bibliography of works published before 1964 can be found in [3.53].

$$\Delta Q_{ss}(t) = \Delta Q_{ss}^0 \exp(-t/\tau_s)^a \quad , \tag{3.7}$$

where τ_s and a are certain parameters.

Further investigations carried out in various gaseous media [3.55,56,59] have demonstrated that Koc's empirical law accurately describes many experimental results. Usually (in dry media and a vacuum) $a \approx 0.3$, and τ_s, at small variations in the potential ($Y_s \leqslant 1$), does not depend upon the magnitude and polarity of the applied field.

It is widely believed that slow states, as opposed to fast states, are found on the outer surface of the semiconductor's oxide film. The corresponding band model (Fig.3.11a) has been discussed in many reviews and books on surface science. This opinion is based on retarded kinetics of relaxation and a strong influence of adsorbed water molecules on the kinetics.

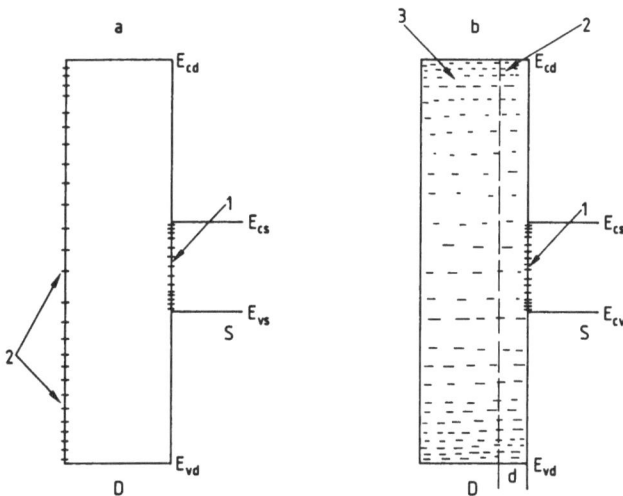

Fig.3.11a,b. Band schemes of real surfaces. (1) Fast states, (2) slow states, (3) superslow states. E_{cs} and E_{cd} are the edges of the conduction bands; E_{vs} and E_{vd} are the valence bands of a semiconductor (S) and a dielectric (D), respectively

One of the first attempts to obtain information on the localization of slow states was made in [3.60]. LASSER et al. observed a sharp increase in the relaxation time of excess conductivity in the field effect during oxidation of germanium at elevated temperatures. That work, however, does not permit unambiguous conclusions, since not only the thickness of the oxide layer, but also its structure and the extent of hydration were subject to changes in the course of oxidation. JÄNTSCH [3.61] challenged these results, reporting that the relaxation rate was independent of film thickness during electrolytic oxidation of silicon.

FROLOV et al. [3.57] conducted a simultaneous study of conductivity and the contact potential difference (CPD) after removal of a transverse field, when there is considerable tension across the oxide layer v_{ox}. They concluded that a slow-state charge localizes on the outer surface of the oxide film. However, the appearance of a "slow" charge could be connected with the diffusion of charged defects in the field [3.62], and not with the recharging of slow states. The data on relaxation during adsorption of water and alcohol, when $v_{ox} = 0$, positively indicate that the relaxing adsorptive charge is found on the oxide-semiconductor interface [3.57].

In order to clarify the location of slow states, experiments were carried out [3.63] in which slow states of the same kind were created via adsorption of ammonia and p-Bq on the surface of germanium specimens covered with a thermally grown oxide film of varying thickness, but having the same extent of hydration. These experiments demonstrated that within the studied range of thickness d (30 to 1000 Å), the kinetic parameters of slow states created during adsorption do not depend on d. This means that in any case the majority of slow states, whose relaxation time is of the order of minutes, are found near the Ge-GeO$_2$ interface (Fig.3.11b).

There now exist numerous experimental data indicating that the presence of an oxide phase is not a necessary condition for the existence of slow states. Slow relaxation has been observed not only with germanium [3.56,63] and silicon [3.64] covered by an oxide film, but also with semiconductors devoid of a surface oxide phase: ZnO [3.65,66], Cu_2O [3.67,68], PbS [3.8,69], and $Pb_{1-x}Sn_xTe$ [3.70]. Slow processes in Cu_2O have been observed in a wide range of phase composition: from Cu_2O to CuO [3.67]. Adsorption of water resulted in a considerable acceleration of relaxation in PbS [3.69] and Cu_2O [3.68]. It is interesting to note that slow relaxation was also observed on germanium free of an oxide film during adsorption of NH_3 and H_2S [3.71,72].

Thorough investigations of the kinetics of slow relaxation carried out by KOZLOV et al. [3.56,63,73], revealed that the kinetic curves $Q_{ss}(t)$, obtained when both donor (H_2O, NH_3) and acceptor (p-Bq, CO_2) molecules were adsorbed on germanium specimens covered by an oxide film of varying thickness (30 to 1000 Å), are nicely linearized in Koc's coordinates, see (3.7), with a = 0.3 and τ_s = 1 to 6 min, and are practically identical with the corresponding relaxation curves obtained with the aid of the field effect (Fig.3.12). For other semiconductors without an oxide film, the kinetics of relaxation is also well

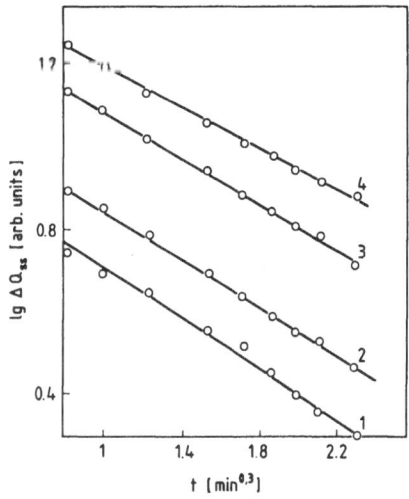

Fig.3.12. Kinetics of slow relaxation after removal of electric field on germanium specimen covered with thin [$d \approx 3$ nm $(1,3)$] and thick [$d \approx 100$ nm $(2,4)$] oxide layers in vapors of p-Bq [$p \approx 10^{-1}$ torr $(1,2)$] and ammonia [$p = 10^{-3}$ torr $(3,4)$]

approximated by (3.7) with $a = 0.3$ for PbS [3.8] and $a = 0.6$ for Cu_2O [3.68].

All these experimental facts do not support the generally accepted model of the real surface (Fig.3.11a). The majority of slow states are found in the immediate proximity of the semiconductor-oxide interface (Fig.3.11b) or in a highly reconstructed surface layer of PbS or Cu_2O. The similarity of the relaxation kinetics in the field effect and in adsorption indicates [3.56,63,64] that the time necessary for the molecules to pass through the oxide layer is quite short. This agrees well with the fact that upon adsorption, the spectrum of fast states, which are found directly on the semiconductor surface, changes in a time period that is small compared to τ_s (Sect.3.1).

Valuable information on the mechanism of slow relaxation can be obtained by analyzing temperature dependences. It was shown in [3.58] that the characteristic relaxation time τ_s varies with temperature as

$$\tau_s \sim \exp(\Delta E_\tau / kT) \quad . \tag{3.8}$$

The quantity ΔE_τ strongly depends on the nature of the environment [3.56, 64]. Since relaxation does not follow an exponential law, see (3.7), the quantity ΔE_τ is not, strictly speaking, the true activation energy of the process and must therefore be treated as a kind of effective parameter. Investigations of slow relaxation at various temperatures have revealed that ΔE_τ exhibits little dependence on temperature. The absence of a noticeable influence of temperature on the rate of slow relaxation was also emphasized in [3.74].

These results imply that the most probable mechanism of charge exchange between slow states and the allowed bands in the semiconductor is the tunnel mechanism [3.73].

Various models of slow relaxation have been proposed in the literature. The choice of a model is determined to a large extent by the choice of the limiting stage of relaxation. In most models the slow states are assumed to be found on the outer surface of the semiconductor's oxide film; in other words, such models are based on the band scheme illustrated in Fig.3.11a. For instance, KOC [3.58] and DORDA [3.59] assumed the limiting stage of slow relaxation to be represented by the diffusion of electrons and holes through the oxide layer ("diffusion mechanism"). FEDOROVICH and VOGEL [3.75] considered the oxide layer of germanium or silicon as an electrolyte (a mixture of various polysilicon or polygermanium acids), and connected slow relaxation with the violation of electron-ion equilibrium in the semiconductor-oxide system. They assumed the limiting stage of the process to be adsorption and desorption, dissociation of H_2O or OH, and reactions between uncharged particles. A similar model of electron-ion processes in the $Si-SiO_2$ system was later discussed in [3.76].

The "field" models of relaxation proposed in [3.55,57] constitute a special case of the model of electron-ion equilibrium. Here the limiting stage is assumed to be either the dissociation of water molecules [3.57], or the diffusion of ions in the oxide layer upon application of a transverse electric field [3.55]. JÄNTSCH [3.53], finally, suggested that the main cause of slow relaxation is the slow transitions between different types of bonding between water molecules and the silicon surface (physical adsorption—coordinative bond —ionization of molecules—formation of hydroxyl groups). In accordance with this, the kinetics of relaxation is determined by four slow processes, and can be approximated by four exponents.

The above-discussed models of slow relaxation do not explain a whole series of experimental facts, namely:

1) The mechanism of relaxation in the field effect (when there is an electric field in the oxide layer) is similar to that in adsorption (when there is no field) [3.77].

2) Similar trends of relaxation are observed not only in adsorption of water, but also in an atmosphere of poorly dissociating substances such as oxygen, iodine, p-Bq, and ammonia. Relaxation also proceeds after partial dehydration of the surface.

3) The temperature dependences of slow relaxation cannot be brought into agreement with either diffusional or ionic relaxation mechanisms.

The most consistent explanation of these experimental facts is provided by the model of electron capture. In it, the limiting stage of the relaxation process is assumed to be the capture of carriers on the surface states, which are separated from the bulk by potential barriers and therefore have small capture cross sections c_p and c_n. MORRISON [3.51] was one of the first to consider slow relaxation on a uniform semiconductor surface as a result of thermal overshooting of free charge carriers across the potential barrier in the oxide. However, strictly speaking, his solution of the simplified kinetic equation is only valid in the case of small perturbations, and cannot account for the nonexponentiality of experimental curves under these conditions. In this connection KINGSTON and McWHORTER [3.74] suggested a model of a heterogeneous surface, also restricting themselves to the region of small perturbations and assuming the tunnel mechanism of charge exchange between semiconductor and oxide.

The most general solution of the kinetic equation describing capture of electrons and holes on slow states (without specifying the mechanism of charge transfer, be it either tunneling or surmounting the potential barrier) was obtained by KOZLOV and NOVOTOTSKY-VLASOV [3.77]. They analyzed the kinetics under the assumption that the relaxation is determined by traps of only one kind which exchange charge carriers with a single allowed energy band of the semiconductor. For a uniform surface, the general solution of the kinetic equation is represented by a function of the form

$$\Delta\sigma_s = \Delta\sigma_0 \frac{\exp(-t/\tau_1)}{\{1 + \alpha\Delta\sigma_0[1 - \exp(-t/\tau_1)]\}} \quad , \tag{3.9}$$

where α and τ_1 are constants, the choice of which depends on the initial concentration of electrons on the surface, their excess in the SCR, and the parameters of the states. Analysis of (3.9) reveals that with surface potential ΔY_s varying within a sufficiently range of values, the relaxation curves $\Delta\sigma_s(t)$ lie between exponents having time constants $1/2\tau_1$ (accumulation layer) and $2\tau_1$ (depletion layer).

To describe a nonuniform surface, the distribution function of surface patches over relaxation times proposed in [3.74] was used: $q(\tau_s) \sim 1/\tau_s$. It was also assumed that the characteristic times fall within the interval $\tau_{min} < \tau_s < \tau_{max}$. In the case of small perturbations ($|\Delta Y_s| \ll 1$), the function $\Delta\sigma_s(t)$ for a nonuniform surface has the form

$$\Delta\sigma_s(t) \simeq \Delta\sigma_0[-E_i(-t/\tau_{max})] \quad , \tag{3.10}$$

where $E_i(-t/\tau_{max})$ is the exponential integral function.

Numerical calculations of this function demonstrated that in the interval $10^{-4} < t / \tau_{max} < 5 \cdot 10^{-2}$, (3.10) is well approximated by Koc's empirical equation (3.7) with $a = 0.3$, and with $a = 0.6$ in the interval $5 \cdot 10^{-2} < t / \tau_{max} < 5 \cdot 10^{-1}$. Thus Koc's empirical law acquired theoretical meaning in the investigation of electron capture on a nonuniform surface.

The distribution over τ_s results from the distribution of barriers, separating the slow states from the bulk, over height W (in the case when the mechanism of surmounting the barrier is the prevailing one) or over thickness d (in the case when charge carriers tunnel through the barrier). KOZLOV and LEVSHIN [3.64,78] assumed the barriers to be distributed according to a natural (Gauss's)law. Then the function $g(\tau_s)$ has the form

$$g(\tau_s) = \frac{1}{\Delta\sqrt{2\pi}} \exp\left(- \frac{\ln(\tau_s / \bar{\tau}_s)}{2\Delta^2}\right) \frac{1}{\tau_s} \quad , \qquad (3.11)$$

where $\bar{\tau}_s = \tau_0 \exp(\bar{u})$, $\tau_s = \tau_0 \exp(u)$, Δ is the mean-square deviation of u, and $\bar{u}$ is the mathematical expectation of u.

If the charge carriers are transferred by the over-the-barrier mechanism, then $u = W / kT$; for the tunnel mechanism $u = d\sqrt{8m^*W/h}$, where m^* is the effective mass. Analysis shows that the greater Δ is, the wider is the interval of τ_s values in which the kinetics of relaxation can be approximated by Koc's functions (3.7). The experimentally observed variations of τ_s can be attributed to variations in the barrier's heights within 0.08 to 0.1 eV or widths within 5 to 7 Å (which is confirmed by experiment).

The majority of researchers describe the process of slow relaxation with just one parameter: relaxation time τ_s. Such a description cannot be considered complete. As we emphasized at the beginning of this section, the equilibrium field effect is hard to measure, since the duration of the relaxation processes (hours) makes it impossible to maintain steady-state conditions on the surface while taking down the capture curve. Attempts to measure the field effect in comparatively short time intervals (see, e.g., [3.10]) cannot be considered reasonable, since such measurements are taken at definitely nonequilibrium conditions. In [3.73], the above-discussed model of electron capture [3.77] was used as the basis for a technique of determining the effective parameters of slow states. The kinetics of slow relaxation was measured at various temperatures with a transverse field being switched on and off.

According to theory [3.73,77] at small deviations from equilibrium ($|\Delta Y_s| < 1$), relaxation obeys the same law after application of a transverse electric field as it does after removal of the field, but the direction of the relaxation is towards the new equilibrium value σ_1 (Fig.3.13a). If the concentration of slow states is large (e.g., in a humid environment), the charge Q_{ind}, induced by the external field, passes almost completely into these states. In this case the relaxation curves with respect to the initial conductivity σ_0 are identical after switching the field both on and off. If the concentration of slow states is low, the relaxation curves $\Delta\sigma'(t) = \sigma - \sigma_1$ (the field is on) and $\Delta\sigma''(t) = \sigma - \sigma_0$ (the field is off) will be identical (Fig.3.13a). It is difficult to determine the value σ_1 directly, because of the long duration of the relaxation process. However, the value of σ_1 can be estimated accurately enough by using the identicalness of dependences $\Delta\sigma'(t)$ and $\Delta\sigma''(t)$, which are represented by parallel straight lines 2 and 3 in coordinates $\lg\Delta\sigma$, $t^{0.3}$ in Fig.3.13b. The value of σ_1 will correspond to a new equilibrium potential Y_{s1}, and to a new charge in fast (Q_{sf}) and slow (Q_{ss}) states and in the SCR (Q_{sc}). The change in potential $Y_s = Y_{s1} - Y_{s0}$, as well as the additional charge ΔQ_{ss} stored in slow states can easily be determined at the new equilibrium values of σ_{s1} and Y_{s1} from data obtained from the field effect using an ac signal (Sect.2.3): $Q_{ss} = Q_{ind} - \Delta(Q_{sf} + Q_{sc})$.

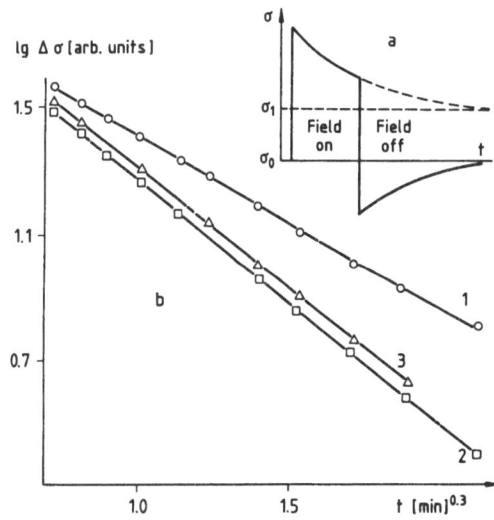

Fig.3.13a,b. The kinetics of slow relaxation after a transverse electric field is switched on and off (a) and in Koc's coordinates (b); σ_0 is the initial conductivity of the specimen; σ_1 is the steady-state conductivity in the presence of the electric field. (1) Relaxation towards σ_0 after application of the field, (2) the same after removal of the field, (3) relaxation towards a new equilibrium state

Let the donor-type states[6] be responsible for the slow relaxation. They can be characterized by a certain effective level $\varepsilon_t = E_t - E_i / kT$, where $E_i = q\psi_0$

---

[6] The type of a slow state of adsorptive origin is determined by the sign of charge buildup on the surface when adsorption takes place.

is the middle of the forbidden band (Fig.2.1), and by concentration N_t. The number of charged states is $N_t^* = \Delta Q_{ss} / q$. The difference between the density of charged states after applying the field (N_{t1}^*) and in the absence of a field (N_{t0}^*) is

$$\Delta N_t^* = N_{t1}^* - N_{t0}^* = \frac{N_{t0}^*[\exp(-Y_s) - 1]}{1 + \exp(\varepsilon_t - Y_{s1} + \ln \lambda)} . \qquad (3.12)$$

Since N_{t0}^* is defined via ε_t and N_t according to Fermi statistics, similarly to (3.1), information about ε_t and N_t can be obtained from measurements of the temperature dependence of the surface charge buildup.

If the surface charge increases with an increase in temperature, the effective energy level of a donor-type slow state lies below the Fermi level. Then (3.12) is simplified

$$N_{t0}^* = \frac{\Delta Q_{ss}}{q[\exp(-\Delta Y_s) - 1]} . \qquad (3.13)$$

If the temperature dependence is of the inverse type, then the corresponding effective level (unoccupied) lies above the Fermi level. Then for the concentration of uncharged states one can write down the expression

$$N_{t0} = \frac{\Delta Q_{ss}}{q[\exp(\Delta Y_s) - 1]} . \qquad (3.14)$$

Having determined N_{t0} and N_{t0}^* at two different temperatures according to (3.13,14), one can calculate the effective parameters of slow states ε_t and N_t. An example is given in Fig.3.14a which shows the temperature dependence of the concentration of charged slow states which emerged when ammonia was adsorbed on a germanium surface. Part b of the figure depicts the energy scheme of a surface state.

A method of estimating the effective cross sections of capture of charge carriers on slow states is proposed in [3.73]. To do this, one must compare the experimental relaxation curves with the theoretical ones [3.77]. With a uniform surface one can immediately obtain the values of c_p and c_n. If the surface is nonuniform, one can estimate the minimal capture cross sections $c_{p\ min}$ and $c_{n\ min}$. For instance, for germanium vacuumized at 500 K, the estimated minimum of c_n was $10^{-26} - 10^{-27}$ cm^2, or almost ten orders of magnitude lower than the corresponding value for fast states (Sect.3.1).

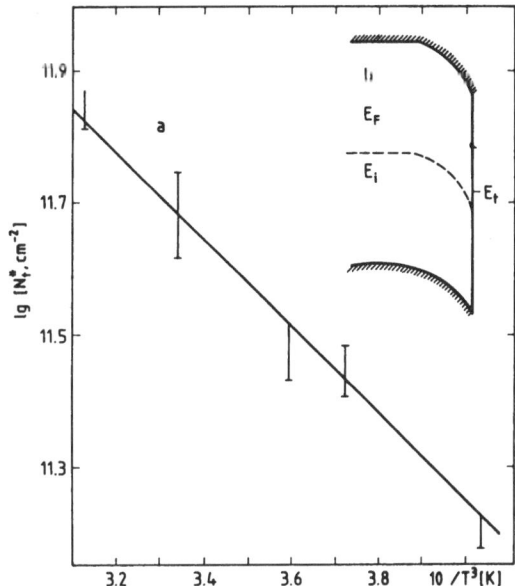

Fig.3.14a,b. The concentration of charged slow states as a function of temperature for the system $Ge_{500}-NH_3$ ($p = 1.7$ torr) (**a**); surface band scheme (**b**)

Many researchers attribute the activation nature of capture on slow states to the existence of a buffer layer which separates these states from the bulk. As indicated above, this is not confirmed by experiment. Slow states, situated in a thin (1 - 1.5 nm) subsurface layer, exchange charges with the allowed bands in the semiconductor mainly via the tunnel mechanism. The probability of tunneling is high enough not to limit the process of capture. Therefore, out of the two phenomenologically introduced distributions of parameters of slow relaxation on a nonuniform surface, see (3.11), the one which is realized is distribution over the height of the barrier W (the mechanism of surmounting the barrier). The considerable dependence of τ_s and ΔE_τ on the structure of the adsorbed molecules indicates that small potential barriers (responsible for a low temperature dependence of relaxation) are not due to conceptual buffer layers, but are connected with the act of capture of charge carriers on the slow state. We shall discuss this in greater detail in Chap.8.

It should be noted that the overwhelming majority of investigations of slow relaxation have been carried out in a comparatively small time interval. By stretching this interval to $5 \cdot 10^4 \leqslant t \leqslant 10^5$ s it was possible to determine the distribution of slow states over the height W of potential barriers, which owe their existence to the specific features of interaction between charge carriers and these states [3.64,78]. The experimental curves of relaxation of slow states turned out to be best approximated by Gaussian distribution.

The concentration of slow states on real semiconductor surfaces varies depending on the conditions of pretreatment, from 10^{11} to 10^{12} cm^{-2} for germanium and silicon, 10^{12} to 10^{13} cm^{-2} for PbS, and is about 10^{13} cm^{-2} for ZnO. The mean distance between these states is thus smaller than the Debye shielding length [3.79]. Therefore their Coulomb fields overlap, just as those of fast states do. All concepts of the theory of disordered systems (Sect.3.2) are entirely applicable to slow states. Their energy spectrum can be described by a single effective level [3.49], and exactly this meaning should be attributed to effective parameters ε_t, c_n, and c_p, obtained from the kinetic curves.

After adsorption the density of slow states is often greater than 10^{14} cm^{-2}. At such concentrations, exchange interactions between electrons in adjacent states are theoretically possible. However, this cannot result in the creation of a surface band (Sect.3.2), since these states are randomly scattered over the surface. All attempts to detect the conduction which would take place in the surface band have so far proved futile, which can be taken as an indirect indication of the validity of our concept of the surface as a disordered system.

So far we have been speaking about systems of states which manifest themselves in slow relaxation of a system which has been driven out of equilibrium by the application (or removal) of transverse electric fields. Such states, however, are not unique. Investigations of photoemission of electrons and holes from semiconductors into oxide layers demonstrate that photo-injected charge carriers are captured on traps in the oxide which originate from numerous defects in the oxide phase. The exchange of charges between the allowed bands in the semiconductor and the traps in the oxide occurs mainly via the over-the-barrier mechanism. At room temperature the charge in these states can be conserved for hours or days. States of this kind are therefore "superslow" (Fig.3.11b) and seem to be responsible for the long tails of the relaxation curve in the field effect. Since these states are connected with the structure of the oxide, they will be discussed in greater detail in Chap.4.

In conclusion, let us emphasize once again that the energy model of the real surface presented in Fig.3.11b must be considered the most reliable one. According to this model, the energy spectrum of semiconductor-oxide heterotransition includes four groups of surface states:

1) fast states, which are in direct contact with the semiconductor;

2) recombination centers;

3) slow states, which penetrate the subsurface oxide layer to a depth of about 1 nm and interact with the semiconductor primarily via the tunnel mechanism;

4) superslow states in the oxide, which exchange charge carriers with the semiconductor mainly via the over-the-barrier mechanism. Naturally, owing to the heterogeneity of the transition, there are no sharp boundaries between regions occupied by states of various kinds.

3.4 The Energy Spectrum of an Atomically Clean Surface

The atomically clean surface is usually considered to be the simplest and most convenient model for investigating the nature and properties of surface electron states. This is true as far as the theoretical aspects of the problem are concerned. The experimenter, however, will reject this opinion. Owing to the high density of surface states, the main technique of investigating the energy spectrum of the surface, the field-effect technique, can provide only scant information about intrinsic states on atomically clean semiconductor surfaces. So far we have only a very general qualitative notion about the character of the spectrum of these states, which cannot match the knowledge we have obtained from experiments with real surfaces of germanium and silicon. For this reason, we will discuss the properties of atomically clean semiconductor surfaces at the end of Chap.3 which deals with the energy spectrum of surface states.

The theoretical techniques of calculating the parameters of intrinsic surface states have been exhaustively treated [3.80-85] and will therefore not be considered here. Instead, we shall restrict ourselves to the physical aspects of the problem. As already emphasized in Sect.2.1, the formation of a surface gives rise to two possible sets of intrinsic states: Shockley states and Tamm states. These states result from different causes.

The main cause of Tamm states is distortion of the Coulomb potential in the elementary surface cell as compared with the potential in the bulk (Fig.3.15a). In principle, Tamm states can appear on any crystalline surface. Shockley states are associated with finite crystals, where there is interaction between states corresponding to adjacent bands. Let us elucidate this point, using the Shockley model of a unidimensional crystal restricted on two sides. When atoms forming a chain are brought from infinity to a certain finite distance, bands are created from terms of isolated atoms (Fig.3.15b). When the distance r_0 between atoms becomes small enough, they develop hybridized chemical bonds. This corresponds to an overlapping of bands (Fig.3.15b). The bonding states from the valence band; the repulsion states form the conduction band [3.86]. Such a situation is typical of covalent crystals (e.g., germanium or silicon).

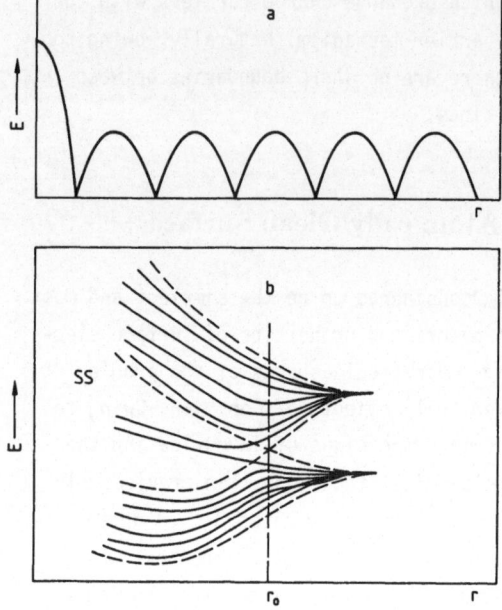

Fig.3.15a,b. The genesis of Tamm (**a**) and Shockley (**b**) levels. E is energy, r is distance, and SS are surface states

Because the chain of atoms has limits, two local levels in the forbidden band are split off from both bands (Shockley levels SS in Fig.3.15b).

As clearly demonstrated by TOMASHEK and KOUTECKY [3.86], the appearance of Shockley states is due to the rupture of a strong localized bond and the creation of an unsaturated valence. The energy levels corresponding to unpaired hybridized orbitals lie in the forbidden band. The location of these levels is very sensitive to any change of potential in the ultimate crystalline cell. A change in potential can drive Shockley levels into the allowed bands and thus destroy them as localized states. For this reason Shockley states, in contrast to Tamm states, exist only at small perturbations of the Coulomb potential.

This leads, even in the case of an ideal surface, to certain ambiguities. The potential in the outermost crystalline cells will depend not only on the nature of the crystal, also on the coordinates of the plane of cross section of the crystal lattice. For example, LIPPMAN [3.87] demonstrated that with a unidimensional lattice, a half-cell shift of the crystal limit under certain conditions can convert Tamm states into Shockley states, and vice versa. External electric fields can also exert a strong influence on the position of surface levels in the forbidden band, and give rise to Tamm levels or eliminate them [3.88]. A qualitative idea of the predominance of either Tamm or Shockley levels on the surface can be obtained in certain cases by comparing thc contributions which the Coulomb and the exchange integrals make to the energy of the chemical bond in the surface cells.

For an ideal three-dimensional crystal, the surface concentration of Tamm and Shockley states is approximately equal to the concentration of atoms on the surface (about 10^{15} cm^{-2}). At such a density (in the case of an ideal lattice), owing to the periodicity of the crystal potential, the corresponding levels are smeared into the surface band.

KOUTECKY [3.89] and TOMASHEK [3.90], using the one-electron approximation within the framework of the MO LCAO technique (Hartree-Fock approximation), evaluated the criteria for the existence of Tamm and Shockley levels on an ideal crystalline surface with a diamond lattice. The surface resulting from the dissection of the elementary cell by the <111> plane behaves like a polyradical, and the surface atoms retain sp^3 hybridization. The resultant band of Shockley states lies within the forbidden zone and is half occupied. If the interaction between sp-orbitals is sufficiently strong, the surface band can partially overlap with the bulk bands. Dissection by the <100> plane results in the rehybridization of two ruptured bonds. The spectrum of Shockley states then consists of a fully occupied band in the forbidden zone and an empty degenerate band overlapping with the conduction band. Similar calculations were carried out for cleavage surfaces of graphite crystal, where the band of surface Shockley states also exists [3.89].

In [3.90], the correlation of electrons in the surface states was taken into account. In this case the half-occupied band of Shockley states on the <111> surface actually consists of two subbands containing electrons with oppositely directed spins. The separation between the subbands depends on the strength of the Coulomb interaction between electrons. At low temperatures, when band-to-band transitions do not take place, the surface should exhibit ferromagnetic properties, due to the presence of bands with two antiparallel spin systems [3.90].

The choice of criteria for the existence of Tamm and Shockley states on the surfaces of heteropolar crystals is much more complicated. Calculations of the energy spectrum of the surface of sphalerite-type crystals were carried out in [3.86,91]. Here the appearance of either Tamm and Shockley states is, to a large extent, determined by the difference in the electronegativities of atoms emerging onto the surface. It should be noted that the appearance of Shockley states excludes the appearance of Tamm states, and vice versa. DAVISON and LEVINE [3.80] in considering the strong influence which the potential in the surface cells has on the location of Tamm and Shockley levels, chose to analyze the ionic states introduced by LEVINE and MARK [3.92] as a result of their investigation of the behavior of Madelung potential on the surface of an ionic crystal. The ionic states lie in pairs in the forbidden band of the crystal, a few electron volts above (or below) the corresponding limits of the allowed bands. A detailed analysis of the energy spectrum of surfaces of $A^{III}B^V$ compounds can be found in [3.84,93].

The literature abounds in theoretical calculations of the energy spectra of surface states for both ideal and reconstructed semiconductor surfaces. Various methods have been used: the strong bond approximation, the resolvent (Green's functions) technique, the method of self-consistent potential, and the method of Mathieu potential [3.84,85]. Reconstruction of the surface not only changes the entire energy spectrum, but also gives rise to new localized states. Due to ambiguities in the interpretation of ELEED data [3.30], theoretical calculations of the energy spectrum of reconstructed surfaces are as yet of a qualitative nature, having the character of a model [3.94,95]. As will be shown later, the presence of domains, steps, and other structural flaws on an atomically clean surface make it more natural to explain a number of physical properties of such surfaces from the standpoint of the theory of disordered semiconductors.

Now let us consider the experimental data on the energy spectrum of intrinsic states. We shall begin with germanium. Numerous investigations of the field effect and surface conductivity [3.1,9,10,96-98] conclusively show that an atomically clean germanium surface always contains, independently of the bulk parameters, a large negative charge (10^{12} electron $\cdot$ cm^{-2}), owing to which the bands are bent upwards [$(12-15)kT/q$], and conductivity in the SCR is p type. GOBELI and ALLEN [3.99] who investigated photoemission of electrons from a Ge cleavage surface and the work function, were among the first to suggest that the energy spectrum of Ge surface states has a two-band structure. But the study of photoemission only yields information on the occupied states. To obtain information on the whole spectrum, the UPS technique is usually combined with ELS, as well as with surface photovoltaic spectroscopy (SPV) [3.100]. The data on the parameters of Ge surface supplied by these techniques [3.85,100, 101] give a qualitative picture of the energy scheme of surface states as an occupied band submerged into the valence band, and an overlapping unoccupied band (Fig.3.16a).

As opposed to the case of germanium, the surface conductivity of silicon is rather low. NESTERENKO and SNITKO [3.102] analyzed the temperature dependence of field-effect kinetics and demonstrated that the value of surface potential Y_s is about 6 kT/q, and that the surface contains a small positive charge ($\sim 10^9$ holes $\cdot$ cm^{-2}). Data on photoemission from Si cleavage surfaces

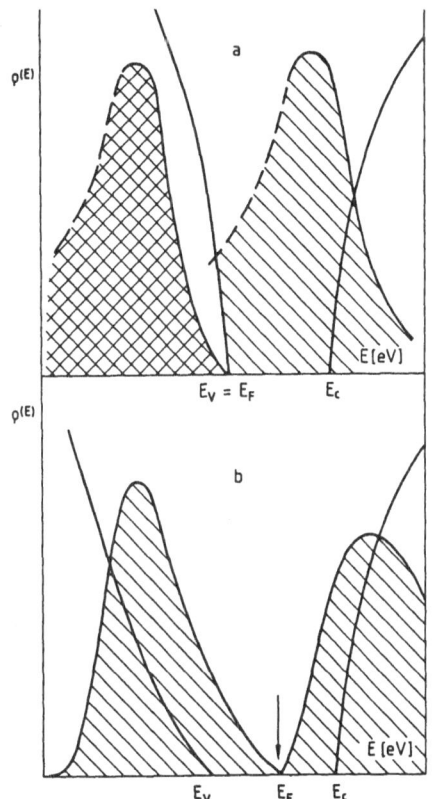

Fig.3.16a,b. Models of the energy spectrum of surface states on atomically clean surfaces of (**a**) Ge (2 × 1) and (**b**) Si (2 × 1) [3.85]

[3.103] and measurements of field-effect kinetics on surfaces cleansed by ion bombardment [3.102] justify the conclusion that the energy spectrum of Si surfaces can be approximated either by two bands (Fig.3.16b) or two discrete effective levels. Within the limits of the discrete spectrum the density of states reaches $10^{14} cm^{-2}$, and the donor-type and acceptor-type states are 0.1 eV away from the Fermi level [3.102]. Owing to the high density of states, the Fermi level is rigidly fixed on the surface. With both germanium and silicon, experiments with the field effect can only provide information on states lying within several kT of the Fermi level. Much more valuable information on the two-band structure of the silicon surface was obtained in subsequent investigations using combined UPS and ELS technique [3.85,104]. Data on the relative location of surface bands with respect to the bulk bands was also confirmed by SPV spectroscopy dates [3.100]. Valuable information on the symmetry of local fields of surface states was obtained with angle-resolved UPS (AR UPS) [3.105] and with a UPS technique employing a synchrotron source [3.106]. These techniques were used to investigate the energy spectrum of $A^{III}B^{V}$ and $A^{II}B^{VI}$ compounds [3.93,100,104].

HENZLER [3.107-109] conducted combined investigations of variations in the structural and electric parameters of atomically clean germanium surfaces during short-term heat treatment of specimens in an ultra-high vacuum. Reconstruction of the <111> surface from (2 × 1) into (2 × 8) [Ref.3.30,Chap.3] was found to induce considerable changes in surface conductivity $\Delta\sigma - \Delta\sigma_0$ and field-effect mobility μ_{fe} (Fig.3.17). HENZLER interpreted the data from the standpoint of the two-band theory, considering two extreme cases: continuous distribution of states, and two discrete levels (acceptor and donor). After reconstruction of the surface, the total concentration of states (in the discrete-levels model) falls from $6 \cdot 10^{13}$ to $3 \cdot 10^{13}$ cm^{-2}. Both groups of states are shifted upwards from the top of the valence band, and the distance between the groups of acceptor and donor states decreases. Similar variations in the electrophysical

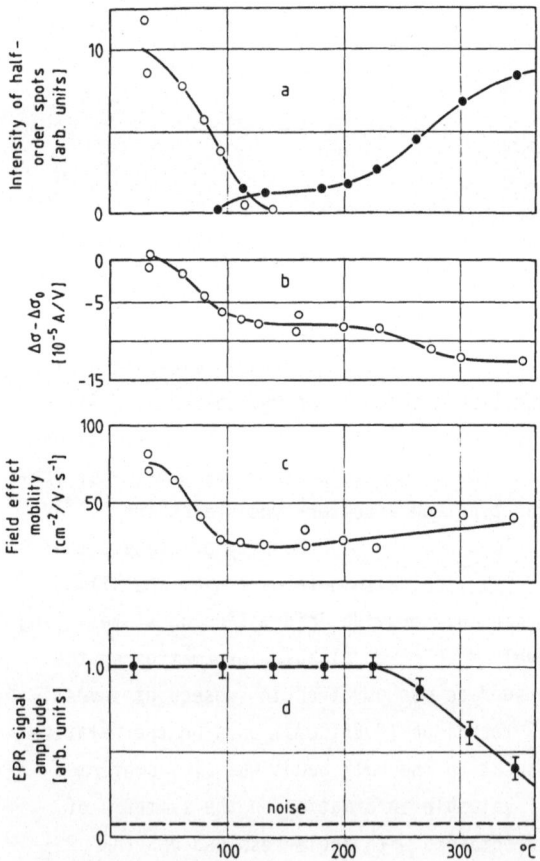

Fig.3.17a-d. Variations in the intensity of diffraction maxima (**a**), surface conductivity (**b**), field-effect mobility (**c**) [3.108], and strength of EPR signal (**d**) [3.113] for a atomically clean germanium surfaces during isochronous annealment of specimen

parameters of atomically clean surfaces were later also observed for the transition of silicon from <111> Si (2 × 1) to <111> Si (7 × 7) at 670 K [3.110, 111]. We see that transitions from one superstructure to another strongly influence the electrophysical parameters of the surface. However, our knowledge of the corresponding changes in the energy spectrum of surface states is still rudimentary.

EPR investigations of atomically clean surfaces are very promising in the study of intrinsic surface states. The most important experimental results obtained with this technique are discussed in [3.30]. Let us now analyze the relationship between the paramagnetic centers observed on the surface, and the Tamm and Shockley types of intrinsic surface states. From the very definition of these states it is clear that what will be observed as paramagnetic centers in the EPR spectra are unoccupied Shockley states (free valences) and Tamm states which have captured one electron. For homopolar lattices of germanium and silicon, the paramagnetic centers A observed in the EPR spectra are most likely empty Shockley states $\geqq Si \rightarrow$ [3.30]. This is in conformity with the theoretical models of the surfaces of these semiconductors [3.89,90]. Two-electron states, i.e., Shockley states which have captured one electron, do not show up in the EPR spectra (centers B, $\geqq Si \updownarrow$ [3.30]). Note that the existence of two-electron states also follows from a consistent analysis of the energy spectrum of a disordered system [3.44]. Since the two-electron traps do not contribute to the EPR signal, we have no direct means of measuring their concentration. There can be no doubt, however, of their existence [Ref.3.30, Sect.3.1.4].

The capture of an electron by a Shockley trap (i.e., the A → B transition) results in a buildup of negative charge on the surface. In the case of germanium the predominant states are of the B type, while for silicon, A-type centers (unoccupied traps) prevail [3.30]. This conclusion is confirmed by the above-mentioned experimental data obtained via electrophysical measurements: the negative charging of the germanium surface is considerably greater than that in the case of silicon. The dependence of the concentration of spin centers upon the temperature of crushing [3.30] attests to the activated character of the A → B transition. For instance, in the case of germanium, crushing at 300 K created a surface that mainly had occupied traps (centers of type B). For silicon, which has a wider forbidden band and a lower concentration of free charge carriers, the filling of Shockley traps occurred at crushing temperatures of about 700 K. As pointed out in [3.30], with narrow-band semiconductors, unoccupied Shockley states were not observed at all in the EPR spectra. This mechanism of trap-filling resembles to some extent the mechanism of occupation of the lower subband of surface Shockley states with parallel spins in the model of the energy spectrum proposed by TOMASHEK [3.90].

A study of EPR signals from atomically clean silicon surfaces carried out by CHUNG [3.112] demonstrated that the concentration of spin centers n_s is influenced very little by bulk doping of Si. The transition from degenerate n type to degenerate p type, when the Fermi level in the bulk travels over the whole width of the forbidden band, caused almost to changes in the value of n_s. Since the potential of an atomically clean silicon surface is about equal to the potential of flat bands, the locations of the Fermi level with respect to the middle of the band differ little on the surface and in the bulk. The absence of any considerable recharging of surface traps in these experiments, as well as data indicating the existence of a high activation barrier, that separates the traps from the allowed bands [3.113,114] should be taken as evidence that at least some of the traps detected by EPR fall within the allowed bands of the semiconductor. The barrier separating these states from the allowed bands is probably connected with the different symmetry of the wave functions. It has been firmly established that localized or resonance states may originate in the allowed bands. If so, then the traps observed by EPR are not the same as the traps located near the top of silicon's valence band and detected by the energy-loss spectra of incident electrons [3.115]; the latter are attributed to dangling silicon bonds on the surface.

In a very interesting work, ASSMAN and MÜNCH [3.116] reported optical transitions between dangling bonds, which they detected in the spectra of photovoltaic effect and photoconductivity (two absorption peaks at 0.23 and 0.5 eV). The extremely slow kinetics of these states (about 10 s) agrees well with our data on isochronous annealing of spin centers on the surface, which exhibited a rather high (about 0.8 eV) activation energy [3.113,114]. We attributed the decrease in the concentration of dangling paramagnetic centers to the occupation of empty resonance states of type A, which converts them into two-electron states of type B.

HANEMAN [3.117] drew a connection between EPR signals from atomically clean silicon surfaces and delocalized electron states which appear on the surface as a result of strong overlapping between the wave functions of unpaired electrons of adjacent atoms. However, as demonstrated in [3.113], the overlapping integral is quite small. It was pointed out in [3.112] that HANEMAN [3.117] overestimated the energy of by nearly two orders of magnitude. Measurements of the temperature dependence of the parameters of the EPR signal also point to the absence of any strong delocalization of electrons in paramagnetic centers. For instance, the width of the FPR signal from Ge did not change in the temperature range from 77 to 220 K.

Should the EPR signals be connected only with delocalized electrons in the surface bands, then the parameters of the signals ought to depend strongly upon reconstruction of the surface during heat treatment of the specimens. As follows from Fig.3.17, the concentration of spin centers undergoes no change in the region of transition from superstructure <111> Ge(2×1) to <111> Ge (2×8), where there are noticeable variations in the electrophysical parameters $\Delta\sigma$ and μ_{fe}. A considerable drop in concentration is detected at much higher temperatures (550 K for Ge and 700 K for Si). These facts are not in accord with the viewpoint of HANEMAN [3.117], who assumed equivalence of spin centers and the surface electron states which govern the electrophysical parameters of the surface. We suggested in [3.113,114] that the energy spectrum of an atomically clean surface includes two groups of states: sufficiently localized Shockley states that are responsible for the EPR signals, and partially delocalized states in the surface bands, as well as fluctuation states responsible for the electrophysical parameters of the surface. This assumption is based on the concept of partial disorderliness of the atomically clean surface and the presence of a rather large number of fluctuation levels in the energy spectrum.

Indeed, cleavage surfaces of Si and Ge exhibit a high concentration of steps and other defects (1 to 20% of the number of surface atoms [3.109]).The presence of steps markedly affects the surface potential, the structure of the energy spectrum, and the dipole component of the work function [3.118,119]. According to ellipsometric data [3.120], surfaces cleansed by ion bombardment, a thin amorphous layer is present whose optical parameters are close to those of amorphous silicon. As we pointed out in [3.30], the reconstructed surface very probably has a domain structure. The boundaries between domains may serve as an additional source of fluctuating fields [3.113].

The assumption about the partial disorderliness of an atomically clean surface [3.113] does not contradict the data obtained with the aid of electron spectroscopy. Investigations of photoemission on <111> Si surfaces carried out in [3.121] revealed that the spectrum of surface states is quasi-continuous and that the density of states increases towards the edge of the valence band. A spectrum of this kind is typical of disordered systems (Sect.3.2). In their most recent work on EPR from silicon surfaces, LEMKE and HANEMAN [3.122] admitted also the possibility of Anderson localization, which is peculiar to disordered systems.

The question of whether there is conduction over, the bands of surface states remains open. Experimental work has so far failed to provide convincing data on conductivity in the surface band [3.123].

As demonstrated in [3.113], the overlapping integral for surface atoms is more than 20 times smaller than the integral of overlapping of atomic orbitals involved in the creation of chemical bonds. Therefore, even is surface bands do appear, they must be rather narrow [3.124-126]. The mobility of charge carriers in narrow bands must be quite low, owing to a large effective mass and to scattering on surface phonons and flaws. On the other hand, if the surface is considered a disordered system, then a band of delocalized states cannot be formed even at a high density of surface states, which, as pointed out in Sect.3.2, is a consequence of the random nature of fluctuating fields.

Adsorption of oxygen on Ge and Si, even when the coverage is far less than a monolayer, results in the elimination of the superstructure and causes a sharp change in surface conductivity σ_s and field-effect mobility μ_{fe} [3.107, 108]. According to [3.127], when oxygen is adsorbed the band of occupied surface states must either disappear or submerge into the valence band. As opposed to the electrophysical parameters, the parameters of the EPR signal remain unchanged for both germanium and silicon. It is not yet clear whether this is due to "geometric" inaccessibility of some of the spin centers (microcracks) [3.122], or to "upturned spins" [3.113]. Whatever the reason, the EPR data so far are in poor agreement with the variations which occur in the energy spectrum upon adsorption [3.114].

The most prominent feature is the highly dissimilar behavior of atomically clean surfaces of Ge and Si towards adsorption of oxygen. As we demonstrated in [3.30], "superdry" oxygen is not adsorbed on Ge surfaces at all, while adsorption of O_2 and Si is accompanied by the release of a large quantity of adsorption heat which is close to the energy of formation of bonds of type $\gtrsim Si = O$. We attribute this to the adsorption of O_2 on occupied Shockley traps [3.30]. The great dissimilarity is most probably due to the different properties of the atomic orbitals of Ge and Si. It is known that 4s electrons of Ge show a reduced participation in chemical bonding. The valence 4 sp electrons form a so-called Sadgwig's inert pair [3.128]. This is reflected in the band structure of germanium. NISHIDA [3.129] demonstrated that the main contribution to dangling bonds on relaxed <111> Ge surfaces comes from s-like back antibonding orbitals while on relaxed <111> Si surfaces it is the p_z orbitals.

4. Electron Processes
in Semiconductor Adsorbents and Catalysts

The main objects of investigation in the study of adsorption and catalysis on nonmetallic surfaces are semiconductor with a rather wide forbidden band, primarily metallic oxides and solid solutions having such oxides as their base. The width E_g of the forbidden band for many useful catalysts is of the order of several electron volts, and under ordinary conditions they are poor conductors of electricity. In the majority of investigations the electron processes on the surfaces of such substances are treated (without any additional assumptions) from the standpoint of the phenomenological theory discussed in Sect.2.2, which was developed primarily for the classical semiconductors, germanium, and silicon.

4.1 Electron Processes in Wide-Band Semiconductors (Dielectrics)

In the formal sense, semiconductors differ from insulators in the width of the energy gap which separates the conduction band from the valence band. At the same temperature, the concentration of free charge carriers will be much higher in a semiconductor's band than in an insulator. Indeed, the probability of filling a state is given by the Boltzmann factor $A \exp(-E_g / kT)$. At room temperature the value of kT is about 0.025 eV. In semiconductors ($E_g \sim 0.1 - 1$ eV) the Boltzmann factor is about 10^{-10}, while in insulators it is about 10^{-30} - 10^{-40}. Since the concentration of states in the conduction band is about 10^{22} cm^{-3}, one can assume that at room temperature the conduction band in an insulator is empty. On the basis of this simplified reasoning it is often assumed that the difference between the electron processes in semiconductors and in dielectrics is just a matter of scale: at high temperatures a dielectric will behave just like a semiconductor at lower temperatures. This is, however, not exactly the case. While the behavior of electrons in the conduction band of both semiconductors and insulators is equally well described by Bloch states, the behavior of a hole in the valence band can differ for different substances.

A number of crystals are better described within the framework of the localized valence states approximation [4.1] than on the basis of band theory.

HARRISON [4.2] proposed that all wide-band semiconductors be classified into two groups, depending on the ionicity of the bonds:

1) substances with low ionicity, which it would be legitimate to consider on the basis of the valence-band concept; this group includes nearly all $A^{III}B^V$ compounds, and

2) substances with a high bond ionicity, which are more rightly considered in terms of localized states. This group consists of all compounds of type $A^{I}B^{VII}$ (e.g., halides), most oxygen-containing compounds, and certain compounds of the $A^{II}B^{VI}$ type. Since in the second group the valence band is mostly characterized by separate atomic bands belonging to the corresponding ions, the valence band is quite narrow. Therefore the effective mass of charge carriers is high, and their mobility is very low. In the extreme case of localized states (realized in van der Waals crystals and in many oxides of transition elements), the band concept loses its meaning. Here the conduction in a crystal takes place through "jumps" of charge carriers from one ionized atom to another. The transition from band-type conduction to jump-type conduction (the so-called Mott transition) is reflected in the temperature dependence of the mobility of charge carriers. In band-type conduction the mobility of electrons falls with an increase in temperature, due to collisions with lattice phonons, while in jump-type conduction the mobility rises with an increase in temperature according to the exponential law $\mu \sim \exp(-E_a / kT)$, E_a being the energy of activation of a jump between the localized states.

Notice yet another peculiar feature of the conductivity of crystals having a high ionicity of bonds. This feature is connected with the interaction of electrons with the optical modes of lattice vibrations. In nonpolar crystals (e.g., germanium and silicon), the linkage of an electron with the lattice field is small and can usually be neglected. An electron in the conduction band is considered a free charge carrier. In ionic crystals, owing to the strong Coulomb interaction between electron and ions, one can no longer neglect the effects of lattice deformation caused by this interaction. With its own field the electron polarizes the lattice, and thus creates the so-called polaron state. In other words, the electron "digs out" its own potential pit. Two situations are then possible. If the displacement of ions is small, and embraces a large number of ions, then the electron, under the influence of the external field, will travel along the crystal, tugging the deformation field behind. The electron carries a polarization "coat" and is called a polaron, a term coined by PEKAR [4.3] who developed a comprehensive theory of the motion of polarons in crystals. The mobility of a polaron will naturally be lower than that of an electron, and its energy will be smaller than the energy of electron excitations. The band of polaron states is below the bottom of the conduction band, and is many times narrower. The temperature dependence of a polaron's mobility can differ from that of an electron on the conduction band.

The second possible situation arises when the displacement of ions caused by an electron is large enough so that the polaron state becomes localized at a lattice site; in other words, the electron is self-captured in some region of the crystal. Such a state is unstable. Owing to the tunnel effect, the wave packet of such an electron can become smeared over the lattice. In this case we can consider its behavior as we would the behavior of a band electron in the polaron band. This process dominates at low temperatures. At high temperatures, owing to intense thermal vibration of the lattice atoms, potential pits are often created which may serve as localization centers for polarons. Most of the time a polaron will be found in the localized state. Upon application of an external field, this polaron will travel in jumps from one localized state to another.

We see that in wide-band semiconductors two entirely different mechanisms of electron transfer are possible: movement of free electrons or polarons over the band, and the jump mechanism. The mobility of charge carriers for each of these mechanisms of conduction has its own temperature dependence. The band mechanism prevails at low temperatures, while at high temperatures the jump mechanism is predominant.

In the general case the conductivity of a crystal also has an ionic component, whose role increases with the specimen's temperature. The ionic conductivity can take place through jumps of interstitial ions (Frenkel defects) or vacancies (Schottky defects) in a crystal. In order to distinguish between the different components of electronic and ionic conductivity, detailed investigations must be made of electroconductivity, magnetic resistance, and the Hall effect, with the applied electric field varying widely. This is not easy, if one considers the low concentration of charge carriers and the low effective mobilities of charge carriers in insulators.

A characteristic feature of the energy spectrum of many wide-band semiconductors is the presence of well-defined levels of exciton states, which are manifested in the absorption spectra at higher temperatures than in classical semiconductors (e.g., germanium). The concept of excitons was first introduced by Frenkel to account for the fact that the absorption of light connected with the formation of electron-hole pairs does not always give rise to photoconduction. Depending on the nature of the crystal, two extreme cases are distinguished:

1) Mott excitons, which arise due to the Coulomb attraction between an electron and a hole belonging to different atoms in the crystal; and

2) Frenkel excitons (low-radius excitons), where the electron and the hole belong to one and the same atom in the lattice at any given instant. The exciton as a whole is an electrically neutral formation; it travels within the crystal without carrying any charge. An exciton is unstable; it may either annihilate by virtue of electron-hole recombination, or break down into an electron and a hole.

4.2 Disordered Wide-Band Semiconductors

So far we have been considering the specific features of electron processes in crystalline wide-band semiconductors. However, for some important applications it is crucial to be able to predict the behavior of electrons and holes in amorphous insulators. Many industrially significant adsorbents and catalysts are amorphous oxide systems. When high-dispersion crystalline catalysts are used, their surface layer, where the electron processes pertinent to catalysis take place, is usually highly destructuralized [4.4]. Amorphous semiconductors are equally important in semiconductor technology: the dielectric layers used in planar integral circuit technology are mostly amorphous ones.

As was shown in Sect.3.2, the band scheme is formally retained when we go from crystalline to amorphous bodies; its parameters, however, acquire statistical meaning. Owing to a blurring of the limits of the allowed bands and the presence of "tails" of the density of states, we must now speak of the limits not of allowed and forbidden bands, but of localized and delocalized states. This kind of limit is often referred to as a Mott limit [4.5]; let us designate it E_c' from the side of the conduction band, and E_v' from the side of the valence band (Fig.4.1). The conductivity at $E > E_c'$ will have the usual drifting character. At $E < E_c'$ the conduction will be due to jumps by electrons between the localized states. It is quite clear that when the boundary E_c' between localized and delocalized states is crossed, there will be a sharp decrease in the mobility of charge carriers μ (Fig.4.1). As estimated by MOTT and DAVIS [4.5], $\mu_{E > E_c'} / \mu_{E < E_c'}$ is about 10^3. In the $E > E_c'$ region the mobility of electrons drops with an increase in temperature roughly as $\mu \sim T^{-1}$, while in the region of jump conduction the mobility increases exponentially with temperature: $\mu \sim \exp(-E_a / kT)$. Additional information on the nature of the conduction can be obtained from measurements of the ac conductivity $\sigma(\omega)$. While the dc ($\omega = 0$)

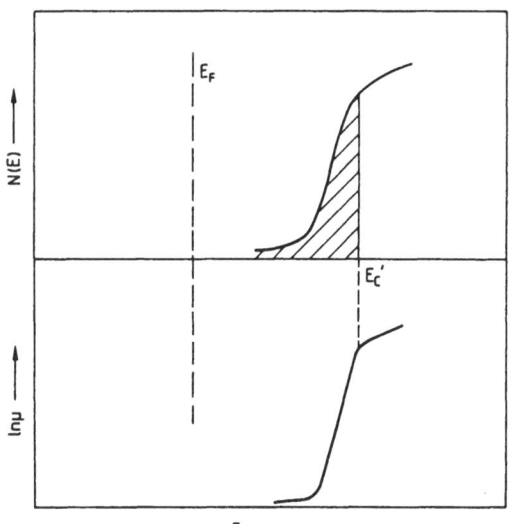

conductivity $\sigma(0)$ tends to zero at $T \rightarrow 0$, this is not the case at sufficiently high frequencies: $\sigma(\omega) \neq 0$ at $T \rightarrow 0$.

For disordered systems, which are characterized by an energy spectrum of Type a or b (Fig.3.10), MOTT and DAVIS [4.5] suggested three distinctive mechanisms of conduction which exhibit different frequency and temperature dependences. For clarity, let us consider the case of hole-type conduction.

1) Conduction over delocalized states. For dc current $\sigma_1(0)$ we have

$$\sigma_1(0) = \sigma_{1,0} \exp[-(E_F - E_v') / kT] \quad . \tag{4.1}$$

The conductivity for ac current $\sigma_1(\omega)$ is practically independent of frequency up to 10^7 or 10^8 Hz.

2) Conduction over localized states having energies E that are a little greater than E_v':

$$\sigma_2(0) = \sigma_{2,0} \exp[-(E_F - E_v' + E_a) / kT] \quad . \tag{4.2}$$

The value of $\sigma_{2,0}$ is 10^2 to 10^4 times lower than $\sigma_{1,0}$ due to the leap in mobility (Fig.4.1) and the reduced density of states. At sufficiently high frequencies $\sigma(\omega) \sim \omega^{0.8}$.

3) Conduction over localized states near the Fermi level E_F. In this case

$$\sigma_3(0) = \sigma_{3,0} \exp[-(E_a' / kT)] \quad , \tag{4.3}$$

where E_a' equals approximately one half of the energy range occupied by the states which lie near the Fermi level. As in Case 2, $\sigma(\omega) \sim \omega^{0.8}$.

Despite the approximate nature of theoretical estimates, measurements of the frequency $\delta(\omega)$, and temperature, $\delta(T)$, dependences of the conductivity yield valuable information on the type of conductivity that prevails under given experimental conditions, and permit qualitative estimates of the concentration of localized states. In amorphous material this concentration can be 10^{19} cm^{-3} or higher.

Considerable difficulties arise when one attempts to determine the sign and mobility of majority carriers in amorphous semiconductors. When dealing with crystals, one can measure the Hall effect and thermal emf, but in high-resistivity amorphous semiconductors the observed Hall effect is very insubstantial, owing to a low mobility of charge carriers, and measuring the Hall emf is very complicated. Because the free parth of charge carriers in amorphous bodies is so small, grave complications are also encountered in interpreting data on thermal emf. Only by comparing the data on electroconductivity, the Hall constants, and thermal emf can one draw conclusions about the mechanism of the kinetic phenomena which take place in these systems. In addition, when dealing with amorphous bodies having a high ionicity of bonds, one must also take polarization effects into account. The concept of polarons and their jumping between localized states is employed more and more often in interpreting the electrophysical properties of noncrystalline materials. Polaronic effects complicate the expressions describing the frequency dependences $\sigma(\omega)$, the Hall constants, and thermal emf.

A striking feature of amorphous semiconductors is their very low sensitivity to doping. An admixture of atoms whose valence differs from that of the semiconductor usually does not cause any significant change in either the type or magnitude of the conductivity. In this respect these systems are quite dissimilar to crystalline semiconductors. There are experimental results which indicate that structural defects in amorphous bodies have more influence on conductivity than impurities.

The theory of disordered semiconductors is currently undergoing a stage of active development. A number of model-type calculations have thrown new light on many peculiarities in the behavior of electrons in disordered systems. Unfortunately, these results, which relate directly to the electronic processes on the surface (which is disordered in the majority of systems of practical interest) have not yet found their way into surface science.

4.3 The Energy Spectrum of Wide-Band Semiconductor Surfaces

Although binary (as well as more complex) semiconductors were the materials
most extensively employed in the early stages of semiconductor technology, we
still know much less about their band structure than we do about that of germanium and silicon. The same can be said about our knowledge of the energy
spectrum of the surface of these compounds. The literature does contain a vast
body of experimental data on the conductivity, work function, thermal emf, and
Hall emf of the polycrystalline powders, films, and ceramics widely used in
catalysis and semiconductor technology, and studies have also been made of the
influence which the adsorption of various molecules has on these parameters,
as well as of the influence of heat treatment. But unfortunately, all these
data, which have been obtained from experiments with polycrystalline substances, give, as will be shown below, very scant, and sometimes even incorrect
information on the energy spectrum and the mechanism of charge-carrier transport in semiconductors.

Only in the past decade has progress been made in this direction. Thanks to
the explosive developments in semiconductor technology acousto- and optoelec-
tronics wide-band semiconductors have found many practical applications. This
has stimulated the development of methods for obtaining flawless monocrystals
and the search for new techniques of investigating electron states in wide-
band semiconductors and dielectrics.

One of the most rapidly developing fields is currently the investigation
of thin dielectric films on semiconductors, laminated MIS (metal-insulator-
semiconductor) and MIM (metal-insulator-metal) structures, which are at the
heart of contemporary microelectronics. Silicon is still the most common base
for integral circuits. For this reason, efforts are concentrated on investigating SiO_2 films and the $Si-SiO_2$ interface [4.6]. A detailed discussion of
the structure and properties of MIS structures would have grown into a separate book, so we shall restrict ourselves here to just a few aspects, analyzing in brief the surface-sensitive effects which occur in the dielectric layers of these laminated structures.

First we should mention that in studies of the optical and emissive properties of the solid, the spectral range investigated has been considerably
broadened. By extending spectral investigations into the region in a vacuum
and employing synchrotron radiation, unique information could be obtained on
the band structure of such dielectrics as SiO_2 and $\gamma-Al_2O_3$, and the position
of the energy levels of defects and other localized electron states in the

bands of these insulators could be determined. For instance, in the case of γ-Al_2O_3 (having a forbidden band $E_g \sim 7.5$ eV) synchrotron radiation spectroscopy made it possible to determine the exciton levels [4.7]. Important advances in the investigation of interfaces were made with the aid of synchrotron spectroscopy by researchers of Stanford Laboratory, to mention only the studies of the Al-GaAs and Si-SiO_2 [4.8,9] interfaces. In recent years the structure and defects of dielectric layers have been intensively studied using X-ray techniques (SSXA and EXAFS) employing monochromatic sources of synchrotron radiation [Ref.4.4,Chap.3]. Insulator-semiconductor structures are now being intensively studied with the modern techniques of electron and ion spectroscopy discussed in [4.4](AES, ELS, XPS, ESCA, UPS, and ISSS). The angle-resolved electron spectroscopy methods (AESAR, UPSAR, ELSAR) are widely used, and nuclear-reaction methods are being applied to the analysis of impurity states in insulator films [4.10].

A study of the location of traps in the energy spectrum of insulators was recently performed with the aid of a variety of thermal activation spectroscopic techniques. The method was based on the observation that charge carriers which are captured in deep traps at low temperatures are released upon heating. At first the concentration of charge carriers increases, and then, having reached a maximum, it begins to decrease, owing to exhaustion of the carriers captured in traps. The charge carriers which are released contribute to conduction (thermally stimulated conductivity or TSC), and in some cases give rise to thermally stimulated luminescence (TSL) [4.11] and thermally stimulated exoelectronic emission (TSE) [4.12,13]. An analysis of the temperature dependence of these effects (where the maximum corresponds to the uppermost concentration of released electrons) can yield valuable information about the activation energy of emptying the deep traps in a wide-band semiconductor and about the concentration of these traps[1]. In some dielectric systems it was possible to observe the emptying of deep traps, which was stimulated by applying strong electric fields to the dielectric [4.14]. The field lowers the barrier that separates the traps from the allowed bands, and thus increases the probability that charge carriers will escape from the traps, owing to the Pool-Frenkel effect [4.15,16]. In another technique commonly employed to study the parameters of traps, photostimulated currents in MIS structures are monitored [4.17]. KREUTZ [4.18] utilized photoconductivity spectra to investigate the energy distribution of traps located in insulator layers adjacent to the semi-

---

[1] Data obtained with the aid of TSC, TSL, and TSE are processed in a manner similar to that adopted in the method of thermally stimulated desorption [4.4].

conductor. Valuable information on the distribution and concentration of traps can be obtained by measuring the contact potential (e.g., with the "vibrating condenser" technique) [4.19], especially in combination with the field-effect technique.

A convenient method of studying traps in thin dielectric layers has been to investigate injection currents in MIM and MIS structures [4.20]. Essentially this consists in measuring the volt-ampere characteristics (VAC) of the current created in an insulator by charge carriers injected by one of the electrodes. The information on the traps inside the insulator is given by the part of the VAC that corresponds to the flow of current bounded by the dielectric's space-charge region (SCRC). While before the injection the equilibrium occupancy of traps in the dielectri is determined by the Fermi level in accordance with (3.2), if a transverse electric field is applied which is not too strong (such a field changes in the parameters of the traps), the balance between the empty and the occupied traps will be shifted because of injected electrons. The space charge thus created will limit the flow of current through the dielectric.

Consider, for example, the experimental VAC of an Al-Al_2O_3-Si system [4.21] with an Al_2O_3 layer about 100 nm thick. As can be seen from Fig.4.2, when the field strength is below 10^4 $V \cdot cm^{-2}$ (corresponding to an applied voltage of 0.1 V), the current is linearly dependent on the voltage, and Ohm's law is valid (Region I of the figure). The same dependence is conserved when the sign

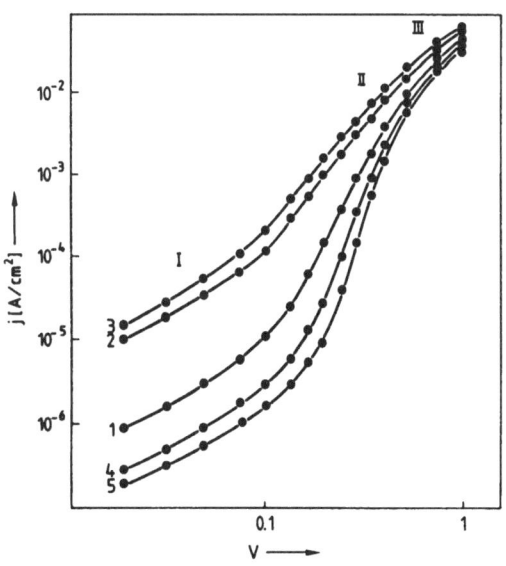

Fig.4.2. Volt-ampere characteristics for the Al-Al_2O_3-Si system after dehydration at 570 K (1) and after adsorption of water at 6 torr (2), NH_3 at 350 torr (3), NO_2 at 1 torr (4), and O_2 at 1 torr (5). Positive potential is applied to metal electrode

of the field is reversed. If the voltage is further increased, the concentration of free electrons exceeds the equilibrium concentration n_0, owing to an injection of electrons from the Si (Region II of Fig.4.2), and the current exhibits the dependence $j \sim V^{k_0+1}$, which is typical for the space-charge region current. According to the theory of SCRC [4.20], a dependence of this kind is characteristic of the exponential distribution of electron traps in the band scheme of an insulator

$$N(\epsilon_t) = \frac{N_t}{kT_t} \exp\left(\frac{E_t - E_c}{kT}\right) \quad , \tag{4.4}$$

where T_t is the characteristic temperature of distribution, N_t is the concentration of levels of discrete traps in the kT_t energy range, and E_c is the energy corresponding to the lower limit of the insulator's conduction band.

The parameter $k_0 = T_t / T$, which enters the function $j(V)$, determines the width of the energy distribution of levels. In Region III of Fig.4.2, with fields stronger than 10^5 V·cm^{-1} (applied voltage V > 1 V), the VAC is described by the Mott-Herni law $j \sim \epsilon \mu_d (V^2 / d^3)$, where μ_d is the drift mobility of charge carriers, and ϵ is the permittivity. The current density here is so great, that the traps have almost no influence on conduction.

As we have already mentioned, the most informative part is Region II of the VAC which corresponds to the SCRC. The value of kT_t estimated for this region (Fig.4.2) is about 0.13 to 0.15 eV, and the concentration of traps [as in (4.4)] is about 10^{15} cm^{-3}. The shape of Region II of the curve corresponds, according to the theory, to a dense distribution of traps around the Fermi level in accordance with the exponential law. This is typical of a disordered system, and amorphous Al_2O_3 is one. By analyzing the temperature dependence of the current in the space-charge region, one can draw conclusions [4.21] on the mechanism of charge-carrier transport pertinent to various temperature ranges, be it tunneling, jumping over the barrier separating the localized states, or drifting over the band of delocalized states.

All of the above-discussed methods of investigating the energy spectrum of the centers of capture of charge carriers deal with bulk traps. The spectrum of surface traps which are of considerably greater concern in adsorption and catalysis, is much more difficult to study. With wide-band semiconductors having a high density of surface states ($\sim 10^{12} - 10^{13}$ cm^{-2}), the vacuum modification of the field-effect technique (Sect.2.3) can no longer be used. Even the highest possible (prebreakdown) transverse fields induce, as a rule, almost negligible changes in surface potential Y_s and surface conductivity σ_s. It is especially difficult to alter the surface potential of a disordered semiconductor.

There is still much controversy [4.22] over whether the field-effect technique can be used to investigate the energy spectrum of amorphous semiconductors such as amorphous silicon. Advances in this direction have been made using the field-effect and voltage-capacitance characteristics techniques combined, and backed by an adequate data processing program [4.23].

The effects of charging the surface can be considerably enhanced if field-effect experiments are conducted in an electrolyte, since the field strength in the Helmholtz layer can be two orders of magnitude greater than in a vacuum, reaching 10^7 V $\cdot$ cm^{-1}. A discussion of investigations of the semiconductor-electrolyte interface would go beyond the scope of this book; the reader can find a detailed treatment [4.24]. HAUFFE and SCHMIDT [4.25], and MORRISON [4.26] proposed the built up of the charge on the surface with the aid of negative ions created in a corona discharge. However, the adsorption of charged particles generally will not only build-up the charge on the surface, but will also change the energy spectrum of the already-existing surface states.

Emission technique appear to be a very promising method of studying surface states, but must be borne in mind that they have certain limitations when those states are of adsorptive origin. The high-energy quanta of the exciting radiation may cause dissociation or excitation of adsorbed molecules. Electron processes in a solid (including the Auger effect) that are stimulated by primary radiation may change the charge state of adsorbed particles, or trigger adsorption and catalysis. For example, in [4.27] it was reported that irradiation of a ZnO crystal surface with electrons having an energy of about 2 or 3 keV caused dissociation of the surface phase and desorption. Auger spectroscopy (AES) often employs electron beams with such energy. To some extent these techniques appear feasible for investigating the spectra of tough chemical complexes (e.g., hydroxyl groups on silicon). The same drawbacks are inherent in various modifications of thermal activation spectroscopy. Heating can result in thermal desorption of adsorbed particles, so this method can only be used to study thermally stable surface states.

Other direct techniques of investigating traps in insulators, including those of adsorptive origin, are the EPR and NMR methods. In [4.28-32], investigations of paramagnetic defects in Si-SiO$_2$ systems helped to reveal vacancy-type defects in SiO$_2$ (centers of type E'). In [4.31,33] the EPR technique was used to investigate electron exchange between such defects in GeO$_2$ and the semiconductor's bands, and thus to detect the influence of adsorbed molecules on E' centers. The NMR technique has as yet been used comparatively rarely to study the structure of dielectric films, owing to its low sensitivity and the low content of nuclei having a magnetic moment in many substances of interest (e.g., SiO_2). It was however, successfully used to investigate the concentration and distribution of protons in amorphous Si films [4.34].

The method of injection currents is also very promising for studying adsorptive surface states. If a dielectric layer in a MIS or MIM structure is sufficiently porous, then a number of centers of capture on the surface of pores responsible for the SCR current (SCRC) can come into direct contact with the environment. In [4.21] it was reported that adsorption of acceptor (O_2, NO_2) and donor (H_2O, NH_3) molecules has a considerable influence on the section of the VAC which describes the flow of SCRC (Fig.4.2). Calculations carried out for donor molecules indicate that the concentration of traps N_t rises after adsorption from 10^{15} to 10^{16} cm^{-3}, the effect being reversible. The width of energy spectrum k_0 decreases from 0.15 to 0.07 eV. Similar results were also obtained for a MIM structure Ti-TiO$_2$-Au [4.21]. In this case it turned out that variations in the electrophysical parameters upon adsorption were in good agreement with the corresponding variations in the conductivity of a rutile monocrystal.

Valuable information on the energy spectrum of electron states in an insulator has been obtained from investigations in which the traps were charged by charge carriers photoinjected to study traps in Ge-GeO$_2$ [4.35-37] and Si-SiO$_2$ [4.38-40] (see also the references cited in [4.35-40]). As an example let us consider the Ge-GeO$_2$ system ("real surface"), which was thoroughly studied in [4.35,37]. In these works the value of the captured "photocharge" and the corresponding change in the germanium surface potential were determined with the aid of the field-effect technique. Four thresholds of photocharging (Fig.4.3) were discovered. The first threshold corresponds to the passing of electrons from germanium's valence bands across the "tails" of localized states in amorphous GeO$_2$, with subsequent localization in deep traps; the threshold value of the energy of the excitant quanta is $h\nu_{thr}^1 = 1.85$ eV. A second sharp threshold, connected with the passing of charge carriers across the band of delocalized states in GeO$_2$, was observed at $h\nu_{thr}^2 = 2.9$ eV. Both these processes lead to negative charging of the dielectric. With a further increase in the energy of the quanta of excitant light ($h\nu > 3.2$ eV), two thresholds of positive charging were observed, which are connected with the passing of electrons from the oxide into the semiconductor (or photoinjection of holes from the semiconductor into the oxide) (Fig.4.3, Transitions 3 and 4).

These investigations indicate that both negative and positive charging of traps in the oxide is sensitive to the processes of adsorption and desorption. For instance, the photosensitivity of a thin dielectric layer to negative charging increased sharply after adsorption of acceptor molecules NO$_2$ and p-Bq.

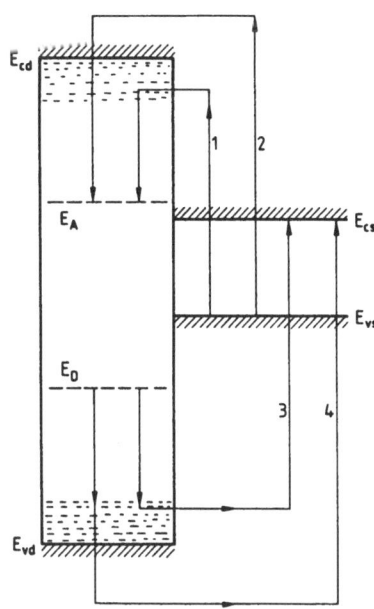

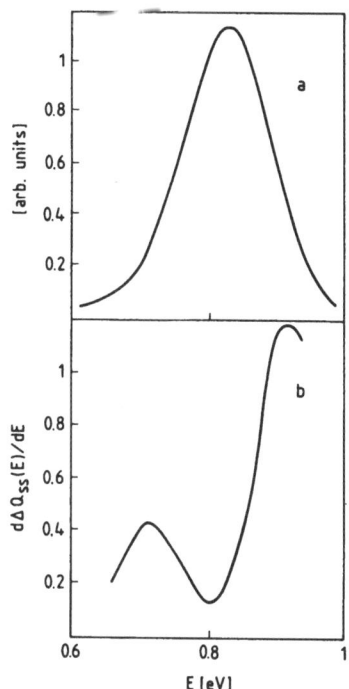

Fig.4.3. Transitions of charge
carriers in Ge-GeO$_2$ heterojunction upon illumination. Designations are the same as in Fig.
3.11

Fig.4.4a,b. Activation energy of thermal
ejection of negative (**a**) and positive (**b**)
charge carriers from traps in the oxide film
in Ge-GeO$_2$ system [4.37]

KASHKAROV and KOZLOV [4.37] studied thermally stimulated ejection of charge
carriers from the deep traps responsible for the charge buildup in the dielectric, and managed to estimate the location of traps in the energy spectrum.
The occupied traps corresponded to blurred maxima of the density of states in
the insulator's band (Fig.4.4), which is typical of the spectrum of a disordered system (Fig.3.10c). Similar blurred maxima were also observed upon optical deexcitation of the excess charge from the insulator's traps [4.35,41]. Adsorption influenced the parameters of the dielectric's traps, the charging
thresholds, and the extension of "tails" of the insulator's localized states
[4.36,37]. The possible nature of the traps will be discussed in Chaps.6,7.
What is important here is that these methods are capable of "sensing" the surface states and the way in which they interact with the environment.

We consider the "sandwich" MIS and MIM structures to be the most convenient
model objects for studying the processes of adsorption and catalysis that take
place in real systems. By combining measurements of the VAC (in the SCRC region)
and CV characteristics (Sect.2.4), and by supplementing these with the photo-

injection and thermal activation methods (depending on the objects of study), one can, using model systems, carry out qualitative investigations of the influence which the processes of adsorption have on the energy spectrum of electron states on the surface of pores in a dielectric layer. MIS and MIM structures provide a more or less accurate model of catalysts which are complex in structure and include metallic and semiconductor phases, together with boundaries between these phases.

Let us conclude this section with a discussion of a possible method of studying the phonon spectra in MIM structures: inelastic electron tunneling spectroscopy (IETS), pioneered by JAKLEVIC and LAMBE [4.42,43]. Essentially this method consists in investigating tunnel currents through a MIM structure at liquid helium temperatures. Consider an $Al-Al_2O_3$-Pb system (Fig.4.5). Lead suggests itself because it has a rather high temperature of transition into the superconducting state. In the band scheme of the MIM structure presented in Fig.4.5, Al is in the ordinary state and Pb is in the superconducting state; Δ is the energy gap of the superconductor, where the Fermi level $E_F = -qV$ is located (V is the applied voltage). The majority of electrons tunnel through the thin dielectric layer without loss of energy (elastic tunneling), as represented in the figure by Transition 1. In this case the VAC is almost linear. Some of the electrons, however, lose part of their energy (inelastic tunneling), as represented by transitions of Type 2. The excitation energy $h\omega = E_1 - E_2$, lost by electrons in the barrier layer, can be expended on the excitation of phonons and vibrational modes of adsorbed molecules.

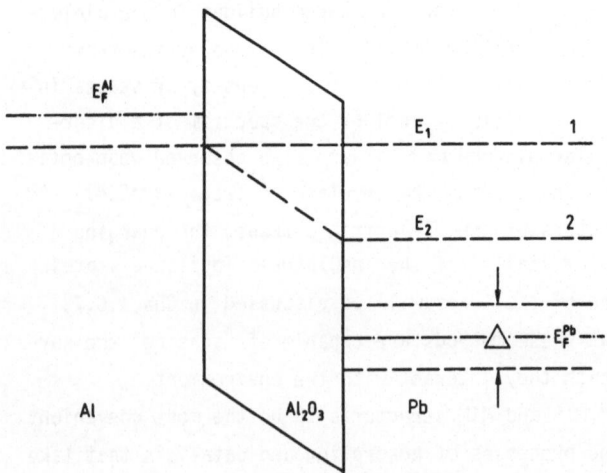

Fig.4.5. Inelastic tunneling spectroscopy: (1) Elastic tunneling, (2) inelastic tunneling

BENNETT et al. [4.44] demonstrated that the second derivative of the VAC of tunnel currents flowing via Channel 2 (Fig.4.5) is proportional to the probability of interaction of tunnel electrons with excitations in the insulator:

$$\frac{d^2j}{dV^2} \sim A(qV)N(qV) \quad , \tag{4.5}$$

where $N(qV)$ is the spectral density of excitations in the insulator, and $A(qV)$ characterizes the dynamics of interaction of electrons with excitations.

The function $(d^2j/dV^2)(V)$ represents the "tunnel spectrum". The IETS technique makes it possible to probe into the spectral range of energies from 0 to 0.5 eV, which corresponds to near IR range in the optical spectrum. Spectral resolution at 1 K can reach 1 cm^{-1} (1 eV = $8.1 \cdot 10^3$ cm^{-1}). The phonon spectra of the metal and the metal-insulator boundary fall within the 0.01 - 0.02 eV range of the IETS spectrum. The frequencies of optical phonons of the insulator occupy a spectral range of up to 0.1 eV [4.45]. The vibrational spectra of many adsorbed molecules extend up to energies of 0.5 eV [4.43-48]. In [4.4] we discussed the results of IETS investigations of the vibrational spectra of a number of simple molecules adsorbed on Al_2O_3.

The IETS technique opens up wide prospects for the study of the phonon spectra of wide-band semiconductors and the influence of adsorption on these spectra. By combining the IETS technique with electrophysical methods of investigating the traps in thin layers, it should be possible to obtain important information on the interconnections between adsorptive, electron, and phonon perturbations on the surface.

4.4 Polycrystalline Semiconductors

When dealing with thin monocrystalline films or microcrystals in disperse powders as opposed to solid single crystals, one must establish the extent to which the SCR theory (Sect.2.2) remains applicable. The band theory of semiconductors implies that when the size of a crystal, ℓ, becomes close to the Debye shielding length L_D for the given substance, the bulk of the crystal is no longer electrically neutral. The surface and the bulk charges have different signs, so with a decrease in ℓ, the Fermi level in the bulk will be shifted with respect to its position in a thick specimen. This violation of electric neutrality makes it necessary to modify all formulas in the SCR theory which define the relationships between surface potential, surface charge, and surface conductivity (Sect.2.2).

KOGAN [4.49] performed a qualitative analysis of the influence which the degree of disperseness has on the concentration of surface states of adsorptive origin. More recently, PESHEV and BLIZNAKOV [4.50,51] considered this problem, taking into account the biographic states on the surface of small particles. PESHEV [4.51] assumed that the charging of the surface is completely determined by the charge arrested in states, whose energy spectrum and concentration do not depend on ℓ. Under these assumptions it turns out that a decrease in ℓ shifts the Fermi level downwards (with respect to the bottom of the conduction band) if $Q_s < 0$, and upwards if $Q_s > 0$. The adsorptive activity of the surface is also altered in a corresponding fashion. It should be noted however, that it is incorrect to assume that the concentration of defects is independent of the particles' size: according to the thermodynamics of small particles, the concentration of defects increases with a reduction in the size of the particles [4.52].

GOODWIN and MARK [4.53] considered the case in which the thickness of the semiconductor film is several Debye shielding lengths. Having obtained the numerical solution of the Poisson equation (2.1) for adsorption of acceptor molecules on the surface of a partially compensated n-type semiconductor, they derived the theoretic dependences of the concentration of charge carriers and the location of the Fermi level in the film upon the temperature and concentration of the surface states. These calculations were found to be in good agreement with the experimental data on variations in the conductivity of monocrystalline CdS films during chemisorption of oxygen. A helpful practical aspect of this work is that it establishes the optimal film thickness and concentration of dopants, that ensure the greatest sensitivity of conductivity to adsorption.

When dealing with small particles, one must take into account changes not only in the equilibrium electronic parameters, but also in the phonon spectrum [4.4]. A number of experimental results on the optical properties and heat capacity of thin films directly indicate that the phonon spectrum for small particles is biased. This affects not only the mobilities of electrons and holes in small particles, but also the characteristic times of capture and recombination of charge carriers, i.e., the kinetic parameters of the relevant states. This side of the problem has not yet received the attention it deserves in the theory of electron processes on the surface.

Another equally important complication encountered in investigations of polycrystalline films and powders arises from ambiguities in the interpretation of the electrophysical parameters measured in these heterogeneous systems. Let us first discuss the electroconductivity σ. With powders and polycrystalline films, the conductivity for dc current, $\sigma(0)$, is to a large extent determined by the nature of the numerous contacts between crystallites. MORRISON [4.54] considered the structure of Zinc oxide sintered at a high temperature as a system of uniform particles, linked together by thin "necks" ("Swiss cheese" structure, Fig.4.6a). He demonstrated that the changes in conductivity during chemisorption are mainly determined by surface phenomena in these necks.

SLATER [4.56] and PETRIZ [4.57] proposed a mechanism of barrier conduction as a phenomenological description of the conduction in such systems. This mechanism has found wide application. According to Slater and Petriz, the density of current flowing through a polycrystalline system is

$$j = Mn_b \exp(-q\varphi/kT)[\exp(q\Delta V_b/kT) - 1] \quad , \qquad (4.6)$$

where M is a coefficient that depends on the shape of the barrier, n_b is the concentration of charge carriers near the barrier, $q\varphi$ is the height of the barrier, and $\Delta V_b = V/N$ is the voltage across a single barrier (V is the voltage across the whole specimen, and N is the number of barriers per unit length of the specimen).

If $V_b \ll kT/q$, then j (and consequently σ) are given by expression of the type

$$\sigma = q\frac{Mn_b}{NkT} \exp(-q\varphi/kT) \quad . \qquad (4.7)$$

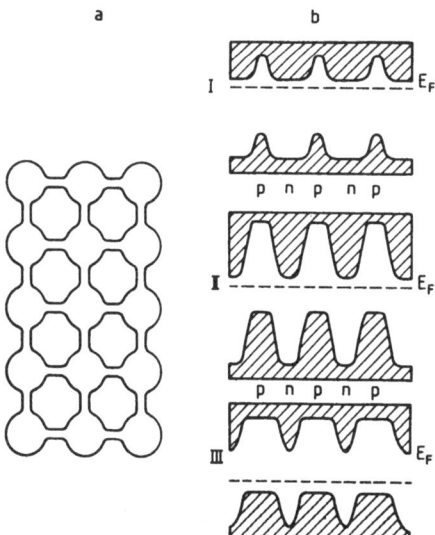

Fig.4.6. (a) Sintered structure and (b) band scheme illustrating the barrier model of PbS [4.51]. (I) Underoxidized, (II) normally oxidized, (III) overoxidized film [4.55]

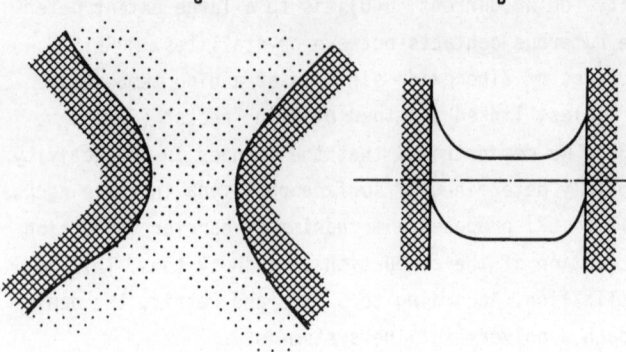

Fig.4.7. (**a**) Model of barrier and (**b**) its energy diagram [4.55]

Customarily, though without sufficient grounds, n_b is assumed to be equal to the bulk concentration of charge carriers, n_v.

To illustrate, let us consider a model of contact between two semiconductor particles sintered together, and the energy diagram of this contact (Fig.4.7) [4.55]. In the general case the stoichiometry in the link between the two particles can be violated, as opposed to the case for the bulk of the particle. PETRIZ [4.57] approximated such contact by a potential pit, with the barrier's height being determined by the surface charge Q_s and (in the case of adsorption of polar molecules) by the dipole component of the work function. As follows from (4.7), if the shape of the barrier does not change, see (4.6), the change in the conductivity during adsorption is wholly determined by the change in the barrier's height $q\varphi$. Since this term stands in the index of the exponent it is this term, and not the preexponential factor Δn_b, that determines the change in conductivity. MORRISON [4.58] proposed a method, based on the temperature dependence of σ, see (4.7), of determining the barrier's height $q\varphi$ and consequently, the position of the Fermi level on the surface of the particles. However, this technique is poorly suited for the study of adsorptive states, owing to the desorption of a number of adsorbed molecules in the course of heating.

The equivalent electric circuit of polycrystalline structures can be presented as a combination of a large number of resistors and capacitors. The value of the conductivity for ac current will increase with an increase in frequency. Reliable information on the conductivity of separate grains (i.e., Δn_v) can be obtained only with sufficiently high frequencies, using VHF techniques [4.59]. Indeed, EROFEICHEV and KURBATOV [4.59] observed a drastic change

84

in conductivity (and its temperature dependence) when they shifted from dc current to ac current with a frequency of about 10^{10} Hz. Similar observations could also be made in the optical frequency range.

It is known [4.60] that the coefficient of optical extinction, α, exhibits a linear dependence on the concentration of free charge carriers in the conduction band

$$\alpha = \frac{q^3}{\pi c^3 n_r} \frac{\lambda^2 n_v}{\mu_n m_n^*} \quad , \tag{4.8}$$

where n_r is the index of refraction, c is the speed of light, λ is the wavelength, m_n^* is the effective mass, and μ_n is the mobility of electrons.

In order to estimate the extent to which changes in the concentration of charge carriers in the grains of powder (Δn_b) and changes in the barriers between particles ($\Delta \varphi$) affect the change in conductivity, $\Delta \sigma$, BURBULYAVICHUS et al. [4.61] measured the dc conductivity for a rutile monocrystal (no barrier effects) and a compact tablet of polycrystalline rutile during adsorption of oxygen (Fig.4.8). The variation in the number of charge carriers in the grains (Δn_v) during adsorption was estimated from the absorption spectra in the IR spectral range (Fig.4.9).

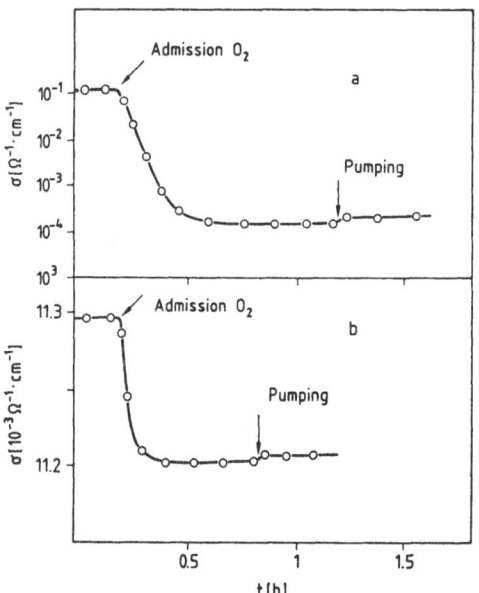

Fig.4.8a,b. Variations in the conductivity of vacuum-reduced polycrystalline (**a**) and monocrystalline (**b**) rutile specimens during adsorption and desorption of oxygen

85

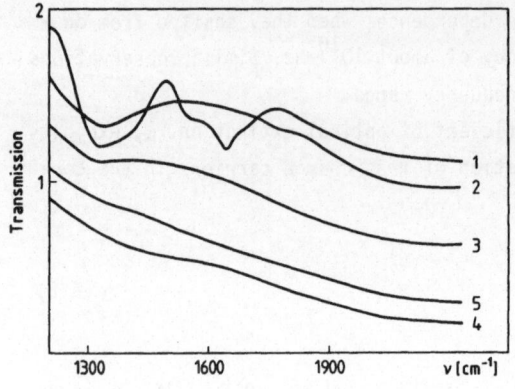

Fig.4.9. Transmission of polycrystalline rutile specimens after vacuum heat treatment at 500 (*1*), 700 (*2*), 750 (*3*), and 850 K (*4*), and after adsorption of oxygen and vacuumization at 300 K for the specimen pretreated at 850 K (*5*)

As seen from Fig.4.8, the injection of oxygen into a polycrystalline specimen reduces its conductivity by nearly three orders of magnitude, as compared with a mere 1% in the case of a monocrystal. This is convincing proof that in the case of a powder, $\Delta\delta$ is entirely determined by the change in the barriers between particles, i.e., by $q\Delta\varphi$ in the Petriz's formula (4.7). Indeed, according to (4.7), the variation of $\Delta\sigma$ by three orders of magnitude is caused by a variation in the surface barrier by $7kT/q$. Direct measurements of the contact potential difference indicate that the change $\Delta\varphi$ may be $5kT/q$ or higher, depending on the degree of reduction of the rutile powder.

Optical measurements demonstrate (Fig.4.9) that with an increase in the temperature of reduction, optical transmission is reduced due to an increase in n_v. Adsorption of oxygen results in an increase in the transmission, which is due to a decrease in the concentration of free electrons owing to negative charging of the surface of the grains of powder. A similar reduction in the transmission of rutile during adsorption was observed earlier by FILIMONOV [4.62]. Notwithstanding the rather qualitative nature of this estimate (recall the Rayleigh scattering and absorption of light by impurities), the value of n_v is about 10%, according to the data presented in Fig.4.9. The change in the concentration of charge carriers, if attributed to the SCR, is about 1% and this value is in good agreement with the data on the change in conductivity of a single crystal. The huge variation (three orders of magnitude) in the conductivity of a polycrystalline specimen (Fig.4.8) is entirely due to the change in the height of barriers between particles. We see that measurements of dc conductivity or conductivity at low frequencies do not yield any direct information about variations in the concentration of charge carriers in grains of powder during adsorption.

Sometimes the apparent fulfillment of Ohm's law is used as an argument against the presence of barriers in powders. However, since the number of particles along the current flow is $N \sim 10^5 - 10^6$, $\Delta V_b \ll kT$ at commonly employed voltages (several hundred volts), and Ohm's law will always be fulfilled. At higher voltages $V_b > kT$, and the VAC will have the exponential shape typical of a diode.

Other electrophysical parameters of polycrystalline objects, e.g., thermal emf, are hardly more informative. In the case of powders, owing to a highly nonuniform thermal field at the points of contact between particles (especially when the particles themselves exhibit anisotropic properties), the data on thermal emf cannot even be used to determine the type of conductivity. For instance, KALASHNIKOV and POKROVSKI [4.63] observed a change in the sign of thermal emf (and thus a change in the type of conductivity) during crushing of germanium. With small particles, when the size of the particle is close to the length of the free path of lattice phonons, the effect of entrainment by phonons can even occur at room temperatures [4.64] which further complicates attempts to interpret data on thermal emf. Similarly, data on the Hall effect can also be ambiguous. The nonuniform conductivity of grains and links between them can cause the Hall fields to be shunted off, making it difficult to interpret the values of Hall constants. New possibilities for investigating parameters of polycrystalline structures are offered by measurements of the temperature dependences of conductivity, thermal emf, and the Hall constant [4.65-67], as well as measurements of Hall mobility in the VHF range [4.68].

Unfortunately, all these specific features of the electrophysical parameters of polycrystalline substances are neglected in the overwhelming majority of theoretical works dealing with electronic and adsorptive phenomena on semiconductor surfaces. Should one dispense with all measurements of the electrophysical properties of surfaces of polycrystalline substances? Certainly not; these substances are of no less practical importance than monocrystals. They form the basis of all catalysts and adsorbents, resistors, photoresistors, thermistors, etc. In many cases it is barrier effects that are responsible for the high sensitivity of polycrystalline structures to external agents, which makes them so useful in many branches of technology (photoresistors, semiconductor gas analyzers). The question of whether to use polycrystalline or monocrystalline objects must be decided on the basis of the experimenter's aims.

Naturally, to check theoretical conclusions and illuminate the mechanisms of interactions between molecular and electronic subsystems on solid surfaces, all measurements ought to be carried out on single crystals. However, the mi-

87

nuteness of their surface area limits the extent to which one can probe the mechanisms of chemisorption. It seems to us that until more sensitive methods are developed for studying adsorption on a wide range of monocrystals[2], it would be more correct to measure electrophysical parameters on monocrystals at varying pressures of gaseous medium [e.g., $\sigma_s(P)$, $\varphi_t(P)$, etc.], and then to convert these values (with the aid of adsorption isotherm $n_a(P)$, measured on a powder under the same conditions) so as to present them as functions of the concentration of adsorbed molecules [$\sigma_s(n_a)$, $\varphi_t(n_a)$, etc.] (Sect.6.2). Of course, the adsorptive properties of a powder and a monocrystal differ, even if they do undergo similar treatment. However, these differences being accounted for, one can estimate the concentration of molecules adsorbed on the surface of a monocrystal with an accuracy of up to 20 or 30% [4.4]. In any case, any errors that arise will be smaller than those made when one attempts to compare results derived from electrophysical measurements on powders with the corresponding theoretical results that are valid for a single crystal.

The same considerations apply to the comparison of electrophysical data with the results of spectroscopic investigations. It must be remembered that electroconductivity is a collective property of a solid, and must be studied on a monocrystal. Spectroscopic data on the nature of adsorption centers and surface complexes mainly reflect local properties. For instance, the EPR signal from a paramagnetic center of a surface is determined primarily by the structure of the closest coordination sphere of the center. We emphasized in [4.4] that the parameters of an EPR signal from dangling bonds of Ge on a high-dispersion powder (particle size $\sim 1\,\mu m$) are in excellent agreement with the corresponding data for a monocrystal. Improved spectroscopic techniques will make it possible in future to correct errors which have crept into the interpretation of experimental data on disperse systems as compared with monocrystals.

The path outlined above does not, of course, preclude investigations of electric properties of polycrystalline structures themselves. Modern technology makes it possible to produce sufficiently flawless epitaxial films of many binary compounds. By means of simultaneous measurements of the temperature and frequency dependences of the electrophysical parameters of monocrystalline and polycrystalline films of the same composition under various conditions of adsorption and stoichiometric disturbances, it will be possible to construct more adequate equivalent circuits for modeling the electron processes in complex heterogeneous semiconductor structures. Let us briefly consider

---

[2] In investigations of monocrystalline films, this can be achieved by the method of piezoelectric balance (Sect.6.2).

one such model structure. This will help us to understand the differences in the interpretation of data on adsorption and conductivity for polycrystalline systems and monocrystals.

Imagine a sintered polycrystalline system ("Swiss cheese structure", Fig.4.6) composed of rather large crystallites (size $d \sim 10^{-3} - 10^{-4}$ cm) linked together by narrow necks ($d \sim 10^{-6}$ cm) [4.54]. It is quite clear that after heat treatment of such a structure in a vacuum (or in oxygen), as well as after surface doping, the concentration of defects in the crystallites and in the necks will be different, owing to the different curvature of the surface and the different surface energy [4.4]. Actually, in the case of such typical n-type semiconductors as ZnO, TiO_2, and the like, the polycrystalline structure can be modeled as a set of a great number of n^+n-diode structures, connected in parallel and in series (p^+p structures in the case of p-type semiconductors). Often the neck will display the type of conductivity opposite to that of the crystallite (e.g., films of deposited and partially oxidated metals, PbS, CdS films, etc.). In this case the film will be presented as a set of pn transitions. The energy scheme of such a structure is illustrated in Fig.4.6b [4.55]. One can observe here the development of barriers in p-type and n-type regions as the oxidation of the polycrystalline PbS film proceeds.

The change in the conductivity of the film during adsorption is almost completely determined by the change in the concentration of carriers in the "necks". Indeed, the dimensions of the necks are close to the Debye shielding length L_D for the given substance. In this case the location of the Fermi level within the volume of the neck becomes a function of the surface charge and thus depends on adsorption. Assuming that the change in mobility is negligible, one can write for the change in the concentration of electrons in the neck: $\Delta n_b \sim \exp(-\Delta\varphi / kT)$; the change in the concentration of electrons in the crystallites, Δn_v, is considerable. When data on adsorption and catalysis are compared with the results of conductivity measurements, the magnitude of adsorption is usually ascribed to the total surface of the film or powder, which is primarily determined by the area of the grains, whereas changes in the concentration of carriers take place in the necks, whose surface area is 4 or 5 orders of magnitude smaller. Roughly speaking, this is that very difference in $\Delta\sigma$ which is observed in Fig.4.8 for a powder and a monocrystal. The data on the Hall effect also apparently reflect a change in the concentration of carriers, Δn_b, in the necks, and thus are similar to data on the conductivity of necks. For instance, adsorption of oxygen on ZnO shifts the conductivity of necks into the n^+ region. Consequently, it is hard to understand the references made in the electron theory of chemisorption (Chap.5) to unambiguous correlations between variations in the conductivity $\Delta\sigma$ of such structures and the concentration of charged adsorptive states on the surface (the total surface). In order to give a theoretical interpretation of the data obtained in experiments with powders, one must consider (at least within the framework of the model of heterojunctions, Fig.4.6) the problem of how adsorption influences the parameters of these junctions. This problem is also of practical interest.

Owing to the large number of barriers, the sensitivity which a polycrystalline structure's conductivity exhibits towards chemisorption is quite high. For this reason such structures have found many applications as semiconductor sensors in gas analyzers [4.69-71]. However, neglect of the peculiar mechanism of conduction in heterojunctions often causes the influence of adsorption on the electric parameters of these structures to be incorrectly interpreted. For instance, the sensitivity of ZnO films to adsorption (i.e., the magnitude of adsorption, which results in a considerable change in conductivity) is estimated in [4.72] to be $10^7 - 10^8$ particles $\cdot$ cm^{-2}, or about 10^{-7} of a monolayer. It is easy to see that since the ratio of the surface area of grains to that of necks is about 10^4 or 10^5, the actual sensitivity is $10^{11} - 10^{13}$ particles $\cdot$ cm^2, if ascribed to the surface area of necks. This figure is close to the number of electrically active surface defects (adsorption centers) on the surface of a single crystal (Sect.6.2).

However, by no means should one conclude that polycrystalline films are of no inherent interest to the researcher. The possibility of creating (under adequate technological conditions) an immense number of n^+n or p^+p junctions in such films can be used to construct detectors with selective sensitivity to the adsorption of mixtures of acceptor and donor molecules. Equally promising are investigations of the role played by these junctions during the acceptor and donor stages of catalytic reactions. Naturally, this necessitates model experiments with isolated junctions, as well as theoretical estimates of the influence of chemisorption on the height of barriers and the currents in these junctions.

The influence of field on the height of intercrystallite barriers and the conductivity of polycrystalline films is discussed in [4.57], but a very important factor is overlooked, namely the recharging of surface states on the boundary between grains when the difference in potentials changes. GOLDMAN and ZHDAN [4.73] published a pheomenological theory of the steady-state and transition electron processes that occur when the occupancy of intercrystallite traps varies. The theory is based on a bicrystal model, and considers what happens when the following are present on the interface:

1) monoenergetic levels,

2) a discrete set of levels, and

3) a continuous distribution of levels of traps [4.73].

The latter case is especially interesting with regard to the real boundaries between grains, which are not always ordered (Sect.3.2). Under certain conditions the theory developed in [4.73] allows one to determine the shape of the distribution function of surface states (including the adsorptive states) in the forbidden band of a semiconductor on the basis of the known shape of experimental volt-ampere characteristics.

Important results were reported by GOLDMAN and GULYAEV [4.74], who for the first time constructed a consistent theory of the field effect in polycrystalline structures which accounts for the modulation of intercrystallite barriers and the filling of surface states located on the interface in bicrystals. This theory makes it possible to use the experimental field dependence of the conductivity of polycrystalline films to determine the most important parameters: density of surface states, mobility of charge carriers, and height of the intercrystallite barriers. Direct information on the height of the barriers can also be obtained by measuring the thermally stimulated discharge of a condenser through a semiconductor having intergranular barriers [4.75].

Due to its comparatively large surface area, a polycrystalline film is a very convenient model for the study of adsorption and catalysis. The theory developed in [4.73-75] opens up new prospects in this direction.

5. The Electron Theory of Chemisorption and Catalysis on Ideal Semiconductor Surfaces

The idea that there might be a link between chemisorption and semiconductor properties was expressed as early as the 1930s by IOFFE [5.1], ROGINSKY and SCHULZ [5.2], WAGNER and HAUFFE [5.3,4], WILSON [5.5], and other researchers. The theoretical foundations of the electron theory of chemisorption and catalysis were laid in the 1950s. In the same period numerous investigations were conducted to discover the effects of adsorption on conductivity and the Hall effect in a number of semiconductors. It was found that changes in the conductivity are connected with changes in the concentration of charge carriers in the space charge region[1]. BRATTAIN and BARDEEN [5.9], KINGSTON [5.10], HEILAND [5.11], LOW [5.12], MORRISON [5.13], and MARK [5.14] established that adsorption changes the surface potential of germanium. This made it possible to develop a method of controlling the surface potential (the Bardeen-Brattain gas cycle). This method involves changing the magnitude and sign of the surface potential with the appropriate choice of adsorbates, whose adsorption results in a positive or negative charging of the semiconductor surface. In some cases this method can be used to locate the minimum surface conductivity and thus (in accordance with Sect.2.3) calculate the surface potential. However, as will be shown below, adsorption not only alters the surface potential, but also changes the whole system of surface states, which makes it difficult to interpret experimental results.

5.1 Electronic and Adsorptive Equilibrium

Any theory which attempts to explain a phenomenon should begin by examining the causes which give rise to it. Thus, the primary task of surface science is to investigate the causes of the charge buildup which occurs on the surface when a given molecule or atom is adsorbed. ZENTER [5.15] was among the first to consider the interaction between a foreign atom and a unidimensional crystal; in 1938 he demonstrated that allowed states are created in the forbidden

---

[1] A bibliography of these works can be found in [5.6-8].

band of the crystal. Later this problem was considered under various assumptions by VOLKENSTEIN [5.16], BONCH-BRUEVICH [5.17], KOUTECKY [5.18], GRIMLEY [5.19], and NAGAEV [5.20]. This was the real inception of the electron theory of chemisorption. However, this theory is still far from able to calculate the parameters of these levels. In some cases it is possible to predict the type of level created (acceptor or donor) from general information on the values of the ionization energy, the energy of the electron affinity of the adsorbed molecules, and the known properties of the semiconductor. More often, however, the required knowledge is obtained experimentally. The lack of data on the energy spectrum of the surface predetermines the qualitative nature of the theory.

Local electron states created by adsorbed particles on the surface of a crystal serve as traps for electrons and holes. The occupancy of these levels at thermodynamic equilibrium is unambiguously described by the Fermi distribution function [see (3.1)]. Consequently, VOLKENSTEIN [5.21] introduced the idea that there are two possible forms of chemisorption: neutral and charged. The neutral form corresponds to empty local levels, the charged form to occupied levels. If n_a is the concentration of chemisorbed particles, then the relative concentrations for various forms of adsorption will be expressed as [5.21]: $\eta^0 = n^0 / n_a$ (neutral chemisorption); $\eta^- = n^- / n_a$ (acceptor levels), and $\eta^+ = n^+ / n_a$ (donor levels) in the case of charged chemisorption. The values and η^+ can be easily calculated from (3.1) if one assumes $N_t = n_a$. Then $n^\pm = Q_{SA} / q$, where Q_{SA} is the charge accumulated via the charged form of chemisorption[2]. Finding $\eta^\pm$ is a standard problem of the population of impurity levels in semiconductor physics. In our case the surface impurity is represented by adsorbed particles.

A somewhat different approach to electron processes is assumed in the boundary layer theory of Hauffe and other researchers. According to this theory [5.22], local levels are created only by the charged form of chemisorption (ionosorption). A local level vanishes when the electron that has been occupying it is delocalized. This viewpoint was criticized in [5.21]. Later HAUFFE [5.23] introduced the concept of a neutral form of adsorption, by which he meant physical adsorption. MANY [5.24] also suggested the possibility that physical adsorption could pass into the charged form of chemisorption.

The controversy between HAUFFE [5.22] and VOLKENSTEIN [5.21] centered on the question of whether or not physical adsorption can create a localized state.

---

[2] The presentation of η^0, $\eta^\pm$ in [5.21] differs from that adopted here (3.1), in [5.21] the energy levels relate to the edges of the allowed bands, while in surface science they are customarily related to the middle of the band.

To a large extent this is a matter of terminology, owing to ambiguities in the experimental criteria for distinguishing between various forms of adsorption [Ref.5.7,Chap.2]. If one uses theoretical criteria (physical adsorption being considered within the framework of the second-order approximation of perturbation theory [5.7]), then physical adsorption does not result in the creation of any levels. It would hardly be worthwhile to draw a connection between the classification of forms of adsorption and the theoretical assumptions. Rather, it would be more natural to attribute the presence or absence of localized states to particular types of adsorptive interactions. In some cases (Sect.6.1), so-called physical adsorption can give rise to localized states on a surface.

In a widely accepted modification of the electron theory, which will henceforth be called the VOLKENSTEIN model [5.21], adsorption centers are represented by electrons and holes in the semiconductor. They are called the "free valences" of the surface. These free valences differ fundamentally from their counterparts in quantum chemistry and from unsaturated valences in the Shockley sense [5.25]. All of the remaining localized states on the surface are assumed to take no part in adsorption ("biographic states"). A nonreal model of a surface which contains no biographic states at all is adopted in [5.26-28]. The TOMASHEK-KOUTECKY model [5.25], which accounts for interactions of adsorbed molecules with localized Shockley and Tamm states seems more justified to us as an approximation of the ideal surface. In the case of a strong chemisorptive bond, a state is classified as a Tamm adsorptive state; if the bond is weak, one speaks of a Shockley adsorptive state.

Within the framework of the free valence model, electron transitions between the charged and the neutral forms of chemisorption are referred to as localization or delocalization of a charge carrier on the adsorbed particle [5.21]. Localization of a charge carrier on a chemisorbed particle strengthens its bond with the surface. If q^0 is the bonding energy of a neutral particle and $q^\pm$ is the energy of bonding of a charged particle with the surface, then $q^\pm = q^0 + v^\pm$, where $v^\pm$ is the position of the local level of the particle with respect to the edges of the allowed bands [5.21]. Consequently, charged chemisorption is called the strong form, and neutral chemisorption the weak one.

At thermodynamic equilibrium, when there is equilibrium between bands and surface states, as well as between the gas phase and the surface, the magnitude of equilibrium adsorption can be expressed as the sum of concentrations of the neutral and the charged forms, which are obtained from the Fermi distribution (3.1). Having carried out the pertinent calculations, VOLKENSTEIN derived the adsorption isotherm, which formally resembles the Langmuir equation

$$\theta = \frac{1}{1 + (b/P)\ \exp(-q^{\pm}/kT)} \quad , \qquad \text{where} \qquad (5.1)$$

$$q^{\pm} = q^0 - kT\ \ln(1 - n^{\pm}) \quad . \qquad (5.2)$$

Here q^0 is the bonding energy of a chemisorbed particle in the neutral state. However, as opposed to physical adsorption, where the coefficient b in the Langmuir equation is a function only of temperature, here the coefficient b $\exp(-q^{\pm}/kT)$, in accordance with (5.2), depends on the location of the Fermi level on the surface, i.e., on Y_s. If the surface charge in the biographic states does not change, then the surface potential will be a function of the occupancy, i.e., $Y_s = Y_s(\theta)$. Thus Volkenstein demonstrated that even with an ideal surface, with all the assumptions underlying the Langmuir adsorption mechanism satisfied, the equation of the adsorption isotherm has a distinctly non-Langmuir form, and its coefficient turns out to be dependent on the occupancy: $b = b(T,\theta)$.

In chemical adsorption the adsorptivity of the surface depends not only on external parameters P and T, but also on the status of the adsorptive system as a whole, which is determined by the location of the Fermi level on the surface. KOGAN and SANDOMIRSKY [5.29] analyzed (5.1) for various curvatures of the bands and different values of the parameter λ, which characterizes the position of the Fermi level in the bulk of the semiconductor, see (2.7). They also demonstrated that the heat of adsorption, q, which in the Langmuir model does not depend on occupancy, will here, in accordance with (5.2), be a function of surface potential and, by virtue of the arguments developed above, a function of occupancy: $q = q(\theta)$. We see that a change in the heat of adsorption, which in the theory of physical adsorption is associated with nonuniformity of the surface or with interactions between adsorbed molecules [5.7], acquires a totally different meaning in the electron theory of adsorption. As occupation proceeds, the heat of adsorption changes even with a uniform surface, as a consequence of the adsorption itself. For this reason the functions of distribution of the energy of adsorption over the surface have an entirely different form here [5.21].

An equally important task of the electron theory is to establish the relationships between various forms of chemisorption and the experimentally obtained values. Naturally, purely adsorptive measurements only supply information about the total number of adsorbed atoms or molecules. In order to draw a distinction between neutral and charged adsorption, the results of electrophysical measurements must be taken into consideration.

Charged chemisorption results in a buildup of charge on the surface as compared to the bulk. Provided the charge in the biographic states, Q_{sB}, is constant, the theory developed in [5.21,22] uses expressions from the SCR theory (Sect.2.2) that are similar to (2.11,12,19)[3] to draw a connection between charge Q_{sA} in adsorptive states (and consequently $n^{\pm}$) and the measured values of surface potential Y_s (or thermionic work function φ_t) or surface conductivity σ_s. As we shall see later, much greater ambiguities are encountered in attempts to estimate the concentration n^0 of neutral adsorption.

Finally the third main task of the electron theory of adsorption is to assess the possibilities of controlling the chemisorptive activity of the surface by modifying the state of the electron-hole gas on the semiconductor surface. As follows from the above discussion, the model proposed in [5.21] centers on the concept of the Fermi level as a regulator of the chemisorptive activity of the surface. Indeed, in the ideal lattice approximation, assuming invariability of the energy spectrum of biographic states, the occupancy of the local levels of chemisorbed particles [i.e., the relation between the charged $(n^{\pm})$ and the neutral (n^0) forms] is uniquely determined by the location of the Fermi level on the surface, $q\varphi_s / kT$, see (3.1). Since $q\varphi_s / kT$, as seen from Fig.2.1, depends on the bending of bands Y_s and on the location of the Fermi level in the semiconductor bulk, $\ln \lambda$, the realtion between $n^{\pm}$ and n^0 can be modified by changing either of these quantities. The position of the Fermi level in the bulk, $\ln \lambda$, can be changed by introducing donor- or acceptor-type impurities into the bulk, i.e., by means of doping.

An alternative way to control the chemisorptive activity of the surface is to vary Y_s with the aid of a transverse electric field. As follows from (3.1), a change in $q\varphi_s / kT$ will result in a change in the occupancy of levels. This will bias the equilibrium between the charged (strong) and neutral (weak) forms of chemisorption on the surface, which, depending on the polarity of the field, will result in extra adsorption or desorption of molecules. This electroadsorptive effect was considered by BARU and VOLKENSTEIN [5.31].

One more way to control the chemisorptive activity of a surface is to expose the semiconductor to photoelectrically active light: the so-called photoadsorption effect [5.32]. The appearance of nonequilibrium charge carriers on the surface changes the relative values of $n^{\pm}$ and n^0. Depending on the re-

---

[3] The form of the equations linking Q_s to Y_s and σ_s in [5.21] differs from that in the GARRETT-BRATTAIN theory [5.30], because the edges of the allowed bands are chosen as the reference points for the energy levels of adsorbed particles.

lative concentration of excess electrons (Δn_s) or holes (Δp_s), additional adsorption (photoadsorption) or desorption (photodesorption) will take place on the surface. The corresponding calculations of Δn_s and Δp_s have been carried out, making it possible to formulate criteria for the existence of photoadsorption and photodesorption effects [5.32].

5.2 The Kinetics of Adsorption and Catalysis

On the basis of the above model of adsorption on "free valences", VOLKENSTEIN and PESHEV [5.21,33] examined the kinetics of adsorption and desorption. Assuming that surface potential Y_s does not change during adsorption ($Q_{SB} \gg Q_{sA}$), the kinetic equations for neutral and charged adsorption have the following form [5.33]

$$n^0(t) = n^0_\infty + C'_1 \exp(-t/\tau_1) + C'_2 \exp(-t/\tau_2) \quad ,$$

$$n^\pm(t) = n^\pm_\infty + C''_1 \exp(-t/\tau_1) + C''_2 \exp(-t/\tau_2) \quad ,$$

$$(5.3)$$

where C_1 are constants, n^0_∞ and $n^\pm_\infty$ are the concentrations of adsorbed particles corresponding to equilibrium ($t \to \infty$), and τ_1 and τ_2 are characteristic times which under certain assumptions can be expressed via the lifetimes of adsorbed particles in the charged and the neutral forms (τ^0, $\tau^\pm$). The corresponding calculations were carried out for low occupancies (Henry region).

The results of the calculations indicate that for arbitrary relative values of the characteristic times, the functions $n^0(t)$ and $n^\pm(t)$ are depicted by monotonic curves (Fig.5.1). If $\tau^- \ll \tau^0$, then neutral adsorption forestalls the creation of the charged form (Curves 1 and 1'). If the slower stage is associated with the kinetics of electron transitions ($\tau^- \gg \tau^0$), then the neutral form prevails in the initial stage (Curves 2 and 2'). In the general case, when Y_s varies during adsorption [$Y_s = Y_s(n^\pm)$], calculations of the kinetic curves were carried out only for the initial stage of adsorption. Here the functions $n^0(t)$ and $n^\pm(t)$ are also represented by monotonic curves.

Analysis of the experimental data reveals that the kinetic curves of conductivity $\Delta\sigma(t)$ and work function $\Delta\varphi_t(t)$, which describe the kinetics of charged adsorption $n^\pm(t)$, are often nonmonotonic with pronounced maxima. This shape cannot be explained within the framework of the surface model used in [5.21, 33]. In [5.34] the kinetic equation of adsorption was solved taking into account a nonzero probability that both an adsorbed molecule and a free charge

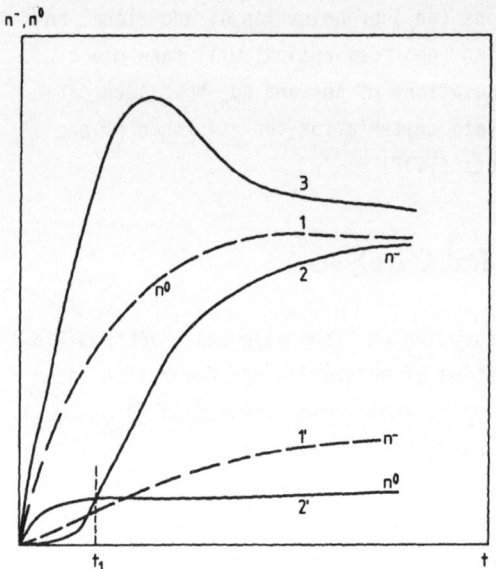

n⁻,n⁰

Fig.5.1. Theoretical curves
of the kinetics of chemi-
sorption in charged [$n^-(t)$]
and neutral [$n^0(t)$] forms,
according to [5.21]
($1,1',2,2'$) and to [5.34]
[$n^+(t)$, (3)]

carrier could be simultaneously captured on an adsorption center. This resulted
in kinetic equations for n^0 and $n^\pm$ that were similar to (5.3). However, in this
case the coefficients C_1 are no longer constant, but depend on the characteristic times and other parameters of the system. With an appropriate choice
of coefficients, the theoretical dependence $n^\pm(t)$ can be made to exhibit a
maximum (Curve 3 in Fig.5.1). Consequently, the kinetic equation obtained in
[5.34] allows a formal description of the nonmonotonic curves of charge buildup
obtained from experiment. This problem is an important one in regard to adsorption kinetics, and we shall return to it in our discussion of processes
that take place on real surfaces.

Experiment shows that in many cases chemisorption is to some extent irreversible (i.e., a number of molecules at a given temperature turn out to be
too strongly bound to the surface to be desorbed). Volkenstein and Peshev
gave two possible reasons for this irreversibility of adsorption:

1) Adsorption is irreversible if the establishment of electron equilibrium
between the set of local levels and the band lags behind the desorption of
neutral particles. In this case, all the neutral particles leave the surface
first, and only then do the charged particles take their turn, leaving the
surface in an amount of time determined by the relaxation of the charged form
of chemisorption. If the relaxation time of the charged chemisorption is sufficiently long, then during the experiment the charged chemisorption will be
seemingly irreversible, thus the term "apparent" irreversibility.

2) Irreversibility can be the result of a reaction between chemisorbed particles and an impurity on the semiconductor surface. Then a strong chemical compound is formed which cannot be removed by evacuation. This is called "true" irreversibility; it is typical at elevated temperatures, when diffusion of impurities becomes considerable.

In the case of "apparent" irreversibility, Volkenstein and Peshev derived the relationship between the ratio of the reversible and irreversible forms of adsorption, n_r / n_i, and the position of the Fermi level with respect to the energy level of the adsorbed particle.

Let us now consider the electronic aspects of catalysis. The main aim of the theory of catalysis is to elucidate the nature of adsorption centers and the reactivity of adsorbed particles. In the VOLKENSTEIN model [5.21], the principal reactable type is considered to be represented by the radical or ion-radical forms of adsorption. If a chemisorbed particle has an unpaired valence electron (e.g., Na, H), then in the neutral state such a particle will be radical, while in a charged state it will be valence saturated. If a particle is valence saturated, it will remain so in neutral adsorption, while in charged adsorption it will be ionized and turn into an ion radical (e.g., H_2O^+, CO^+). In the former case, determining the concentration of the radical form will be equivalent to determining n^0, while in the latter case it will be equivalent to determining $n^\pm$ using the already known Fermi distribution (3.1). In catalysis the neutral valence-saturated form is assumed to be inactive — in other words, the theory considers only the radical reactions from the whole variety. Since the ratio between n^0 and $n^\pm$ determines the surface charge Q_{sA}, the catalytic activity can be connected with conductivity $\Delta\sigma_s$ and work function $\Delta\varphi_t$ in a way similar to that outlined in Sect.5.1.

The main characteristic of any catalytic reaction is the reaction rate v, which was shown in [5.7] to depend on the surface area of the catalyst particles in contact with the reaction components, and on the concentration of reactants c_i in the surface phase. For the model of an idealized surface, VOLKENSTEIN [5.21] established the relationship between the reaction rate constant and the location of the Fermi level ε_s on the surface: $k \sim \exp[(E_a + \varepsilon_s)/kT]$. This expression is a function of the temperature and pressure of the gaseous phase.

All heterogeneous catalytic reactions were divided by GARNER [5.35] and later by VOLKENSTEIN [5.21] into two classes: acceptor-type reactions, which are accelerated by an increase in the concentration of electrons in the SCR, and donor-type reactions, which are boosted by the presence of holes. Consider, for example, catalytic decomposition of ethyl alcohol, which can proceed either

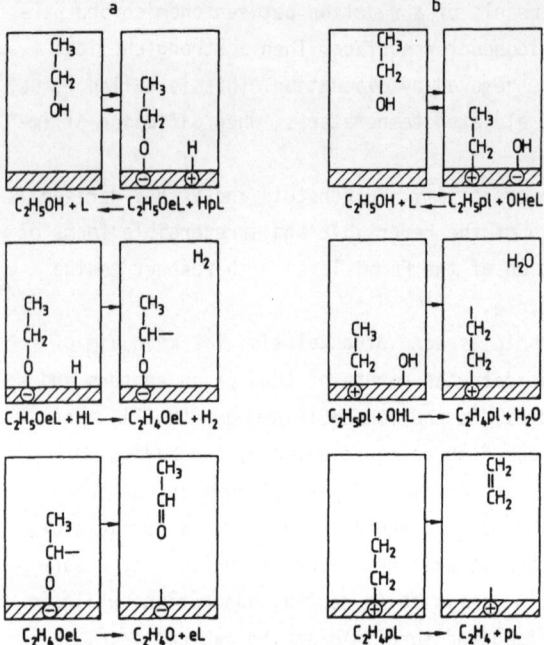

Fig.5.2a,b. Dehydrogenation (**a**) and dehydration (**b**) of alcohol according to [5.21]. L denotes lattice

by dehydration or dehydrogenation. According to [5.21], dehydrogenation is an acceptor-type reaction (Fig.5.2a), while dehydration is a donor-type reaction (Fig.5.2b). By controlling the position of the Fermi level on the surface, one can boost one direction of the reaction and slow down the other. We see that the selectivity of a catalyst is also associated with the location of the Fermi level on the surface. Hence there are various possible ways of controlling the catalytic activity by changing the position of the Fermi level. We have already mentioned them: introduction of impurities (doping), application of transverse fields, and illumination. It is worth noting that HAUFFE [5.22] considered a different electron mechanism of catalytic decomposition of alcohol, and came up with an exactly converse dependence of reaction type on the position of the Fermi level on the surface.

In the Volkenstein model the Fermi level acts as a universal regulator of adsorptive activity. The role of the parameters of surface states is not explicitly analyzed in that theory, though they make a tacit appearance in the problem of neutral chemisorption. MORRISON [5.28] considered the choice of the Fermi level as the independent variable in this model to be one of its major shortcomings. Indeed, the location of the Fermi level on the surface depends

on a great number of often uncontrollable agents: the temperature and pressure
of the ambient gas, the concentration of surface dopants, the concentration
and parameters of biographic states, etc. In contrast to the model proposed
in [5.21], other authors, in describing adsorptive [5.8,36-38] and catalytic
[5.28,39,40] processes, prefer to consider the variations in the energy spectrum during adsorption. MORRISON [5.28,40], in the case of an idealized surface
devoid of biographic states, chose as independent variables the energy levels
of adsorbed acceptor and donor molecules.

It should be noted that in the overwhelming majority of works on kinetics
and catalysis utilizing the model proposed in [5.21], a unique Fermi level is
used as the principal parameter of the electronic subsystem. This is incorrect
[5.41]. When electron equilibrium is violated, which is certainly the case in
catalysis and chemisorption, the system is characterized by Fermi quasi levels
for electrons and holes (2.5). The majority of expressions obtained for the
kinetics of adsorption, catalysis, and photoadsorption [5.21,32,33] are valid
only for very low levels of injection [$\delta \to 0$; cf. (2.4)], an assumption never
realized in experiments.

When considering nonequilibrium processes, one must take into account the
kinetics of electron transitions between surface states and the semiconductor's
bulk. For instance, GARRETT [5.26] and LEE [5.27] considered a trivial reaction
of acceptor (A) and donor (D) molecules on the surface: $A + D = AD$. They assumed
that adsorbed molecules act as centers of capture for electrons and holes with
these centers being characterized by stationary concentrations n_A and n_D, po-
sitions in the band ε_A and ε_D, and capture cross sections c_{nA} and c_{pD}. Desorption takes place only through removal of neutral molecules AD. Biographic states
are assumed to be absent on the surface. Under these assumptions and using the
statistics of Shockley and Reed electron transitions (Sect.2.4), the authors
obtained expressions for the rate of exchange between acceptor (v_A) and donor
(v_D) traps and the allowed bands of the semiconductor.

If the limiting stage of the reaction leading to the creation of AD is the
transfer of charge carriers from bands to the surface states, then [5.27]

$$v_A = c_{nA} n_A n_s^* \quad ; \qquad v_D = c_{pD} n_D p_s^* \qquad (5.4)$$

The steady-state values of surface concentrations of charge carriers n_s
and p_s are determined, in accordance with (2.5), by the positions of the Fermi
quasi levels for electrons ($q\varphi_n$) and holes ($q\varphi_p$). If the limiting stage is the
reaction of A^- and D^+, then the reaction rate is

$$v = kn_A n_D [exp -(\varepsilon_D - \varepsilon_A - \Delta)/kT] \quad , \tag{5.5}$$

where $\Delta = q(\varphi_n - \varphi_p)$; k is the reaction rate constant.

It follows from (5.5) that the greater the separation of quasi levels φ_n and φ_p (Fig.2.1), i.e., the higher the injection level δ is, the higher the reaction rate is. This makes the expression for the reaction rate fundamentally different from the results reported in [5.21], where v is assumed to depend only on the position of the equilibrium Fermi level and the location of the surface state in the energy band.

Another interesting conclusion also follows from [5.27]. The activation energy of the reaction, E_a, at $\delta \neq 0(\Delta \neq 0)$ depends on the width of the forbidden band of the semiconductor, E_g, i.e.,

$$E_a = 1/2(E_g - \Delta) + f(c_{nA}, c_{pD}, n_A, n_D, \varepsilon_A, \varepsilon_D) \quad . \tag{5.6}$$

It follows from (5.6) that a decrease in E_g will result in a decrease in the activation energy and an increase in catalytic activity. The activation energy will also decrease with an increase in Δ, i.e., with an increase in injection level δ.

This result has a clear physical meaning. If in the course of catalysis the semiconductor is operating in the region of intrinsic conductivity, then the rate of capture of charge carriers on the adsorptive surface states is determined by the flow of electrons and holes through the SCR and depends on E_g and δ. If we set $\delta = 0$ and consider (in the case of a catalytic reaction) a rare situation of complete thermodynamic equilibrium in the electronic subsystem [5.21], then the value of E_g will not enter the expression for catalytic activity. Earlier KRYLOV [5.42] used a large number of catalysts of some typical catalytic reactions to establish a clear correlation between catalytic activity and the value of E_g (see also [5.7]): the activity increases with a decrease in E_g, which agrees well with (5.6)[4].

Rejecting the assumption that $\delta \neq 0$ also makes it possible to observe other types of regularities in the changes which occur in the reaction activation energy when transverse electric fields are applied [5.27], as compared to those observed at thermodynamic equilibrium [5.31]. Despite a number of important

---

[4] A different standpoint was adopted in [5.20]. Since both the adsorptive properties of a crystal and the width of its forbidden band depend on the ionicity of the bonds (i.e., on the effective charges of the atoms), they must exhibit a certain correlation. It was demonstrated in [5.20] that the energy of dissociation of adsorbed molecules increases with an increase in E_g, while catalytic activity is reduced.

simplifying assumptions made in Lec's model (connected with neglect of biographic states, nonequilibrium capture, and recombination of charge carriers
on these states), the kinetic approach to describing catalysis is more legitimate than attempts to consider it under the assumption of almost complete
electron equilibrium in the system.

5.3 Chemisorption: Electron Theory and Experiment

Let us briefly analyze the current state of the electron theory of chemisorption and its relationship to other theoretical approaches in semiconductor
surface science. The most important task of the electron theory is to investigate the elementary act of interaction between adsorbed molecule and surface,
and to calculate the parameters of the emerging electron state in the band
scheme of the crystal. In VOLKENSTEIN's language [5.21] this is the problem of
neutral adsorption. The treatment of this problem in [5.21] is based on a
quantum-chemical calculation carried out by the author in 1947 for the adsorption of a univalent atom on an ideal unidimensional ionic lattice. This
calculation only proves the fact that a level is created. As pointed out in
[5.43], this model is far from real, and does not permit one to probe the properties of the state that is created.

However, the literature of the past three decades abounds with calculations
of local interactions between molecules and crystals, carried out in various
approximations. There is obviously a wide gap between calculations of local
interactions and the concepts of band theory. This gap, however, is narrowed
if cluster models of the solid are used. Calculations for large clusters are
already in good agreement with those based on band theory [5.40,44], and calculations for surface clusters can yield the parameters of both localized and
delocalized states, which are the counterparts of allowed bands and surface
states [5.45]. It is very important that theoretical calculations of the band
schemes, even in the case of wide-band semiconductors, can be checked and cor-
rected, thanks to highly developed experimental techniques (Sect.4.3). But unfortunately, all advances in modern quantum surface chemistry still do not reach
the scope of the model concepts developed in [5.21]. An urgent task of the
electron theory of the ideal surface is to combine these two approaches in surface physics. So far the theory only introduces the concept of surface states
phenomenologically and does not even suggest experimental approaches to determining their parameters.

The second task of the electron theory is to establish the relationship between charged (n±) and neutral (n^0) adsorption and variations in the electrophysical parameters of the surface. The theory postulates the parameters of the adsorptive surface state; its further object is to calculate the occupancy of such states on the basis of Fermi statistics. Assuming that all other (biographic) states are either absent [5.26-28] or insensitive to adsorption [5.21, 22], relationships are sought between surface charge Q_s, surface potential Y_s, thermionic work function φ_t, surface conductivity σ_s, etc. This second task of the electron theory is similar to that of the SCR theory (Sect.2.2) developed by GARRETT and BRATTAIN [5.30]. Since the latter is a phenomenological theory, it is not concerned with the origin of the surface states responsible for charge buildup on the surface.

Since the 1950s, two equivalent phenomenological approaches have been pursued in semiconductor physics and in the theory of chemisorption. These theories establish the relationship between the surface charge and the phenomena taking place in the SCR. Physicists concerned with semiconductor surfaces use a well-developed theory of the SCR. When phenomenological studies of this kind give rise to questions related to the nature and parameters of adsorptive states, the researchers turn to the electron theory of chemisorption. However, this theory has no answers, since it also introduces the parameters of states phenomenologically.

Chemists investigating electron processes in chemisorption and catalysis have often employed the model concepts developed in [5.21]. That model, however, implies a rather rare situation: almost complete electron equilibrium in the system ($\delta \to 0$). At the same time, the SCR theory offers a well-developed apparatus capable of establishing, in some cases, the quantitative relationships between nonequilibrium experimental values (photoconductivity, photo emf, nonequilibrium field effect, etc.) and the kinetic and energetic parameters of surface states. These states can, of course, be represented by adsorption states. Based on the SCR theory surface physicists have proposed a number of novel experimental techniques for studying both equilibrium and nonequilibrium surface processes, techniques of which specialists in adsorption and catalysis are often unaware.

The isolated development of two theoretical approaches which actually have the same objective has led to the existence of different terms, designations, and representations for essentially the same things, a circumstance which irritate and confuse the reader. Here we shall stick to the more convenient terminology and presentation adopted in the SCR theory whose basic concepts were outlined in Sect.2.2.

The enticing prospect predicted in [5.21], of being able to control catalytic activity and selectivity brought forth a great number of experimental
works attempting to elucidate the role of electron phenomena in catalysis.
These works analyzed the correlations between the catalytic properties of the
surface, electroconductivity, and the work function of semiconductor catalysts.
However, the examination of all these experimental data [5.8,41,42] leads to
rather pessimistic conclusions. In many cases experiment either fails to detect
any correlations between the location of the Fermi level on the surface and
the surface's catalytic properties, or suggests correlations that are opposite
to those predicted by the theory. This has given rise to a sceptical and dis-
illusioned attitude towards the usefulness of the electron theory. Some researchers have come to the conclusion that the "collective" properties of the
surface, which are determined by the position of the Fermi level, do not play
any significant role in catalysis. According to some rival hypotheses, the
predominant role in catalysis is played by local interactions.

As we shall see later, the main reason for the discrepancies between theory
and experiment lies in the striking dissimilarity between the idealized model
of the surface which underlies the theory, and the real surface with which
the experimenter must deal. The concept of "collective" surface properties
for real systems is quite different from that adopted in the idealized models
of the electron theory.

In most cases, comparisons of theory and experiment are based on a modification of VOLKENSTEIN's theory [5.21], which is constructed under the following
three assumptions:

a) Chemisorption proceeds on the regular atoms of the lattice; free electrons
and holes act as centers of localization for chemisorbed particles.

b) The energy spectrum of a surface (of nonadsorptive origin) does not
change during adsorption.

c) Interaction between a chemisorbed particle and the solid is considered
in the one-electron approximation (i.e., within the framework of the band theory
of solids).

The first approximation in solid-state physics is that of a flawless ideal
lattice. In this regard, matters are much more complicated in surface science:
it is impossible even to conceive of a surface without surface states, since
the surface itself is a defect of the crystal structure. Surface states arise
due to termination of the periodic lattice, i.e., owing to the mere existence
of a surface. Be it an atomically clean surface or just any real surface with
which we must deal in experiments there is always a spectrum of states of nonadsorptive origin. This immediately violates the first two assumptions. First,

surface states are themselves active adsorption centers [5.25]. Adsorption on
a defect-ridden surface will occur on defects first. Second, the adsorbed particles, interacting with defects, will modify their parameters, i.e., will
change the spectrum of nonadsorptive origin. For instance, on an atomically
clean surface, adsorption takes place directly on surface states; it causes
rehybridization of the chemical bonds of surface atoms and irreversibly re-
constructs the entire energy spectrum of the surface [5.7].

The third assumption presented above is also frequently violated under real
conditions. A high density of surface states results in blurring of surface
local levels. This makes it necessary to use the theory of highly doped semiconductors, which as yet has not found its way into surface science. For this
reason we shall consider electron processes on surfaces only within the domain
of applicability of the band model.

VOLKENSTEIN [5.21] pointed out two main corollaries which can be experimentally verified. The first is that various forms of adsorption exist on the
surface (charged and neutral, reversible and irreversible) which influence the
electrophysical parameters of the semiconductor. This is the central point in
Volkenstein's model.

Indeed, according to the Garrett-Brattain theory, measurements of the surface potential and surface conductivity make it possible to calculate the total surface charge $Q_s = Q_{sA} + Q_{sB}$. However, the density of biographic states is
large on any real surface (including atomically clean surfaces). The total
charge in biographic states (Sect.2.3) is composed of the charge in fast
(Q_{sf}) and slow (Q_{ss}) states $(Q_{sB} = Q_{sf} + Q_{ss})$, which have entirely different
relaxation times. During adsorption the value $Q_{sB} = qn_B$ will change for two
reasons. First, the occupancy of biographic traps changes owing to a change
in the bending of bands. Second, the parameters of these traps are modified
when adsorption occurs on or near a defect which forms the basis of a biographic
state. In the overwhelming majority of systems considered in chemisorption
and catalysis, $qn_B \geqslant qn_A$. Within the framework of measurements of integral characteristics, such as variations in $\Delta\varphi_t$ and $\Delta\sigma_s$ during chemisorption and catalysis, we cannot separate the total surface charge into qn_B and qn_A, owing
to the different characteristic relaxation times of the individual states.
Despite the abundance of reported data on $\Delta\varphi_t$ and $\Delta\sigma_s$ these data cannot be used
to derive even qualitative estimates of $n^\pm$. Therefore if the first two assumptions are not fulfilled, experiments of this kind cannot be used to determine
the main parameter of the Volkenstein model.

It is worth noting that the expressions derived in [5.21] which relate $n^\pm$
to the position of the Fermi level $q\varphi_s / kT$ on the semiconductor surface are

assumed to be valid not only for low occupancies θ, but also for high ones. However, in the latter case, one cannot neglect Coulomb interactions between charged adsorptive states. Here the density of charged surface states is no longer a simple function of $q\varphi_s / kT$, as assumed in [5.21]; it also depends on θ. A self-consistent calculation of n^+ as a function of $q\varphi_s / kT$, taking into account the Coulomb interactions between donor centers, was carried out in [5.46]. The dependence derived differs qualitatively from that assumed in [5.21].

It is equally impossible to determine separately the concentration of neutral adsorption n^0. According to [5.21], neutral adsorption ought to modify the dipole component of work function $\Delta\varphi_D$ [cf. (2.13)], which in principle can be determined from combined measurements of thermionic work function $\Delta\varphi_t$ and surface potential ΔY_s, see (2.12) [5.36], or from measurements of photoelectronic work function $\Delta\varphi_{ph} = E_g + \varkappa + \varphi_D$. During adsorption $\Delta\varphi_{ph} = \Delta\varphi_D$. However, in order to estimate n^0, one must know the dipole value $\mu_\perp$, see (2.13), which involves quantum-mechanical model-type calculations. Note that physical adsorption, especially of polar molecules, can result in a considerable change in φ_D. For instance, VILESOV adn TERENIN [5.47], while studying the change in the photoelectronic work function during adsorption of benzene on ZnO, discovered a strong polarization of benzene molecules, with the negative end of the dipole being turned towards the crystal. Also of interest are simultaneous measurements of $\Delta\varphi_t$ and $\Delta\sigma_s$ during chemisorption of CO_2 and a ZnO monocrystal in the initial stage of occupation θ [5.48]. It was demonstrated that in order to explain the observed values of $\Delta\varphi_t$ and $\Delta\sigma_s$, it is necessary for adopt extraordinarily large values for the dipole moment of adsorbed CO_2 molecules ($\mu \sim 15$ D). GÖPEL [5.48] assumed that reconstruction of the surface takes place at an early stage of adsorption. In general, this will induce changes in $\varkappa$, which enters (2.12). It must be concluded that the proposed techniques do not allow quantitative separation of the neutral form of chemisorption from physical adsorption.

Let us now investigate the possibilities of experimentally verifying the second principal implication indicated by VOLKENSTEIN [5.21], i.e., that chemisorptive equilibrium can be controlled by changing the concentration of electron (hole) gas on the semiconductor surface.

The first way to achieve such control is to introduce impurities into the semiconductor ("doping"). This, however, leads to ambiguities when one compares theoretical and experimental results. For one thing, the experimenter never knows the distribution of impurities between the bulk and the surface, especially with the high-dispersion powders used in investigations of adsorption

and catalysis. In many cases the dopants are introduced in concentrations of
about 10^{20} cm^{-3} or higher, which even at equilibrium distribution corresponds
to a surface concentration of about 10^{13} cm^{-2}, and usually even higher [5.49]
At the same time, according to estimates made by MORRISON [5.50], at surface
concentrations of impurities of about 10^{12} cm^{-2} the Fermi level on the surface
is stabilized.

It should be noted that in many cases, studies of chemisorption are carried
out with powders whose surface layer is considerably amorphized. As was pointed
out in Sect.4.2, doping of disordered systems does not influence the position
of the Fermi level in the bulk. A higher concentration of impurities
($> 10^{20}$ cm^{-3}) can result in aggregation of impurities and the formation of
solid solutions. The latter has been observed experimentally for both powders
and monocrystals [5.51,52]. This changes the bulk properties of the semiconductor, and any correlations with the electron theory are out of the question.

The second reason for the ambiguities resulting from doping is that the
impurity atoms on the surface can themselves act as centers of chemisorption.
In this case the mechanism of adsorption is no longer described by the theory
constructed for ideal surfaces. Furthermore, the impurity on the surface can
change the spectrum of biographic surface states, as indicated by numerous
data from experiments on doping of germanium, silicon, and oxide surfaces
[5.53-55]. Moreover, in [5.56] it was discovered that bulk doping of germanium, which according to [5.21] should only shift the Fermi level of the specimen, actually led to a considerable increase in the density of surface states
on an atomically clean Ge surface.

Earlier NAGAEV [5.20] demonstrated that with partially ionic crystals the
appearance of impurity atoms on the surface must lead to changes in the effective charges of adjacent regular lattice ions. This alters the electron density distribution and, in turn, the chemisorptive and catalytic activity of the
surface. The effect which a defect has on adsorptive activity depends primarily on its charge status, i.e., on the location of its local level with respect
to the Fermi level. Thus a connection is established between adsorptive activity and extrinsic conductivity within the framework of a model differing from
that offered by the electron theory [5.21]. Owing to possible manifestations
of different mechanisms, through which the impurity ions influence the electrophysical and chemisorptive characteristics of the semiconductor surface,
it is still hard to give an unambiguous interpretation of the numerous data
on surface doping.

Experiments on photoadsorption might seem able to verify the effects which
bulk properties of the semiconductor have on its chemisorptive activity in a

less adulterated form. According to [5.21], what is responsible for photodesorption or photoadsorption are the free charge carriers. What, then, will change in the mechanism if we replace an ideal surface with a real one, in other words, if we reject the first two assumptions? The presence of defects on the surface gives rise to a set of centers of recombination and "sticking" of nonequilibrium carriers, which will immediately affect the processes of generation, recombination, and bipolar diffusion. Nonequilibrium carriers can be captured on surface defects, which will result in increased adsorptive activity of the defects. We see that the existence of a complex surface energy spectrum will influence the characteristics of the photoadsorption process, and this spectrum will also be subjected to variations during adsorption. For all these reasons the simple correlation predicted in [5.21] between the sign of the photoadsorption effect and the position of the Fermi level in the specimen prior to illumination will be completely disguised.

There is one more factor which severly limits the possibilities of verifying the theory. As we have already pointed out, investigations of photoadsorption almost always consider the extreme case of almost complete thermodynamic equilibrium (with a very low level of injection) [5.32], which is actually never realized in experiments[5]. This restriction was recently lifted for the case of photocatalysis [5.58], but it was replaced by the equally strong assumption that the surface potential varies little at high injection levels. This assumption is almost never satisfied at high densities of surface states, as confirmed by numerous measurements of capacitor photo emf during adsorption.

A comparatively large number of investigations have concerned the influence of violated stoichiometry and doping on the sign of the photoadsorption effect [5.58-61]. However, as we have already noted, doping not only shifts the position of the Fermi level, but also modifies the energy spectrum of the surface, thus preventing verification of theoretical predictions. KWAN [5.59] and MOLINARE [5.60] emphasized the nonelementary nature of photoadsorption and its connection with surface defects. The same conclusions were drawn by BASOV et al. [5.61] on the basis of thorough investigations. It was convincingly demonstrated in [5.60,61] that certain distinctive features of photoadsorption cannot be explained by the electron theory within the framework of the idealized model, e.g., the wide spectral range, going beyond the limits of the intrinsic absorption band of the crystal; the irreversibility of photo-

---

[5] A more general consideration of photoadsorption at arbitrary injection levels was carried out by LEVINE [5.57].

adsorption; "memory" with regard to illumination prior to adsorption; and similarities between the changes in the conductivity of ZnO during photoadsorption of O_2 (as well as H_2 and CH_4) and photodesorption of O_2. However, these facts can be satisfactorily explained if one considers electron transitions between adsorbed molecules and surface defects.

The same objections can be raised as regards the criteria for the electroadsorption effect. The correlations indicated [5.31] between the sign of the effect and the location of the Fermi level on the surface will not exist if the first two assumptions are violated. Because the redistribution of carriers between fast and slow surface states has been neglected, and because it is so difficult to distinguish between Q_{sA} and Q_{sB}, we are still unable to verify the regularities in electroadsorption predicted by the theory.

All of the above considerations lead to rather pessimistic conclusions regarding the possibility of checking the main elements of VOLKENSTEIN's theory [5.21]. The chemisorptive properties of the surface, which according to [5.21] are determined by the relation between charged ($n\pm$) and neutral (n^0) chemisorption, depend on several factors:

(1) the position of the Fermi level on the surface,

(2) the energy levels of the adsorbed particles and their concentration,

(3) the energy spectrum of the surface (of nonadsorptive origin) and changes in it during adsorption. The theory considers only the first of these, assuming the rest to be constant. This lends great versatility to the theory: should experiment confirm it, then the first factor is assumed to be the decisive one; if theory and experiment are at variance, this is ascribed to the prevalence of the third factor. Naturally, it would be hard to invent any way of experimentally verifying such a theory.

Another equally important problem with the electron theory of catalysis has to do with investigations of the kinetics of the elementary act of adsorption and the charging of the surface. In the idealized model proposed in [5.21], the kinetics of chemisorption and catalysis is considered under conditions close to quasi equilibrium (Sect.5.2), which are never actually realized. The kinetic equations of charged and neutral adsorption (5.3) are to a large extent speculative since the constants entering these equations have not been compared with experimental data already reported for certain simple semiconductors. The range of characteristic times in (5.3), τ^0, τ^-, expands, as indicated in [5.21], from 10^{-9} to 10^5 s. This enormous range of fourteen orders of magnitude offers immense choice to the theoretician seeking the limiting stage of the process, but is of little help to the experimenter endeavoring to clarify the mechanism of the actual processes.

This is an important point. The adsorptive surface states, which according to the electron theory govern the catalytic process, are usually slow ones (Chap.6). As pointed out in Sects.3.1-3, these states are characterized by very low cross sections of capture of free charge carriers. For example, the values of the cross sections of capture for electrons and holes, c_n and c_p, for germanium at room temperature are $10^{-26} - 10^{-27}$ cm^2, which is about ten orders of magnitude lower than the corresponding values for fast states. The capture coefficients $\alpha_n = c_n v_t$ and $\alpha_p = c_p v_t$ (where v_t is the thermal velocity of the charge carriers) are correspondingly small, and the characteristic relaxation times τ_s are large. At low levels of injection the values of τ_s for germanium (having adsorptive surface states) are in the range of tens and hundreds of seconds.

From the above discussion it follows that the fast stages of catalytic reactions cannot be associated with the buildup of charge in slow adsorptive states. Then the fast adsorptive states could be assumed to be responsible for catalysis. However, this is disproved (at least for germanium) by the experimental data. Experimental investigations on the effects of adsorption on the energy spectrum of the surface (Sect.3.1) reveal that the most important and the most active centers of adsorption and charging surface are the surface defects, which form the basis of slow states (coordination-unsaturated Ge atoms). We shall demonstrate in Sect.9.3 that fast states ($\tau_s \sim 10^{-5} - 10^{-7}$ s) do not interact directly with the gaseous phase.

A very slow rate of charge accumulation on the surface during adsorption is not just typical of germanium. Comparably small capture cross sections ($10^{-21} - 10^{-26}$ cm^2) have been observed with the wide-band semiconductors CdS and ZnO (Sect.6.3). ZnO is often employed in model catalytic experiments. In wide-band semiconductors, further limitations on the rate of charge accumulation in the surface states are imposed by slowed transport of charge carriers from the bulk towards the surface [5.62]. For this reason it is unclear how experiments conducted with dielectrics (SiO$_2$, Al$_2$O$_3$) at moderate temperatures can be supposed to offer confirmation of the electron theory [Ref.5.58,pp.115, 121].

Many catalytic reactions are attributed by the authors of [5.21,58] to charged radical adsorption. This only aggravates the situation. The capture of an electron onto the orbitals of an adsorbed molecule weakens the electron's bond with lattice phonons [5.63], thereby reducing the value of the constant of electron-phonon interactions and causing a sharp decrease in the capture cross section. We shall discuss this in more detail in Sect.8.2.

It follows from the above arguments that electron transitions in a solid which are stimulated by chemisorption, illumination, and other external agents participate in the catalytic process in a much more subtle way than is assumed in the simplified models of the existing electron theory of catalysis. The discrepancy between the characteristic times of electron capture and of the catalytic process casts strong doubt on the validity of purely "electronic" explanations of the mechanism of basic catalytic processes such as hydrogenation, dehydrogenation, dehydration, and oxidation [5.64,65].

Most of the experiments cited in VOLKENSTEIN's theory [5.21] were carried out on powdered adsorbents and catalysis. As we demonstrated in Sect.4.4, electrophysical measurements on powders yield little information about electron processes on the surface, and do not permit distinctions between the various forms of adsorption ($n^{\pm}$, n^0). Such distinctions can be made to some extent in experiments on the field effect on monocrystals after appropriate treatment of the surface, which reduces the density of surface states to 10^{11} - 10^{12} cm^{-2} (Sect.2.3). A rather large quantity of data has been accumulated on the influence of chemisorption on the electron properties of the surfaces of single crystals of Ge, Si, CdS, PbS, ZnO, etc. (the data are discussed in the next chapter). For these crystals in most cases qualitative relations have been obtained between surface conductivity, surface charge and potential and the density of fast and slow states on the one hand, and the properties of adsorbed molecules on the other hand. It would seem logical to use these measurements as a basis for verifying the implications of the theory, but VOLKENSTEIN prefers to use electrophysical measurements performed on powders, mostly in the 1950s, even in his latest monographs [5.21,58].

6. The Effects of Adsorption on the Electrophysical Parameters of Real Semiconductor Surfaces

In order to construct a consistent mechanism of the charge buildup which occurs during real surface processes one must first establish correlations between the nature of the emerging adsorptive bonds and the corresponding electrophysical manifestations [6.1,2]. Only then can one hope to unravel the complex interconnections between the molecular and electron processes taking place on the surface. With this end in view, let us analyze the effects which the various forms of adsorption discussed in [6.3] have on the basic electrophysical parameters of semiconductors. We shall start our analysis with the simplest case: physical adsorption of inert gases.

6.1 Physical Adsorption

FIGUROVSKAYA and KISELEV [6.4] investigated the influence of the adsorption of inert gases on the dc conductivity of disperse, slightly reduced rutile ($\underline{n}$-TiO$_2$). It turned out that adsorption of Ar and Xe at room temperature results in a reduction of σ(Fig.6.1), while evacuation at 300 K completely restores the initial value of σ_0. The relative variation σ/σ_0 with Xe is somewhat lower than with Ar, due to the difference in the atoms' size and the different values of adsorption. This effect was proved to bear no connection with oxygen contamination of inert gases or with variations in the specimen's temperature. In the case of a polydisperse specimen the decrease in σ is connected with an increase in the barriers between particles (Sect.4.4), due either to negative charging of the surface or to a change in the dipole component of the work function, φ_D, see (2.13). Taking into account that physical adsorption of Ar and Xe at room temperature is quite low ($\sim 10^9$ cm^{-2} [6.3]), it is hardly possible to attribute the change in σ entirely to the increase in φ_D.

The drop in σ is most probably associated with the accumulation of negative charge in a number of surface traps. Physical adsorption results in polarization of adsorbed atoms and generates induced dipole moments $\mu_i = \alpha \cdot E$. Accord-

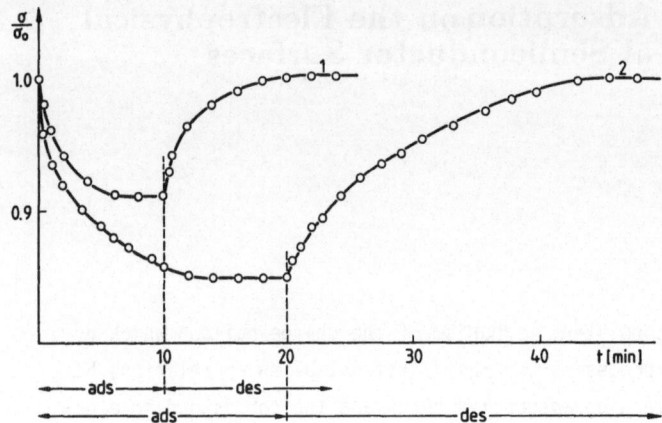

Fig.6.1. Kinetics of variations in relative conductivity of TiO_2 during adsorption of Xe (1) and Ar (2) at T = 300 K, p = 0.1 torr, $\sigma_0 = 1 \cdot 10^{-5}$ ohm^{-1}

ing to [6.5], the average field strength over a rutile surface is $\xi \sim 3 \cdot 10^5$ cgse units, which in the case of Ar and Xe leads to values of $\mu_i \sim (0.5-1) \cdot 10^{-18}$ cgse units. Owing to the high permittivity of rutile, the local field of the emerging dipole can alter the parameters of a defect adjacent to the adsorption center, which can turn it into the center of capture of an electron. Because of the disorderly nature of the surface, the emerging dipole can also change the depth of a level of fluctuative origin. Since the conductivity of rutile is most probably of the polaron kind, the shallow potential pit created by virtue of the interaction between the adsorbed molecule and the defect will become the site of localization of a polaron.

The captured carrier will interact with the adsorbed molecule, strengthening its bond with the surface. Measurements of the differential heat of adsorption of Ar [6.6] indicate that with rutile the initial value for the heat of adsorption (15 kJ $\cdot$ mol^{-1}) is considerably greater than with other adsorbents having lower values of permittivity (8.8 $\pm$ 0.8 kJ $\cdot$ mol^{-1}). This confirms what was surmised regarding the strengthening of the bond between polarized atom and the surface when a charge carrier becomes trapped nearby. The higher the concentration of defects the greater the effect of charging will be. And indeed, as the reduction of rutile proceeded, the effect of surface charging became more pronounced [6.4].

The effects of charging of a semiconductor surface during adsorption of inert gases are rather small. Apparently, the reason they could be detected in [6.4] was that high-dispersion specimens and an adsorbent with a high permitivity[1] were used. Since according to (4.7) the conductivity of such a system is proportional to the number of barriers, which in this case is quite high ($\sim 10^5 - 10^6\,cm^{-1}$), small changes in individual barriers result in detectable variations in σ.

Even with an ideal lattice, the polarization of the lattice during physical adsorption can cause shallow surface states to arise. The capture of charge carriers on these traps will lead on the other hand to charging of the surface (ΔQ_{sA}) and on the other hand to a strengthening of the adsorptive bond. Indeed ROSENBERG [6.9] pointed out that owing to the presence of electrons in the SCR, the value of the heat of adsorption of Kr on germanium is much greater than the corresponding value calculated on the basis of the theory of van der Waals forces [6.3]. An exhaustive analysis of the interaction of physisorbed molecules of Ar, Kr, Xe, N_2, and CO with germanium <111> surfaces was carried out in [6.10]. If the electron subsystem of the semiconductor is taken into account in calculations of dispersive interactions, there is a significant increase in the heat of adsorption (e.g., $16.7 - 18.8\ kJ \cdot mol^{-1}$ instead of $8.4\ kJ \cdot mol^{-1}$ in the case of adsorption of N_2 on germanium).

On a defect-ridden surface, the physisorbed molecules can, by virtue of polarization, change the kinetic parameters of defects and their position in the energy spectrum, causing a change in the surface charge ΔQ_{sB}. This kind of local interaction apparently takes place during reversible low-temperature physical adsorption of O_2 on PbS surfaces leading to a reversible change in the photoelectromotive force (the dark conductivity remains the same) [6.11]. The authors of [6.12] discovered that noble-gas atoms implanted into silicon modify the energy spectrum of vacancy-type traps. The extended Hückel method was used to estimate the distortion of the molecular orbitals of a Si atom caused by nearby implantation of a noble-gas atom. Finally, physical adsorption, by virtue of polarization-type interactions, can lead to some reconstruction of the surface, thus changing the energy spectrum of biographic states. A theoretical analysis of the possibility that surface states are reconstructed during adsorption was made by SPARNAAY [6.13]. As mentioned above, the possibility of surface reconstruction is indicated by experimental data for the ZnO-CO_2 system [6.14].

---

[1] FIGUROVSKAYA and KISELEV [6.4] observed a noticeable change in σ during adsorption of Ar on ferroelectric-barium titanate. The effects of adsorption on the ferroelectric properties of crystals are reported in [6.7]. The connection between the ferroelectric and catalytic properties of adsorbents was investigated in [6.8].

6.2 Adsorption of Donor Molecules

Let us now turn to stronger forms of adsorption, accompanied by the creation of hydrogen and coordination bonds on the surface. It is well established that adsorption of water, alcohols, carbonic acids, and amines at room temperature results in an accumulation of positive charge on the surface of certain semi-conductors. Data has been obtained for both monocrystalline and polydisperse specimens of oxides of zinc, copper, titanium, tin, cadmium, etc., as well as for sulfides and selenides PbS, Sb_2Se_3, and MoS_2 [6.2,15-21]. According to Volkenstein's model [6.22], the buildup of positive charge on the surface during the adsorption of valence-saturated molecules is associated with their ionization. However, the formation of H_2O^+ or ROH^+ ions at such temperatures seems hardly plausible. There is no support for this in the experimental re-sults, and it does not agree with the high ionization potential (greater than 10 eV) of these molecules. As we should in [6.3], adsorption of water and spi-rits on hydrated oxide surfaces is completely reversible, and the heats of ad-sorption do not exceed 1 eV [Ref.6.3,Fig.4.14]. Adsorption proceeds with quite a low activation energy.

It would be still more difficult to explain the formation of H_2O^+ or ROH^+ during the adsorption of such substances on single crystals of germanium, si-licon, or intermetallic compounds, which under ordinary conditions are covered with a hydrated oxide film. Adsorption of water, alcohols, carbonic acids, and amines on these semiconductors always leads to positive charging of the surface [6.21,23-35]. In most works concerned with the influence of adsorption on elec-trophysical parameters, the question of what forms of adsorption are respon-sible for the charging has not received due attention. As a rule, measurements of the surface charge have been carried out separately from measurements of the adsorptive characteristics of the surface.

A consistent series of studies of the mechanism of adsorption and charging of germanium surfaces during adsorption of donor-type water and ammonia mole-cules was conducted in [6.36-39] starting in 1966. Electrophysical measure-ments using the field-effect technique were carried out on Ge monocrystals pretreated in a peroxide etchant, and the adsorptive properties were measured on disaggregated powders prepared by crushing the same monocrystals. The pow-ders and the crystals were treated and handled under exactly the same condi-tions. Of course, the surface properties of a single crystal and a powder dif-fer. The difference, however, can be accounted for, and a comparison of the data yields valuable information on the relationship between the electron and molecular processes occurring on the surface.

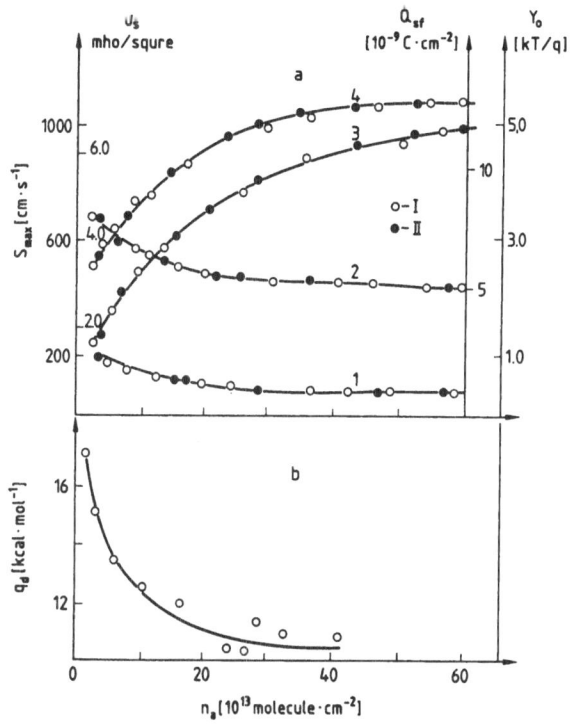

Fig.6.2a,b. Electrophysical parameters (a) and differential heat of adsorption q_d of water on GeO_2 (b) as functions of the concentration of adsorbed molecules n_a. (1) Maximum rate of surface recombination S_{max}, (2) charge accumulated in fast states Q_{sf} at $Y_s = 6kT/q$, (3) surface conductivity σ_s, (4) surface potential Y_{s0} for hydrated germanium surface, (I) adsorption, (II) desorption

Let us first consider the data in [6.37] on changes in the electrophysical parameters during reversible adsorption of water on a hydrated germanium surface, the specimen having been vacuumized at 300 K. As can be seen from Fig.6.2, the surface conductivity σ_s increases with an increase in the concentration of adsorbed molecules n_a. Having estimated the initial value of the surface potential from field-effect measurements ($Y_s = 2.6$ kT), one can use the conductivity figures to compute the function $Y_{s0}(n_a)$(at zero field) (Fig.6.2). The curves $\sigma_s(n_a)$ and $Y_{s0}(n_a)$ are in good agreement with the data in [6.24,33], which we have presented as functions of the concentration of adsorbed molecules n_a with the aid of adsorption isotherms [6.40]. For comparison, Fig.6.2 also displays the data on the differential heat of adsorption of water on germanium dioxide. PRUDNIKOV [6.40] demonstrated that the value for adsorption of water on GeO_2 is close to the corresponding figures for real germanium surfaces. The greatest variations in Y_0 and σ_s are observed at occupancies of about 10^{13} to 10^{14} molecules $\cdot$ cm^{-2}, corresponding to the highest values of adsorption heat. The same region is characterized by a slight decrease in the maximum values of the surface recombination rate S_{max}, as well as the charge in fast states Q_{sf}. The total charge of the surface Q_s and the surface potential Y_{s0} are determined mainly by the charge in slow states, whose concentration after such treatment ($T_{cal} = 300$ K) is an order of magnitude greater than the concentration of fast states [6.39]. Investigations of the kinetics of slow relaxation at different temperatures and calculations of the parameters of slow states in accordance with the technique outlined in Sect.3.3 indicate that at occupancies up to 10^{13} cm^{-2} the concentration of slow states n_{ss} is increased by nearly an order of magnitude [6.39].

In our analysis of the mechanism of water adsorption on hydrated oxide surfaces we demonstrated [6.3] that the coordinative bond mechanism dominates for occupancies up to 10^{13} or 10^{14} molecules $\cdot$ cm^{-2}, while at higher occupancies the hydrogen bond mechanism prevails. On these grounds one can assume that it is the coordinative (donor-acceptor [6.3]) bonds that are responsible for changes in the electrophysical parameters.

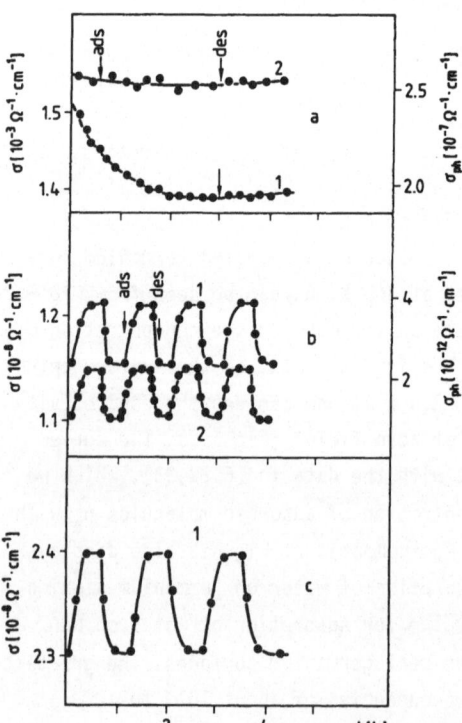

Fig.6.3a-c. Variations in conductivity σ (1) and photoconductivity $\Delta\sigma_{ph}$ (2) of rutile monocrystal in adsorption-desorption cycling of water (**a,b**) and ammonia (**c**) on oxidized and hydrated (**b,c**) and reduced (**a**) specimens. First admission at $t = 0$

A similar effect of reversible positive charging of the surface and a change in dark conductivity σ were observed during adsorption of water on hydrated surfaces of rutile monocrystal [6.41] (Fig.6.3), as well as on polydisperse rutile specimens. Adsorption of water also influenced the steady-state photoconductivity $\Delta\sigma_{ph}$ in a reversible fashion. This is most probably associated with an increase in the effective cross section of capture of minority carriers (holes) by surface centers in the local fields of coordination-bound water molecules. The formation of strong coordinative bonds in this system was also proved by spectroscopic investigations [6.3]. We shall not go into a discussion here of the numerous data from investigations of photoadsorption indicating the effects of adsorption on the photoconductivity of polycrystalline films. The high photosensitivity of such films was proved in some cases to be connected with changes in the intercrystallite barriers due to illumination [6.42].

On hydrated surfaces of oxides and of germanium and silicon, vacuumized at 300 K, the majority of coordination-unsaturated centers are blocked by water molecules. According to spectroscopic data [6.3], the coordination-bound water molecules are completely removed only after heat treatment at $T_{cal} = 500$ or 600 K for germanium and at $T_{cal} = 700$ K for rutile. In this connection it would seem advisable to analyze the variations in the electrophysical parameters during dehydration of the surface. Let us first consider the more reliable data for germanium [6.37,38].

As Fig.6.4b,c shows, in the range $T_{cal} = 300 - 500$ K the values of S_{max} and Q_{sf} increase sharply. At the same time, the surface potential Y_{s0} (Curve 1) varies little, which agrees well with the data cited in RZHANOV's monograph [6.32]. These small variations in Y_{s0} are attributed in [6.32] to sign-reversible changes of charge in fast (Q_{sf}) and slow (Q_{ss}) states. All measurements in [6.31,35,37-39] were performed in vacuums of 10^{-5} to 10^{-6} torr. Having improved the vacuum to 10^{-11} torr, SOCHANSKI and GATOS [6.43] observed a considerable negative charging of the Ge surface ($T_{cal} = 520 - 560$ K), which they reported without attempting to analyze the causes. MATVEEV and PRUDNIKOV [6.44] demonstrated that the function $Y_{s0}(T_{cal})$ in vacuums of 10^{-8} to 10^{-9} torr differs qualitatively from that found earlier (Fig.6.4, Curves 1 and 2). These differences are associated with readsorption of water vapor on the cooling specimen in a poor vacuum. The quality of the vacuum has less effect on Q_{sf}(Curves 1 and 2).

As we showed in Sect.3.1 (Fig.3.6), adsorption of water on a specimen vacuumized at $T_{cal} = 500$ K, starting from occupancies of about 10^7 molecules $\cdot cm^{-2}$, leads to a strong shift of the surface potential in the positive direction and to the appearance of slow relaxation in field-effect experiments. The latter

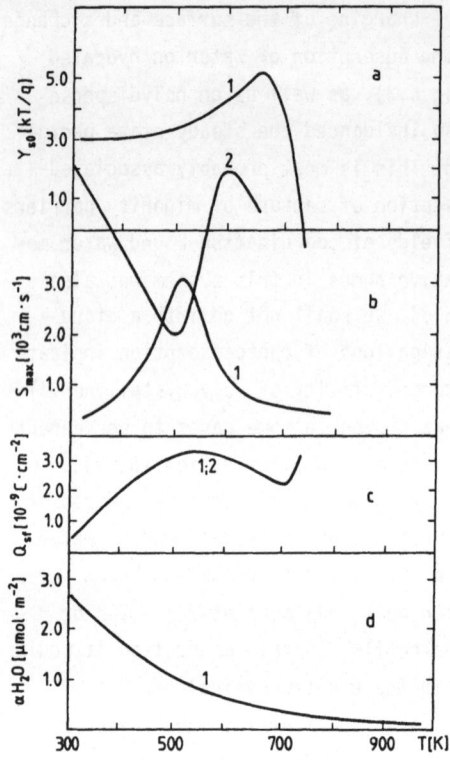

Fig.6.4a-d. Surface potential at zero external field Y_{s0} (**a**), maximum rate of surface recombination S_{max} (**b**), charge in fast states Q_{sf} at $Y_s = 6kT/q$ for germanium monocrystal surface (**c**), extent of surface hydration α_{H_2O} for disaggregated germanium powders (**d**) as functions of vacuumization temperature in "greasy" vacuum of 10^{-6} torr (*1*) and "grease-free" vacuum of 10^{-8} torr (*2*)

indicates the emergence of slow states on the surface. The charge in fast states Q_{sf} and the rate of surface recombination remain the same up to $n_a \sim 10^{11}$ molecules $\cdot$ cm^{-2} (Fig.3.6). This fact, together with the conclusion that the charge in slow states is more than an order of magnitude greater than the charge in fast states, implies that transformations of fast states into slow states and vice versa are out of the question. Most probably, each of these types of states owes its existence to a particular group of defects. Adsorption and subsequent dehydration at T_{cal} = 500 K are accompanied by entirely reversible changes in Y_{s0}, S, and Q_{sf}. All this is evidence of a lack of any noticeable structural changes in the oxide film during vacuum heat treatment of Ge samples at 500 K.

It is an entirely different picture when the temperature of heat treatment is raised above 500 K. As is shown in [Ref.6.3,Chap.3], at T_{cal} > 500 - 600 K the predominant role is transferred to dehydration, which is associated with the removal of structural OH groups. Let us first consider the case of germanium. Figure 6.4 shows that at T_{cal} > 500 K the values of S_{max} and Q_{sf} decrease considerably. At T_{cal} > 600 K the surface potential shifts into the region of negative values. The process of slow relaxation becomes so decelerated that the parameters of slow states can no longer be assessed. Subsequent rehydration in saturated water vapor does not restore the initial adsorptive properties of the surface. As can be seen from Fig.6.4, the value of α_{H_2O} for rehydrated specimens is lower than that for the initial specimen ($T_{cal} = 300$ K). Similar irreversibility is observed for Q_{sf} and S_{max}. At these temperatures the processes of dehydration cause irreversible reconstruction of the film structure, accompanied by the formation of tetragonal GeO_2 elements [6.3,39]. The appearance of phase elements with octahedral coordination of Ge atoms results in a decrease in the number of coordination-unsaturated Ge atoms. This is in good agreement with the observed decrease in the number of slow traps on the $Ge-GeO_2$ interface. The value of Q_{sf} increases sharply at $T_{cal} > 700$ K [6.37]. At these temperatures the destruction of the oxide film begins and Ge atoms start to diffuse through the film. The concentration of defects on the $Ge-GeO_2$ interface increases sharply.

As demonstrated in Sect.3.1, at these T_{cal} the bell-shaped component of the surface recombination rate completely disappears. The recombination is entirely determined by the monotonic component S_m (Fig.3.5). Adsorption of water up to occupancies of $n_a \sim 10^{12}$ molecules $\cdot$ cm^{-2} is accompanied by an increase in the monotonic components S_m; the bell-shaped component S_b appears only at comparatively high occupancies ($n_a > 10^{13}$ molecules $\cdot$ cm^{-2}). It was demonstrated in [6.45] that this pattern of variation in S is connected with irreversible dissociative adsorption of water, accompanied by reconstruction of the surface and the appearance of hydroxyl and hydride groups on it. The latter is confirmed by IR spectroscopy [6.46]. As a result, a number of states drop out from the quasi-continuous spectrum of fast states, and the bell-shaped component of surface recombination S_b appears, which depends on Y_s (Sect.3.1;Fig.3.9).

Adsorption of water on highly dehydrated surfaces [Ref.6.3,Chap.3] proceeds according to a dissociative mechanism, which implies the formation of OH groups. This results in partial restoration of the hydroxyl cover. Figure 6.3a,b shows that the sign of variation in σ during adsorption on a rutile monocrystal (dehydrated and partially reduced at $T_{cal} \sim 850$ K) is reversed, which corresponds to negative charging of the surface during adsorption of the same kind of molecules (H_2O) depends considerably on the conditions of treatment and the extent of surface dehydration.

To obtain more detailed information on the mechanism by which adsorbed molecules influence the electrophysical parameters of germanium surfaces, we performed a series of experiments with adsorption of molecules other than water [6.47-49]. Let us briefly discuss the experiments with hydrated Ge specimens vacuumized at T_{cal} = 500 K [6.47]. As with water, the adsorption of ammonia, pyridine, acetone, and acetylacetone, starting with the lowest occupancies ($n_a \sim 10^7$ molecules $\cdot$ cm^{-2}), leads to positive charging of the surface and to an increase in the density of slow surface states Q_{ss} (Figs.9.6-9.8). Adsorption of all these molecules on hydrated surfaces of oxides and monatomic semiconductors was shown to follow the coordinative mechanism [6.3]. In the case of NH_3 and pyridine the undivided pair of electrons is supplied by nitrogen atoms; with acetone and acetylacetone the donors are the oxygen atoms (this is also true of CO or nitrobenzene, whose adsorption results in positive charging of the surface). Simultaneous measurements of the charging of the germanium surface and the differential heat of adsorption of NH_3 [6.48] indicate that the greatest variations in Q_{ss} correspond to the region where the values for the heat of adsorption are the highest (~ 1 eV). This region of occupancies ($\sim 10^{13}$ molecules $\cdot$ cm^{-2}) corresponds to the formation of coordinative bonds [6.3]. As opposed to the adsorption of water, adsorption of NH_3, Py, and acetone does not influence Q_{sf} and S, which is a strong argument against the dipole mechanism of neutralization of fast states and recombination centers (see Sect.3.1). These questions will be discussed in greater detail in Chap.9.

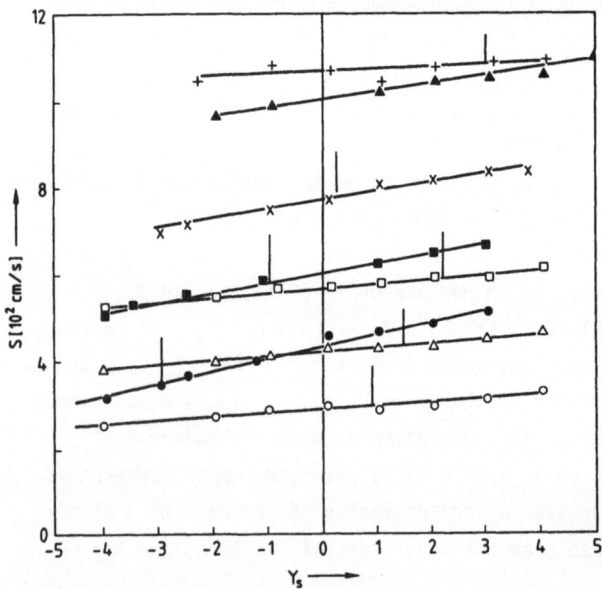

Fig.6.5. Variations in the monotonic component of the rate of surface recombination S(Y_s) for germanium during adsorption of CH_3OH and NH_3.
o - T_{cal} =700 K, □ - CH_3OH, 10^{-2} torr, ▲ - CH_3OH, 10^{-1} torr, Δ - T_{cal} =500 K after adsorption of CH_3OH, ● - T_{cal} =700 K, ■ - NH_3, 10^{-3} torr, x - NH_3, 10^{-2} torr, + - NH_3, 10^{-1} torr. Vertical dashes indicate surface potential at zero field

Let us now consider the case of adsorption on a highly dehydrated germanium surface (T_{cal} = 700 K). Adsorption of NH_3 and CH_3OH, like adsorption of water, resulted in irreversible positive charging of the surface and an increase in the monotonic component of the rate of surface recombination S_m (Fig.6.5). The latter is apparently connected with dissociative adsorption of these molecules accompanied by the formation of amine and methyl groups on the surface. This notion is also supported by spectroscopy [6.50]. In contrast to adsorption of NH_3 and CH_3OH, molecular nondissociative adsorption of Py had no effect on S_m (Fig.3.7). In none of these cases did the curves of capture on fast states show any changes during adsorption. Apparently, recombination and fast capture take place on different group of defects.

An attempt to verify these conclusions was made in [6.51] in experiments on the adsorption of hydrogen atoms on germanium surfaces dehydrated at 750 K. The hydrogen atoms were obtained either by pyrolysis of H_2 on a hot tungsten filament, or in high-frequency discharge. Adsorption of molecular hydrogen did not alter the electrophysical parameters of the surface of Ge_{700} specimens. This fact is in agreement with mass-spectroscopic data [6.52], from which it follows that the sticking coefficient for H_2 is less than 10^{-8}. Adsorption of H atoms resulted in a shift of Y_{s0} by 1.5 to 3 kT/q in the positive direction (vertical dashes in Fig.6.6). The variation in Y_{s0} was accompanied by an increase in monotonic component S_m, with about the same kinetic parameters (Fig.6.6). The steady-state values of Y_{s0} and S_m were attained after the specimen had been exposed to atomic hydrogen for about 40 minutes. The curves of capture did not undergo any changes, which confirms the above conclusion that the system of centers responsible for recombination is independent of the system of fast surface states. The experiments [6.49,51], in which a charge-laden grid was inserted between the sample and the source of H atoms, reveal that it is the neutral H atoms, and not the H^+ ions, that are responsible for the variations in Y_{s0} and S_m. Experiments on adsorption of proton-acceptor pyridine molecules on hydrogenated Ge(H) surfaces also offer evidence against the presence of H^+ ions. As Fig.6.6 shows, adsorption of Py had no effect on S_m, although the potential shifted towards more positive values of Y_{s0}. Vaccumization of the specimen at 300 K after adsorption did not restore the initial values of Y_{s0} and S_m. When, however, T_{cal} was increased to 600 K, these parameters assumed values close to those before adsorption (Fig.6.6), in agreement with the data on desorption of hydrogen [6.52]. Most probably, chemisorption of H$\cdot$ atoms results in the rupture of valence-saturated $\geqq$ Ge - Ge $\leqq$ and $\geqq$ Ge - 0 - Ge $\leqq$ bonds on the Ge - GeO_2 interface and causes the formation of hydride GeH groups, as is also confirmed by IR and UV spectroscopy [6.53].

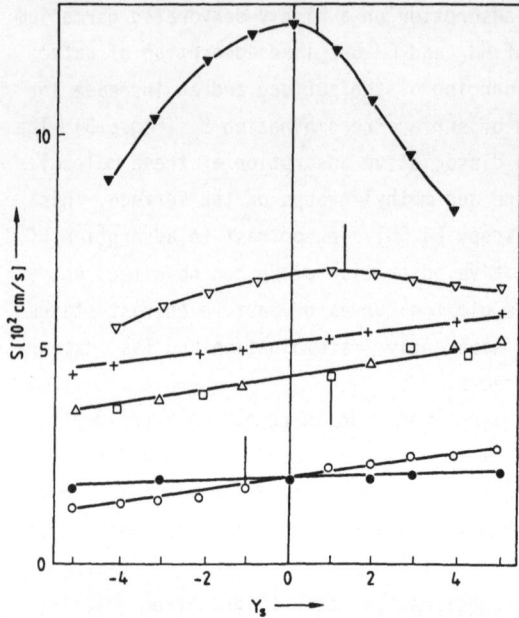

Fig.6.6. Variation in rate of surface recombination $S(Y_s)$ for germanium during adsorption of hydrogen atoms: $\circ$ - T_{cal} =700 K, $\triangle$ - adsorption of H (40 min), $\bullet$ - T_{cal} =600 K after adsorption of H, + - adsorption of H with grid negative, $\square$ - adsorption of pyridine (10^{-1} torr) on hydrogenated Ge(H) surface, ∇ - adsorption of water vapor (8 torr) on Ge(H), $\blacktriangle$ - adsorption of O_2 at 400 K on Ge(H). Vertical dashes indicate surface potential at zero field

This leads to reconstruction of the boundary layer and an increase in the density of surface states[2]. The interface we are dealing with here is different from the interface between Ge and an electrolyte, where the adsorption of hydrogen is accompanied by a symbatic increase in S_b and Q_{sf} [6.55].

The exposure of a hydrogenated specimen to water vapor at about 8 torr for two hours resulted in a certain distortion of the function $S(Y_s)$. Heat treatment of hydrogenated Ge(H) specimens in dry oxygen at 400 K led to the appearance of the bell-shaped component S_b (Fig.6.6). Mass spectra taken under similar conditions registered the release of H_2O molecules [6.52]. Adsorption of O_2 results in partial destruction of hydride groups and further oxidation of the germanium surface.

From the above discussion it follows that slow, fast, and recombination states represent three separate groups of surface states. Coordination complexes, having emerged in the early stages of adsorption, act as slow surface states. The fast centers and the recombination centers at this stage are not engaged in direct interaction with adsorbed molecules. In adsorption of water

---

[2] Adsorption of H atoms on an atomically clean Ge surface, as opposed to a real surface, proceeds on dangling $\geqq$ Ge$\cdot$ bonds which results in negative charging of the surface. ELEED and UPS data point to the formation of hydride groups and reconstruction of the surface (see [6.54] and references therein).

vapor, as we emphasized earlier, the neutralization of these centers takes place at high occupancies and is associated with proton processes (Chap.9). Variations in the monotonic component of the rate of surface recombination S_m, observed with dehydrated specimens, are connected with dissociative forms of adsorption. The bell-shaped component S_b is clearly manifested only during adsorption of water vapor at high occupancies, and when H_2O molecules form on the surface during adsorption of O_2 on a hydrogenated Ge surface. Apparently the appearance of S_b is associated with deep-penetration hydration processes and the resulting reconstruction of the surface phase of germanium[3].

All the above arguments hold, if only in a qualitative way, for real silicon surfaces. Investigations carried out recently at Moscow University [6.58] revealed that the bell-shaped component S_b persists after heat treatment of Si specimen at temperatures up to 800 K. However, because the function has a rather complicated form, it becomes difficult to split S into S_b and S_m components unambiguously. The persistence of S_b up to quite high values of T_{cal} and its immutability with respect to various adsorptive factors are apparently associated with the peculiarities of hybrid wave functions of Si as compared to Ge (the more substantial role is played by p-states in Si-O bonds; see below).

Surface doping with metals is often used to modify the electrophysical properties of semiconductor surfaces. Adsorption of metals can proceed either from the liquid phase (solutions and etchants) [6.59,60], or from molecular beams [6.61]. Depending on the magnitude of the electrochemical potential of the impurities as compared to that of the semiconductor, either donor-type or or acceptor-type states can be created on the surface. Note that the sign and magnitude of the charge accumulated on the surface as a result of doping depends strongly on the method of doping, the concentration of dopant, and the condition of the surface. Most works do not provide a convincing explanation of the mechanism of adsorption and charge buildup. Many researchers assume that metallic atoms can replace hydrogen in the hydroxyl groups that are always

---

[3] NEIZVESTNY et al [6.56] assumed that the coordination bond between H_2O molecules and surface atoms of Ge is responsible for the appearance of S_b. Pyridine, like nitrobenzene [6.56], also produces coordination bonds, but this does not give rise to S_b (Fig.6.6). The formation of a coordination bond results in an increase in the capture cross section of hole c_p on the adsorptive center [6.20,57], and not c_n, as required by the dipole model of recombination centers [6.31](Sect.3.1). Most probably the variations in S_m are due to dissociative forms of adsorption, accompanied by reconstruction of the surface and the creation of new surface chemical groups.

present on most semiconductor surfaces (see, e.g., [6.62]). Such conclusions are often derived from conductivity measurements during adsorption of ions from solutions, which, of course, have nothing to say about the forms of emerging adsorptive bonds [6.63,64]. The mechanism by which surface silicides form and the way in which they influence the electrophysical characteristics of silicon interfaces are discussed in [6.65].

6.3 Adsorption of Acceptor Molecules

One system used to verify the implications of the electron theory of adsorption is that of oxide and oxygen. Oxide semiconductors have served as the basis for numerous investigations on the effects of oxygen adsorption on conductivity, the work function, the Hall constant, and other parameters [6.16, 17,18,66,67]. It has been demonstrated that at moderate temperatures oxygen is chemisorbed on a dehydrated oxide surface as an acceptor, which results in negative charging of the semiconductor surface. At higher temperatures the chemisorption is interfered with by the exchange between gaseous oxygen and impurities or over-stoichiometric metal atoms on the surface. This causes diffusion of oxygen and ionized impurity atoms from the bulk towards the surface and back. The relationship between chemisorption and the electrophysical properties of the semiconductor becomes quite complicated.

Oxide semiconductors are often believed to be the most convenient object for investigating electronic phenomena during chemisorption and catalysis. Unfortunately, this is not the case for several reasons. First, there is the question of the cleanliness of the surface. The surface of every oxide is inhabited by hydroxyl groups which can only be removed at sufficiently high temperatures [Ref.6.3,Chap.3]. Apart from these innate "impurities", various adsorptive organic contaminants are also always present on the surface (contaminants adsorbed from the gaseous environment, vapors of vacuum grease, etc.). These contaminants are "carbonized" in the course of vacuum heat treatment, and EPR spectra usually reveal a strong signal coming from carbon contaminants actively engaged in interactions with oxygen [6.68]. This important feature has been neglected in the majority of investigations on the mechanism of chemisorption. Only recently has the development of Auger spectroscopy, radio spectroscopy, and mass spectroscopy of the gaseous phase over the adsorbent furnished the means for studying the actual composition of the oxide surface [6.69].

A second reason why oxides are not ideal for such investigations is the ambiguity of stoichiometric composition. It is often assumed that heat treatment leads to a loss of oxygen and thus to the appearance of an over-stoichiometric metal. The surface is then assumed to display electron-type conductivity. However, under certain conditions the surface may lose metallic atoms [6.3], as confirmed by mass spectroscopy [6.70] and indirect information derived from measurements of the conductivity of semiconductor films [6.71]. The surface can then display p-type conductivity [6.72]. Also, the oxide is often subjected to prolonged heat treatment in oxygen to cleanse the surface. This leads to the appearance on the surface of firmly bound forms of chemisorbed oxygen, which can appear in various charge states. All this makes it quite difficult to ascertain the initial state of the surface when carrying out electrophysical measurements.

Finally, there is a third important obstacle in the interpretation of experimental data obtained for oxides. As pointed out in Sect.4.1, owing to the substantial contribution of the ionic component in a number of oxides, the polaronic mechanism of conduction begins to play a decisive role.

Adsorption of oxygen on oxides may lead to the formation of a number of ionic forms according to the following scheme:

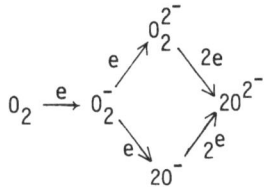

These forms have been experimentally observed in many studies of the effects of temperature and oxygen pressure p_{O_2} on the conductivity of oxides. An especially large number of investigations have dealt with disperse powders of TiO_2 and ZnO [6.18,66,73-83]. For example, the conductivity σ of ZnO at 720 K in an atmosphere of oxygen ($p_{O_2} > 10^{-4}$ torr) varies according to the law

$$\sigma = K p_{O_2}^{1/m} \quad ,$$

where $m = 4$. This corresponds, according to [6.80], to the reaction $1/2\ O_2 \rightarrow O^{2-}$ (lat.), the oxygen in the gaseous phase being at equilibrium with the lattice oxygen. The conductivity of V_2O_5 in the temperature range 400 - 700 K in a similar fashion [6.81]. For oxides with n-type conductivity, the value of σ increase in the oxygen pressure p_{O_2}; for p-type semiconductor oxides the conductivity decreases with an increase in the oxygen pressure. Except at high temperatures, when the direct transition $O_2 \to O^{2-}$ (lat.) is possible, the relationship between conductivity and the pressure of oxygen is usually more complicated, and m is not an integer, since adsorption is accompanied by the creation of several chemisorbed forms which differ in their charge status. For this reason, several techniques are usually used simultaneously in investigations of the mechanism of chemisorption. In attempting to identify the stable charged forms of chemisorbed oxygen, one must bear in mind that an affinity towards electrons is displayed only by neutral atoms and molecules of oxygen: $A_{O_2} = 0.44$ eV; $A_O = 1.47$ eV. The reaction $O^- + e \to 2O^-$ ($E = 2.62$ eV) and $O_2 + 2e \to 2O^-$ ($E = 2.19$ eV). The formation of a doubly charged O^{2-} ion on the surface is not possible. Even if the O^{2-} ion becomes a regular lattice ion and its stability is argumented owing to Madelung energy and the energy of the crystal field, the effective charge of the oxygen ion (as opposed to its formel charge) is much less than 2.

Due to the great variety of forms in which oxygen can appear, the energy spectrum of chemisorbed atoms can no longer be represented as a set of independent levels. As follows from the theory of multiple-charge impurity centers, the very existence of a given level depends on whether the underlying level is occupied or not. All this will modify the statistic of recombination and capture of charge carriers on surface states of this kind [6.31].

A detailed mechanism for the adsorption of oxygen cannot be derived directly from electrophysical investigations. The majority of researchers concerned with analyzing the relationship between conductivity and the pressure of oxygen have come to the conclusion that chemisorption of oxygen involves, apart from free electrons of the oxide, the surface states, which help to change the charge status of adsorbed forms of oxygen. Consequently, there is no experimental proof that the thermally stimulated processes

$$O^-(\text{ads.}) + e \to O^{2-}(\text{ads.}) \quad ,$$

$$O_2^-(\text{ads.}) + e \to O_2^{2-}(\text{ads.}) \to 2O^-(\text{ads.})$$

involve only conduction electrons. As regards the regular lattice atoms, the first process is prevented by high endothermicity, in contrast to the corresponding processes taking place during construction of the lattice. The second process, even though it consumes much less energy ($E = 2.1$ eV), is not realized either, owing to weakness of bonding between O_2^- and the surface. In order for these processes to occur with a sufficiently low activation energy, covalent bonds must be formed between O^- or O_2^- and the surface, which demands the presence of donor-type centers (metallic cations, Lewis acids, anion vacancies) on the surface.

The diversity of forms of chemisorbed oxygen and defects which come into being during heat treatment of oxides, together with the active processes of diffusion of anion-type and cation-type vacancies from the bulk towards the surface (and vice versa), gives rise to an intricate dependence of the electrophysical parameters of the surface on temperature and on the biography of the samples. For instance, preliminary investigations of contact potential difference $\Delta\varphi_t$ and conductivity $\Delta\sigma$ during chemisorption of O_2 [6.74] revealed that both strongly and weakly bound molecules of oxygen are present on the surface; these molecules are responsible for reversible and irreversible variations of $\Delta\varphi_t$ and $\Delta\sigma$ in the adsorption-desorption cycle. Naturally, it would be impossible to elucidate the nature of these forms of adsorption within the framework of the barrier mechanism, since heat treatment and adsorption modify the biographic surface spectrum strongly and in a complicated fashion. One must turn to the study of single crystals.

In the first investigations of surface properties of ZnO monocrystals, carried out by HEILAND et al. [6.76,77], it was demonstrated that both the dark conductivity and photoconductivity of a crystal include a long-term component. The density of biographic states on the surface amounts to about 10^{13} cm^{-2}.

Using the field-effect technique, MANY [6.84] succeeded in demonstrating that a rich accumulation layer is always present on etched surfaces of ZnO monocrystals. He attributed this to the positive charge of ionized superstoichiometric Zn atoms on the surface (about 10^{14} cm^{-2}). Chemisorption of oxygen resulted in a decrease in surface electron-type conductivity; a weak accumulation layer, however, still persisted. Investigations of the kinetics of relaxation of the surface charge (in the presence of oxygen) suggest very low rates and capture cross sections ($c_n \sim 10^{-26}$ cm^2). HEILAND and KUNSTMANN [6.76] also reported highly retarded kinetics of chemisorption of O_2 on ZnO. In order to explain the smallness of the capture cross sections, Many [6.84] assumed the existence of a buffer layer, which separates the adsorbed oxygen from the bulk of the crystal. The exchange of charge carriers through this barrier takes place via a tunnel mechanism. The nature of this buffer layer was not analyzed by Many. It is interesting to note that very low values of c_n were observed in [6.79] for both an atomically clean cleavage surface of ZnO and for a real ZnO surface obtained by etching. As will be shown in Chap.8, the slow processes can be explained without the concept of an intermediate phase. The low values for the capture cross sections can be attributed to the largeness of the constant of electron-phonon interaction on the surface. Many was right to emphasize that the existence of well-developed accumulation layers having a rather narrow barrier makes ZnO a convenient object for the study of oscillations and the analysis of the mobility of charge carriers [6.85]. A detailed analysis of electronic processes in accumulation layers of this kind was performed by HEILAND and KOHL [6.86]. With high-dispersion powders, when the particle size becomes equal by an order of magnitude to the Broglie wavelength, one must take into account dimensional quantum effects, which result in oscillations of certain electrophysical parameters of the crystal. PESHEV [6.87], within the framework of Volkenstein's surface model, considered the role of these effects in the scattering of charge carriers on adsorptive states and their influence on the occupancy of levels of adsorbed molecules.

Owing to the high density of surface states in ZnO, the field-effect technique only makes it possible to study states close to the Fermi level (Sect.2.3). HAUFFE and SCHMIDT [6.88] and MORRISON [6.89] suggested using corona discharge in order to charge the surface more effectively. Morrison analyzed the temperature dependence of the kinetics of relaxation of the surface charge for a ZnO single crystal and obtained an estimate for the effective energy levels ε_t of a number of ions and molecules created after adsorption on the surface: 0.15 eV for J^-, 0.15 eV for $[Fe(CN)_6]^{4-}$, 1.3 eV for CH_3OH, 0.5 eV for C_2H_5OH (ε_t is measured from the lower edge of the conduction band). Unfortunately, the role of the spectrum of biographic states and its variations during adsorption remains unclear, especially as regards bombardment of the surface by very active particles generated in corona discharge.

Another monocrystalline object commonly employed in studies of the chemisorption of oxygen is CdS. Here we shall only mention two series of investigations. MANY et al. [6.84] studied chemisorption on an n-type semiconductor CdS specimen. By combining measurements of the field effect and the photovoltaic effect, they determined the change in the surface potential during adsorption. Assuming incorrectly that the concentration of levels of adsorbed particles is equal to the density of physically adsorbed molecules (Sect.5.1), MANY used the results of measurements of field-effect relaxation to estimate the electron capture cross section for adsorbed O_2 molecules ($c_n \sim 10^{-23}$ cm^2)[4].

---

[4] The magnitude of physical adsorption on CdS at room temperature and $p = 1$ atm which is assumed in [6.84] ($n_a = 10^{15} cm^{-2}$) is overestimated by two orders of magnitude [Ref.6.3,Chap.2]. Consequently, even under Many's assumptions the value of c_n ought to be taken as 10^{-21}, and not 10^{-23}. A similar mistake in estimating the value of n_a is made in [6.79].

He also drew an implausible conclusion regarding the absence of intrinsic surface states on the initial CdS surface, which contradicts both theoretical [6.90] and experimental investigations. For instance, in [6.91] it was shown (with the aid of the field-effect technique) that a CdS surface has at least 10^{12} cm^{-2} fast capture centers. Surface states on CdS surfaces were also detected by surface photovoltaic spectroscopy (SPS) in [6.92]. Many was most likely dealing with compensated surfaces. Furthermore, in some cases the CdS surface has been found to have a rather large number of slow states. MURTAZIN and ZARIF'YANTS [6.93] demonstrated, on the basis of investigations of the field effect in monocrystalline PbS films, that chemisorption of oxygen creates a high density of slow states. A continuous oxide phase did not exist in this case. Similar results were obtained with CdS films.

Investigations on chemisorption of oxygen on n-type CdS were also carried out by MARK et al. [6.67,94-96]. MARK [6.94] analyzed the effects of chemisorption of oxygen on photostimulated conductivity, as well as the temperature dependence of dark current relaxation time, and demonstrated that oxygen adsorbed on a CdS surface creates a level in the forbidden band which lies 0.91 eV below the bottom of the conduction band. The release of an electron from this level results in desorption of oxygen, with the frequency factor ν at desorption being equal to 10^{11}s^{-1}. The author assumed this level to be created by ion radical O_2^-. Up to 468 K the oxygen is adsorbed irreversibly; at 468 K the adsorption is reversible.

Using various models for the distribution of levels of adsorbed oxygen in the forbidden band of thin monocrystalline films of CdS (film thickness: several Debye screening lengths), MARK et al. [6.67,95,96] computed conductivity versus temperature. The curves were found to match the experimental results most closely if an exponential law for the distribution of capture centers was assumed. It is noteworthy that the authors of [6.67,96] were dealing with a disordered surface structure (no reflexes in the ELEED spectra). The exponential dependence $\sigma(E)$ is in good agreement with the theory of disordered systems (Sect.3.2).

So far we have been speaking of adsorption of acceptor-type molecules on binary semiconductors. Similar investigations using the field-effect technique were reported in [6.39] for germanium. The experiments with germanium specimens dehydrated at $T_{cal} > 500$ K indicate that adsorption of acceptor-type molecules such as NO, CO_2, J_2, and p-Bq results in a high concentration of slow states, and consequently in a reduction of relaxation time τ_s in the field effect. Adsorption of p-Bq, as emphasized in Sect.3.1 (Fig.3.1), increases the density of fast states in the upper half of the forbidden band.

Unfortunately, all of the above-mentioned experiments dealing with the effects of adsorption of acceptor-type molecules on the electrophysical parameters of monocrystalline surfaces were performed separately from direct measurements of adsorption, and thus there is no basis for quantitative comparisons. This problem was approached for the first time by BAZHANOVA and ZARIF'-YANTS [6.97], who utilized the extreme sensitivity of a piezoresonance quartz balance (about 10^{-10} g) to carry out simultaneous measurements of adsorption n_a and conductivity σ of an n-type monocrystalline PbS film during adsorption of acceptor-type NO_2 molecules.

As Fig.6.7 shows, the function $\Delta\sigma(n_a)$ is linear up to occupancies of about $\sim 6 \cdot 10^{14}$ molecules $\cdot$ cm^{-2}. The authors of [6.97] used field-effect measurements to estimate the quantity of charge ΔQ_s per adsorbed molecule. The ratio $\Delta Q_s / n_a$ in the linear region was found to be about $5 \cdot 10^{-2}$, which is slightly greater than in the case of germanium (Sect.6.4). In the investigated pressure range (up to 1 torr), physical adsorption, as assessed according to De Boor's formula (see [6.3]), constitutes just several percent of n_a. The concentration of unoccupied levels of NO_2 is also small, since the effective level (as estimated from data on slow relaxation) lies close to the Fermi level. There is no alternative but to assume that as the charge accumulates on the PbS surface, a number of biographic states (whose density is rather high) lose their charge (ΔQ_{sB}), owing to a downward shift of the Fermi level for the surface. As a result, $\Delta Q_s = \Delta Q_{sA} - \Delta Q_{sB} \ll qn_a$. This assumption, as will be shown later, is confirmed by data on the kinetics of adsorption and charge accumulation.

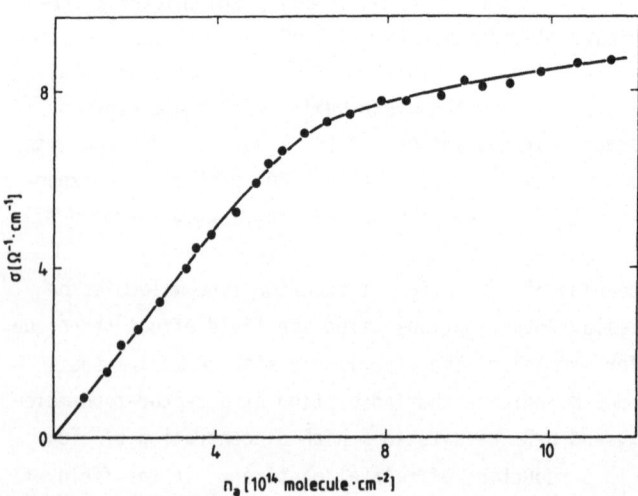

Fig.6.7. Dependence of conductivity of monocrystalline PbS film $\Delta\sigma$ on concentration of adsorbed NO_2 molecules

The effects of chemisorption are not limited to charging of the semiconductor surface. It also strongly affects various excited surface states, including exciton states. Experiments by GROSS and SHEKHMETEV [6.98] and PIMENOV [6.99] suggest a strong influence of adsorption on the exciton luminescence spectrum in CuJ, CdJ$_2$, and HgJ$_2$. The effects of a transverse field and of adsorption of hydrogen ions on exciton luminescence in Cu$_2$O were reported in [6.100]. Also noteworthy are PIMENOV's experiments [6.101] on the influence of adsorption of acceptor-type and donor-type molecules on the long-wave edge of intrinsic absorption of crystals, which Pimenov considered to be a consequence of two possible effects:

1) a manifestation of the Franz-Keldysh effect (a shift to a longer wavelength in the spectrum transmitted by a semiconductor when a uniform electric field is applied, here represented by the field created by adsorbed molecules), and

2) a change in the parameters of surface states (or the generation of new states) during adsorption; these alter the mechanism of annihilation of excitons. Further investigations of these effects would help in determining the intensity of surface electric fields and in elucidating the mechanism of the elementary act of adsorption.

6.4 Slow Adsorptive States

From the standpoint of adsorption and catalysis, slow states are the most interesting. The majority of slow states on semiconductor surfaces owe their existence primarily to adsorption. Almost all of the surface charge during adsorption (at least for germanium [6.39]) is concentrated in slow states. Data on the kinetics of slow relaxation and the field effect were used to calculate (according to the technique outlined in Sect.3.3) the charge in slow states ΔQ_{ss} and the overall concentration of states n_{ss} as functions of the concentrations of water and ammonia molecules adsorbed on germanium [6.39,102] and O$_2$ adsorbed on PbS [6.93]. Also estimated were the effective energy levels of these molecules. The results are presented in Table 6.1. Analysis of these results leads to the following important conclusions.

Table 6.1. Dependence of electronic parameters of Ge and PbS surfaces on the concentration of adsorbed molecules of ammonia, and oxygen

Ge(T_{cal} = 500 K)

n_a [mol·cm^{-2}]	n_{ss} [cm^{-2}]	Q_{ss} [electr·cm^{-2}]
$H_2O(\varepsilon_t \approx -1)$		
Vacuum	$<10^{11}$	--
10^{14}	10^{14}	$1.8 \cdot 10^{11}$
$2.5 \cdot 10^{14}$	$5 \cdot 10^{14}$	$2.3 \cdot 10^{11}$
$NH_3(\varepsilon_t \approx -1)$		
Vacuum	$<10^{11}$	--
10^{13}	$4 \cdot 10^{13}$	$2.8 \cdot 10^{11}$
$1.5 \cdot 10^{14}$	$1.5 \cdot 10^{14}$	$3.6 \cdot 10^{11}$
$4 \cdot 10^{14}$	$2 \cdot 10^{14}$	$3.8 \cdot 10^{11}$

Ge(T_{cal} = 700 K)

n_a [mol·cm^{-2}]	n_{ss} [cm^{-2}]	Q_{ss} [electr·cm^{-2}]
$H_2O(\varepsilon_t \approx -5)$		
Vacuum	$<10^{10}$	--
$2.5 \cdot 10^{13}$	$4 \cdot 10^{11}$	$1.9 \cdot 10^{11}$
$1.5 \cdot 10^{14}$		$2.3 \cdot 10^{11}$
$NH_3(\varepsilon_t \approx -5)$		
Vacuum	$<10^{10}$	--
10^{13}	$3 \cdot 10^{11}$	$2.8 \cdot 10^{11}$
$1.5 \cdot 10^{14}$	$6 \cdot 10^{11}$	$3.7 \cdot 10^{11}$
$4 \cdot 10^{14}$	10^{12}	$4.3 \cdot 10^{11}$

PbS(T_{cal} = 500 K)

n_a [mol·cm^{-2}]	n_{ss} [cm^{-2}]	Q_{ss} [electr·cm^{-2}]
$O_2(\varepsilon_t \approx -5)$		
Vacuum	--	--
10^{12}	10^{13}	10^{12}
--	--	--

Note: The value $\varepsilon_t = (E_t - E_i)kT$ is the location of effective level ε_t with respect to the middle of the forbidden band E_i.

6.4.1 Concentration of Slow States

It has been noted more than once that the surface charge per adsorbed molecule $(\Delta Q_{ss} / n_a$, ΔQ_{ss} being expressed in electron charge units q) is very small, $\sim 10^{-3}$ for both H_2O and NH_3. Its magnitude is determined by the occupancy of levels of adsorptive origin and depends on their position with respect to the Fermi level (3.1). A more interesting value is obtained by comparing the total number of slow states n_{ss} with the number of adsorbed molecules n_a. It can be seen from Table 6.1 that in the initial range of occupancies $n_{ss} / n_a \sim 1$; at higher occupancies $n_{ss} / n_a < 1$ and decreases with an increase in n_a. According to estimates based on adsorption and spectroscopic measurements, the concentration of coordination-unsaturated centers on the surfaces of oxides of Si and Ge [6.3] is several units times 10^{13} cm^{-2}, which is close to n_{ss}.

In its initial stage the adsorption follows the coordination mechanism, and almost every adsorbed molecule of H_2O or NH_3 creates a slow state. At higher occupancies the hydrogen-bond mechanism prevails [6.3]. It was demonstrated in [6.103] that the accumulation of charge on the surface does not correspond either in sign or in magnitude to displacement of electron density in complexes of type $Ge - OH \cdots O\overset{H}{\underset{H}{\diagdown}}$ [6.104-106]. This implies that what is responsible for the charging are coordination bonds with defective coordination unsaturated surface atoms, whose number is almost two orders of magnitude smaller than that of regular lattice atoms. In other words, the accumulation of charge is caused only by those molecules that interact with defects.

6.4.2 Parameters of Slow States

From Table 6.1 it follows that when adsorption of either H_2O or NH_3 proceeds on similarly pretreated surfaces, the effective energy levels ε_t of emerging slow states have similar values, despite the fact that these molecules have different values of ionization potential and basicity. For hydrated surfaces the effective level lies about 1 kT below the middle of the forbidden band. Dehydration shifts the effective level downwards. The value of ε_t for one and the same molecule will depend on the extent of hydration of the surface. This means that the role of charge traps is played not by the adsorbed molecules alone, but by the whole system, which includes the molecule, the adsorption center, and the immediate chemical environment of the center. The parameters of the effective level can be controlled by chemical modification of the surface [6.58]. We shall return to this important practical implication later.

6.5 The Kinetics of Adsorption and Charge Accumulation on Real Surfaces

The first systematic studies of the kinetics of charge buildup on surfaces of single crystals of cuprous oxide, germanium, cadmium sulfide, and other compounds were carried out in Kiev by LITOVCHENKO, LYASHENKO et al. [6.21,27] as early as the 1950s. Their main concern was with the kinetics of change in the work function during adsorption of various polar and nonpolar molecules. They discovered general regularities in the changes in kinetic parameters as functions of the pressure of the adsorbate and the temperature. A large number of kinetic investigations concerned with charge buildup on the surface have been carried out on wide-band semiconductors (mostly single crystals of ZnO and CdS). These investigations are exhaustively reviewed in [6.67,84,107,108] (CdS) and [6.84,109] (ZnO). Their main concern was with the kinetics of variations in the surface conductivity and surface potential during adsorption of oxygen.

Most of these investigations yielded monotonic kinetic curves $\Delta\sigma(t)$ and $\Delta Y_s(t)$, which were interpreted within the framework of the electronic theory of chemisorption (Sect.5.2) under the assumption that the charge in biographic states is either absent or constant. At the same time, the literature contains many reports of experiments in which not only monotonic functions $\Delta\sigma(t)$ and $\Delta Y_s(t)$ were observed, but also nonmonotonic charging curves with definite maxima. These extrema cannot be explained within the framework of the model considered in [6.22,110].

The following reasons have been given for the existence of extrema in kinetic curves $\Delta\sigma(t)$ or $\Delta\varphi_t(t)$:

1) Alteration of the charge status of atoms or molecules adsorbed on the surface, e.g., O_2 on ZnO [6.111], O_2 on TiO_2 [6.75], HCl on Si [6.112].

2) Dissociation of adsorbed molecules, e.g., hydrazine and water on ZnO and TiO_2 [6.113,114].

3) Reactions of adsorbed molecules with ionized surface impurities, e.g., O_2 with Zn^+ on ZnO [6.115], O_2 with Ni^{2+} on NiO [6.116].

4) Inversion of conductivity type during adsorption [6.117].

Kinetic curves $\Delta\sigma(t)$ exhibiting an extremum were also obtained during a catalytic process (dehydrogenation of an alcohol on ZnO film [6.71]). The existence of maxima was attributed to dissociation of adsorbed molecules and recombination of the formed radicals.

These explanations of the extrema are rather speculative, and lack experimental support. Furthermore, all these kinetic curves were obtained in experiments with polydisperse systems. As follows from Sects.4.4, 5.3, it is not possible to use $\Delta\sigma(t)$ and $\Delta\varphi_t(t)$ as a basis for conclusions regarding the causes of extrema on charge curves.

KOZLOV et al. [6.39,118] observed extrema on the kinetic curves of charging for adsorption of CO_2 on real germanium surfaces with n-type conductivity (Fig.6.8a). They were the first to attribute the presence of a maximum on the curve $\Delta Q_{ss}(t)$ to recharging of the set of slow biographic states driven out of equilibrium during chemisorption. As Fig.6.8a shows, during dehydration of germanium, when n_{sB} (according to Table 6.1) ranges from 10^{13} (T_{cal} = 300 K) to 10^{11} (T_{cal} = 500 K), the curve $\Delta Q_{ss}(t)$ with a maximum passes into a monotonic curve. In the first case $n_a \leqslant n_{sB}$, while in the second case $n_a \gg n_{sB}$.

Unfortunately, the kinetic investigations carried out on single crystals were not backed up by direct measurements of adsorption. This task was first undertaken by BAZHANOVA et al. [6.97,119], who utilized the high sensitivity of a piezoresonance balance. They performed simultaneous measurements of the kinetics of variation in total surface charge ΔQ_s of monocrystalline PbS films (using the field-effect technique) and the kinetics of chemisorption of NO_2 (Fig.6.8b). In the case of PbS, the slow biographic states were created by

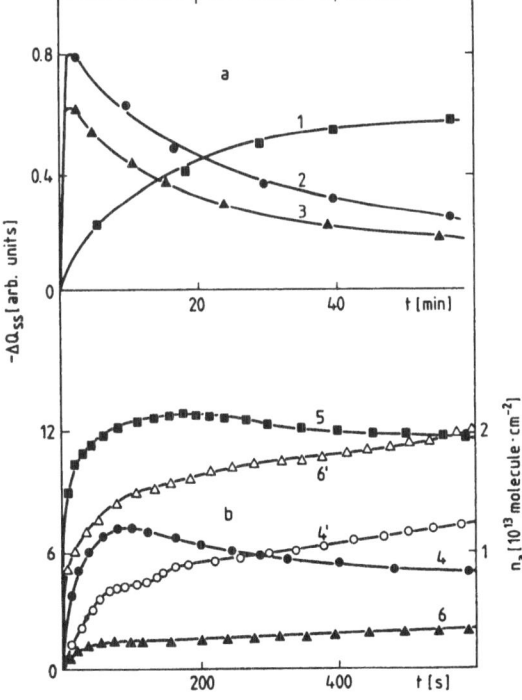

Fig.6.8. (a) Effects of concentration of slow biographic states n_{sB} on the kinetics of charging of germanium surfaces during adsorption of CO_2. (1) 22 torr, (2) 0.4 torr, (3) 50 torr. (b) The same for adsorption of NO_2 on monocrystalline PbS films. (4,5,6) 0.03 torr. (4',6') the kinetics of chemisorption of NO_2. The concentration n_{sB} in cm^{-2} is: (1) 10^{11}, (2,3) 10^{13}, (6,6') 10^{11}, (5) 10^{12}, (4,4') 10^{14}

prior chemisorption of oxygen, their concentration being 10^{11} to 10^{13} cm^{-2}. From Fig.6.8 it is evident that, as in the case of Ge, as n_{sB} decreases (other conditions being equal), the curves $\Delta Q_{ss}(t)$ with a maximum pass into monotonic curves. It turned out that the absolute values of n_a and ΔQ_{ss} also depend on n_{sB}. More recently, similar nonmonotonic curves $\Delta\sigma(t)$ were obtained for adsorption of oxygen and atomic nitrogen on ZnO and CdSe [6.120,121]. Slow relaxation of conductivity on monocrystalline ZnO films was reported in [6.122].

Direct experimental proof that slow biographic states are recharged during chemisorption was obtained from simultaneous investigations of charge relaxation in slow states on Ge during chemisorption and in the field effect. As demonstrated in Sect.3.3, in both cases the relaxation curves of excess surface charge are well linearized in Koc's coordinates (3.7), with the parameters τ_s and a having the same values. A similar result was obtained for the system of PbS and NO_2. The similarity between the kinetics of the declining sections of the charging curves (Fig.6.8) and the kinetics of $\Delta Q_{ss}(t)$ after removal of the transverse electric field is an indication that both cases involve recharging of slow states driven out of balance with energy bands during charging of the surface. These regularities are not peculiar to slow states of a given chemical nature. As pointed out in Sect.6.4, it is coordination-bound water molecules that are responsible for the existence of slow states on germanium surfaces (Fig.6.8a) while in the case of PbS it is oxygen complexes (Fig.6.8b). With PbS the same phenomena was observed when slow states were created by chemisorption of water.

None of the above-discussed possible causes of the extrema on the kinetic curves of charging can be applied to the systems under consideration. It is clear that the maxima cannot be attributed to dissociation of adsorbed molecules or alteration of their charge status. Maxima were observed during adsorption of quite different molecules: O_2, NO_2, p-Bq. The existence of maxima is not connected with inversion of conductivity type in the SCR, since this would be immediately detected by the field-effect technique. A reaction with ionized impurities also could not be of any significance, because prior to adsorption measurements the surface had been exposed to oxygen or water vapor to create the slow states.

There is one other factor which could restrict the rate of capture of charge carriers on adsorptive states: the limited rate at which charge carriers pass from the bulk to the surface. Qualitative estimates based on approximative calculations [6.123] indicate that this phenomenon while it may be of some significance in wide-band semiconductors of ZnO type is negligible in the case of PbS and germanium.

From the above discussion it follows that the most general factor determining the shape of kinetic curves of surface charging during adsorption is the re-charging of biographic states. Unfortunately this factor is overlooked in the majority of theoretical works cited in Sect.5.2. Even in the model problem of adsorption on charged defects (real surface), VOLKENSTEIN's theory [6.22] fails to take into account the change of charge in biographic states.

From the standpoint of formal kinetics, surface charging during chemisorption is due to two simultaneous processes: induction of excess surface charge qN_{SA}^* owing to adsorption, and recharging of biographic slow surface states (BSS) n_{SB}^* caused by additional bending of bands. This problem was theoretically considered by KOZLOV [6.124], who assumed that in adsorption on a uniform surface (deviation from equilibrium being small), the law of change in the concentration of charged adsorptive slow states (ASS) $N_{SA}^*(t)$ can be approximated by an exponent:

$$N_{SA}^*(t) = N_{SA}^0[1 - \exp(-t/\tau_1)] \quad , \tag{6.1}$$

where N_{SA}^0 is the equilibrium concentration of charged adsorptive states, and τ_1 is the time constant which characterizes the rate of creation and charging of these states.

This equation is equivalent to the one in the theory developed in [6.110] (Sect.3.3) under the assumption that one characteristic time is greater than the other[5] (this assumption if often realized in practice).

The change in the charge in adsorptive states $Q_{SA}(t) = qN_{SA}^*$ causes a change in the charge in biographic states $Q_{SB}(t) = qn_{SB}^*$. The kinetics of recharging of surface biographic states can be expressed via the cross sections of capture of charge carriers and their surface concentrations, as was done in Sect.3.3. By combining equation (6.1) with the equation for electron neutrality of the crystal, the author [6.124] obtained the equation for the kinetics of change in the excess concentration of electrons in the SCR (accumulation case):

$$\Delta\Gamma_n(t) = \frac{N_{SA}^0}{k'(1 + r)} \left[1 + \frac{\tau_2 r}{\tau_2 - \tau_1} \exp(-t/\tau_2) + \left(1 + \frac{\tau_2 r}{\tau_2 - \tau_1}\right) \exp(-t/\tau_1)\right] \quad , \tag{6.2}$$

---

[5] If the limiting stage is the formation of the adsorptive state, then τ_1 depends on the gas pressure, the sticking coefficient, and the temperature. If the slowest process is capture on an adsorptive state, then τ_1 depends on the concentration of charge carriers on the surface and the parameters of these states.

where τ_2 is the relaxation time of slow biographic states, r is a constant that can be expressed explicitly (for a real germanium surface $r \geqslant 2$), and $k' > 0$ is a constant.

Analysis of (6.2) reveals that depending on the relative values of τ_1 and τ_2, this equation can define both monotonic curves and curves having a maximum (Figs.6.9,10). From Fig.6.9 it is clear that assuming the parameters of BSS to be constant, an increase in τ_1 shifts the maximum towards greater values of t. The extreme position of the maximum

$$\tau_{max} = \frac{\tau_1\tau_2}{\tau_2 - \tau_1} \ln \frac{\tau_2(r + 1) - \tau_1}{\tau_1 r} \tag{6.3}$$

is determined by the relative values of τ_1 and τ_2. As will be demonstrated below, (6.2,3) accurately describe the principal regularities observed in experiments on adsorption and charge buildup on PbS films.

Equation (6.2) is formally similar to (5.3). However, the parameters now assume a new meaning. Although τ_1 determines the limiting stage in the creation of an adsorptive state, as it did in (5.3), τ_2 in (6.2) describes the change in biographic charge, which in [6.22] is neglected altogether. In fact the coefficients C_1 and C_2 in (5.3) are no constant but represent explicit characteristics of the molecular and electron subsystems.

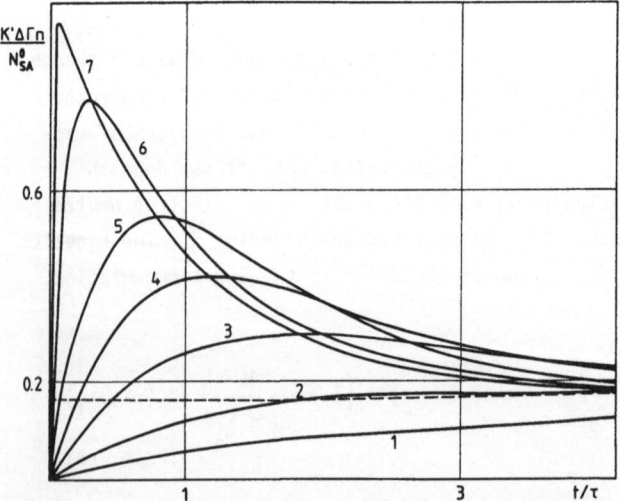

Fig.6.9. Kinetic curves of charging of semiconductor surface during adsorption at constant parameters of biographic slow states ($\tau_2 = \tau$, $r = 5$) and varying relaxation times of adsorptive slow states τ_1 [6.124]. The values of τ_1 are: (1) 10τ, (2) 5τ, (3) 2τ, (4) τ, (5) 0.5τ, (6) 0.1τ, (7) 0.01τ. Dashed line indicates the equilibrium value $k'\Delta\Gamma_n/N_{SA}^0 = 1/r + 1$

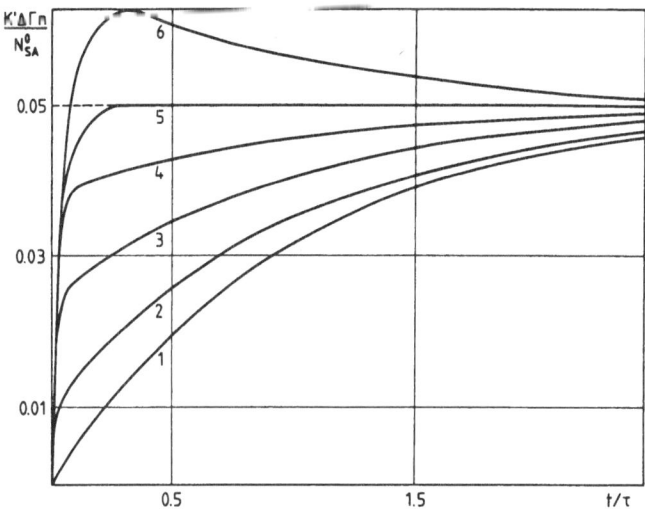

Fig.6.10. Kinetic curves of charging of semiconductor surface during adsorption at constant relaxation time of adsorptive slow states τ_1 and varying relaxation times of biographic slow states τ_2 (at r =19) [6.124]. The values of τ_2 are: (1) 0.001r, (2) 0.01r, (3) 0.025r, (4) 0.033r, (5) 0.05r, (6) 0.066 r. Dashed line indicates the equilibrium value $k'\Delta\Gamma_n/N_{SA}^0 = 1/r + 1$

If $N_{SA}^* \gg n_{SB}^*$, the second term in (6.2) can be neglected, and the function $\Delta Q_{ss}(t)$ will be depicted by a monotonic curve (Sect.3.3). If $N_{SA}^* \leqslant n_{SB}^*$, the curve $\Delta\Gamma_n(t)$ is identical with curves $\Delta Q_s(t)$ and $\Delta\sigma(t)$ [6.124]. The shape of the curve will now be determined by the values of τ_1 and τ_2 relative to each other. If $\tau_2 \ll \tau_1$, at $t \geqslant \tau_2$ the kinetics of charging will follow the law of the creation of charged adsorptive states. If $\tau_2 \gg \tau_1$, then the initially fast charging of the surface (at $t \leqslant \tau_1$) will be followed by a slow decrease in charge, owing to recharging of slow states, which takes time τ_2 (Fig.6.8a).

If the limiting stage in the creation and charging of adsorptive states is the process of adsorption, then the value of τ_1 depends on pressure p. Let us consider the case in which n_{SB} = const, T = const, and τ_2 = const. As is seen from Fig.6.11b, with a decrease in p the value of τ_2 goes up, and in concordance with the theory the maximum on the curve $\Delta Q_{ss}(t)$ shifts towards greater values of t (Curves 1 - 3). The rate of chemisorption is then reduced (Curves 1' - 3'). If τ_1 is sufficiently great, the induced charge has a chance of being captured in biographic states even in the initial stage of adsorption. At low pressure (large values of τ_1) the maximum on the curve $\Delta Q_{ss}(t)$ becomes less pronounced. Charging of the surface proceeds at almost complete equilibrium between surface states of all types and the allowed bands of the semiconductor.

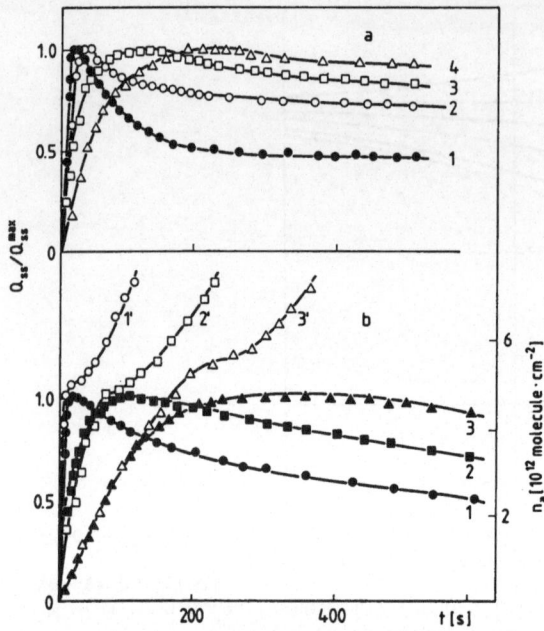

Fig.6.11a,b. Temperature (a) and pressure (b) dependence of kinetics of surface charging (*1-4*) and chemisorption of NO_2 (*1'-3'*) on PbS. (**a**): (*1*) 350 K, (*2*) 294 K, (*3*) 287 K, (*4*) 273 K, p = 0.07 torr; (**b**): (*1,1'*) 1.2 torr, (*2,2'*) $3 \cdot 10^{-2}$ torr, (*3,3'*) $6 \cdot 10^{-4}$ torr, T = 294 K

The values of τ_1 and τ_2 depend on temperature as $\tau_{1,2} = \tau_{1,2}^0 \exp(E_{1,2} / kT)$, where $E_{1,2}$ are the activation energies of processes whose characteristic times are τ_1 and τ_2 respectively. As is seen from Fig.6.11a, with an increase in T, both the ascending and descending branches of curves $\Delta Q_{ss}(t)$ become less steep, and the maximum shifts towards greater values of t. It is interesting to recall here the data on the kinetics of dehydration of alcohol on ZnO [6.71]. As the reaction temperature was lowered, the author observed a considerable flattening of the maximum on the curve $\Delta\sigma(t)$, which also shifted towards greater values of t. This situation is quite similar to that represented in Fig.6.11a. The extremum on the curve $\Delta\sigma(t)$ [6.71] is most likely connected with recharging of slow states (whose concentration on the surface of ZnO amounts to about 10^{13} cm^{-2}), and not with the specifics of the catalytic process, as assumed by the author. We see that neglecting the role of biographic states can lead to ambiguous conclusions about the mechanism of adsorption and catalysis made on the basis of the known kinetics of charging. The same can be said about information obtained using solid-state detectors of chemisorbed particles [6.71].

So far we have been speaking mainly of surface charging. Equally valuable information about the initial stages of adsorption has been obtained with a fast-response piezoelectric balance. For instance, in [6.125] the sticking

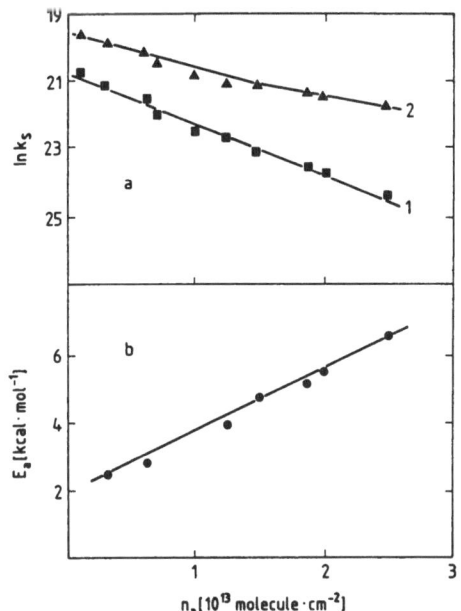

Fig.6.12a,b. Sticking coefficient of oxygen k_S (**a**) and activation energy E_a (**b**) functions of the occupancy of a PbS surface at $T = 298$ K (*1*) and $T = 370$ K (*2*)

coefficients k_S and activation energy E_a for chemisorption of O_2 were defined as functions of the concentration of adsorbed molecules n_a (Fig.6.12). From Fig.6.12a it follows that the function $k_S(\theta)$ is satisfactorily approximated as $\ln k_S = \text{const} \cdot (1 - \theta)$. A similar function $k_S(\theta)$ was derived by AGEEV and IONOV [6.126] for nondissociative adsorption on metals. Starting with the lowest occupancies ($\theta < 0.002$), the activation energy E_a increases linearly with θ.

Simultaneous measurements of $\Delta\sigma(t)$ and $n_a(t)$ indicate (Fig.6.13a) that beginning with the lowest occupancies ($\theta < 2 \cdot 10^{-3}$ of a monolayer), both the $\Delta Q_{ss}(t)$ and $n_a(t)$ curves are entirely symbatic. The kinetics of capture of electrons on adsorptive states does not restrict the process of adsorption. As θ increases and the Fermi level on the surface shifts downwards, some of the charged defects, which act as centers of adsorption, are emptied, and the activation energy increases. It is interesting to note that the curve of adsorption $n_a(t)$ exhibits inflections (Fig.6.13b) simultaneously with the maximum on the curve $\Delta\sigma(t)$ (for specimens with a sufficiently high concentration of biographic states n_{sB}). This implies that the rate of adsorption changes when the sign of derivative $\partial\Delta Q_{ss}(t) / \partial t$ is reversed owing to recharging of a number of traps. This is also evidence in support of adsorption on charged defects.

143

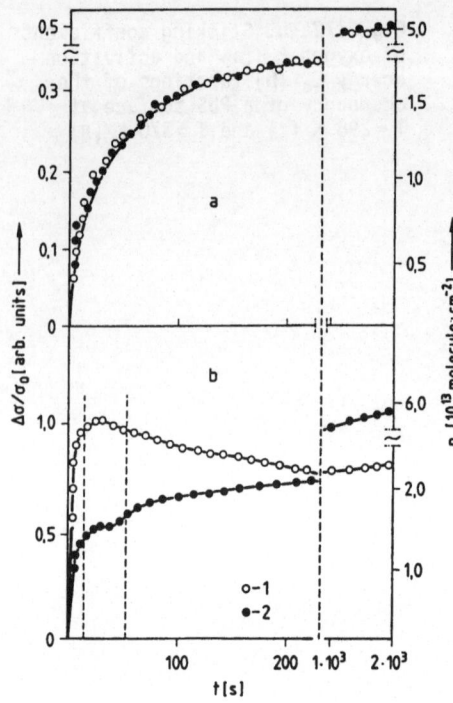

Fig.6.13a,b. Kinetics of variations in conductivity $\Delta\sigma$ (*1*) and chemisorption n_a of NO_2 at $p = 0.7$ torr (*2*) on a PbS surface. (a) $n_{sB} \simeq 10^{11} cm^{-2}$, (b) $n_{sB} \simeq 10^{12} cm^{-2}$

6.6 The Donor-Acceptor Mechanism of Adsorption and Charging

An analysis of individual forms of adsorption on the surfaces of certain semi-conductors reveals that it is often donor-acceptor-type bonds (Sect.6.2) that are responsible for charge accumulation on the surface. One of the authors constructed a convenient working model of the elementary act of adsorption, based on the donor-acceptor mechanism of adsorption and charge buildup on semiconductor surfaces [6.1,2,57,103].

It has been already emphasized [6.3] that the term "donor-acceptor bond" refers only to the origin of the electrons taking part in the creation of a bond; it does not say anything about the nature of the resulting chemical complex. This "genealogic" definition is sufficient as long as we are engaged in a general consideration of the mechanism of charging. However, when investigating the causes of the reactivity of adsorbed molecules, we need details on the structure of the donor-acceptor complex. Depending on the spectroscopic data available for each case, we shall be speaking of coordinative (donor-acceptor), dative, or valence bonds.

The donor-acceptor mechanism treated here is based on the assumption that adsorption (which results in an accumulation of charge on the surface) proceeds mainly on "defective" surface atoms. This assumption is supported by all of the vast experimental data on chemisorption [6.3] and surface charging (Sects. 6.2,3). As a rule, the concentration of strongly adsorbed molecules and the number of generated adsorptive electron states are much smaller than the total density of surface atoms. On the basis of spectroscopic information [Ref.6.3, Chap.3], adsorption centers can be conventionally classified in a first approximation as electron-acceptor (EA) and electron-donor (ED) centers. The former are usually represented by coordination-unsaturated M atoms on the surface, whose effective charge q_{ef} differs from that of regular lattice atoms in the bulk. The role of ED centers is often played by vacancy-type defects (e.g., F centers). Transition-metal atoms on the surface can exhibit either donor or acceptor properties, and in certain cases, donor properties can also be exhibited by oxygen ions on oxide surfaces. Each type of adsorptive center is characterized by its own set of energy levels in the crystal's band scheme.

As was demonstrated in [6.3], adsorption of an acceptor-type molecule A on an ED center results in the formation of a donor-acceptor complex: $A^{-\delta} \cdot M^{q_{ef}+\delta}$ (for a donor-type molecule: $D^{+\delta} \cdot M^{q_{ef}-\delta}$), where δ is excess charge[6]. The emerging complex is a new surface state characterized by a new position in the energy band ε_t and new kinetic parameters c_n and c_p. Depending on the relative values of the parameters, this adsorptive state can act either as a recombination center or as a capture center. It is not necessary that charge carriers be localized directly on the orbitals of the $A - M$ (or $D - M$) complex; the maximum of its wave function can be located, for example, on a neighboring defect [6.58].

According to MULLIKEN and PERSON [6.127], depending on the relative positions of the energy levels of electrons belonging to atoms (or molecules) that $A - M$ or $D - M$ complexes form as well as on the separation between them, the interaction can take any conceivable form, from weak ion-dipole interactions to partially covalent and ionic ones. Sufficiently high excitation may cause an electron to pass completely from orbitals of an adsorptive center over to an M atom, and vice versa (in other words, a charge transfer complex [CTC] is created). In this case $\delta \approx q$, or, in the usage of chemists, $\delta = 1$. The creation of CTCs during adsorption of certain acceptor-type and donor-type molecules on surfaces of semiconductors and dielectrics has been proved in numerous experiments using optical spectroscopy and EPR [Ref.6.3,Chap.3].

---

[6] This notation, as opposed to the form commonly accepted by chemists, i.e., $A^{-\delta} \cdot M^{n+\delta}$ (or $D^{+\delta} \cdot M^{n-\delta}$), where n is an integer, underscores the fact that the effective charge q_{ef} of surface atoms differs from that of regular lattice atoms.

In many experimentally observed cases the transition from $\delta < q$ to $\delta = q$ (CTC) is of an activated nature. This was discovered in investigations of adsorption of acceptor-type p-Bq molecules on real surfaces of germanium [6.128] and rutile [6.129]. For TiO_2 the activation energy of the transition ranged from 0.05 to 1 eV depending on the temperature. Activated transitions of donor-type molecules of H_2O and NH_3 in the state of a firm coordination bond have been observed using IR and NMR spectroscopy [6.3]. A general theoretical analysis is given in [6.130] of the causes which give rise to an activation barrier when a donor-acceptor bond is established.

The extent of charge transfer between the adsorption center and the adsorbed molecules (i.e., the magnitude of δ) depends not only on the electronic properties of the constituents which join to form a complex, but also on the whole ligand surrounding of the center. In particular, the values of q_{ef} and δ for a great many oxide and elementary semiconductors depend on the extent of hydration, i.e., on the number of OH groups linked to the adsorption center. The replacement of OH groups by more electronegative groups should result in an increase in positive charge q_{eff} and, in the case of adsorption of donor-type molecules ($D^{+\delta} \cdot M^{qef-\delta}$), in a strengthening of the bond and deformation of the D molecule. This is confirmed by IR and NMR spectroscopy, as well as by mass spectrometry [6.3], for adsorption of water, ammonia, and alcohols on the surfaces of oxides of Si, Al, and Ti.

EPR measurements give valuable information about the effects of the ligand surrounding on the structure of surface complexes. Such investigations have been carried out for rutile-water and rutile-ammonia systems [6.41]. The parameters of EPR signals from paramagnetic Ti^{3+} ions on reduced rutile surfaces are presented in Table 6.2 [6.41,131]. As reported in [6.41,131], the Ti^{3+} ion is surrounded by a distorted square-base pyramid. The replacement of some of the oxygen atoms with fluorine in the nearest coordination sphere of Ti^{3+} by fluorinating the rutile surface caused considerable narrowing of the EPR line (Table 6.2), which indicates a change in the coordination. Adsorption of water ($p/p_s = 0.01$) and ammonia (at 200 torr) on the initial (nonfluorinated) specimen was accompanied by a slight narrowing of the EPR line, which was attributed to a change in the symmetry of the crystalline field in which the Ti^{3+} ion was found, and to an increase in the spin-lattice relaxation time. A similar phenomenon was observed with isolated Ti^{3+} ions contained in a silicate matrix, where electron exchange with the lattice is not possible [6.132]. In both cases the concentration of spin centers n_s remained constant.

When adsorption proceeds at the same pressures on fluorinated specimens, one observes, in addition to a narrowing of the line, an abrupt drop in n_s

Table 6.2. Parameters of EPR signals from Ti^{3+} ions on rutile surfaces

Parameter of EPR signal	Prior to admission of adsorbate		After admission of H_2O at 300 K		After admission of NH_3 at 300 K	
	Initial specimen	Fluorinated specimen	Initial specimen	Fluorinated specimen	Initial specimen	Fluorinated specimen
$g_\perp$	1.956	1.961	1.970	1.971	1.968	1.967
$g_\parallel$	1.945	1.945	1.942	1.946	1.94	1.95
$\Delta H [G]$	50	31	44	28	40	25
$n_S \cdot 10^{12}$ [spin $\cdot$ cm^{-2}]	2	1	2	0.01	2	0.2[a]

[a] After admission of NH_3 at 500 K, the value of n_S became equal to $0.03 \cdot 10^{12}$ spin $\cdot$ cm^{-2}

(Table 6.2). For NH_3 the value of n_S is reduced by a factor of 5; however, if adsorption is made to proceed at 500 K, with subsequent cooling of the specimen to 300 K, then the strength of the signal is reduced by nearly two orders of magnitude which is an indication of adsorption of an activated nature. During adsorption of donor-type molecules of H_2O and NH_3, the complex $D^{+\delta} \cdot Ti^{3-\delta}$ is formed. For nonfluorinated specimens, $\delta < q$, and the change in the EPR signal is due only to an alternation in the symmetry of the crystalline field. When some of the oxygen atoms are replaced by more electronegative atoms of fluorine, the effective charge of the Ti^{3+} ion increases, as does the extent of pullover of an undivided pair of electrons of H_2O and NH_3 to Ti^{3+}, and δ assumes a greater value. This causes a weakening of the intramolecular bonds, which may lead to dissociation of the molecule and, as a result, to a decrease in n_S. This is confirmed by mass spectrometry [Ref.6.3, Chap.4]. However, one cannot totally exclude the possibility that some of the molecules (at sufficiently high excitation) pass into ion radicals, transferring an electron to a Ti^{3+} ion. The EPR spectra from Ti^{2+} (in d^2 configuration) can only be observed under special conditions [6.133]. Data on the formation of cation radicals during adsorption of perilene on reduced TiO_2 suggest the possibility of a $Ti^{3+} \rightarrow Ti^{2+}$ transition [6.134]. This possibility is also indicated by electron energy loss spectroscopy (ELS) [6.135].

The above discussion leads to an important conclusion: by chemically modifying the surface, one can exercise control over the magnitude of the effective charge of defect M atoms and, consequently, the value of δ in the donor-acceptor complex. It is thereby possible to control the strength of the bond between the adsorbed molecule and the adsorption site and thus to control the reactivity of the molecule.

A relevant example is provided by the above-cited EPR data obtained for hydrated and fluorinated surfaces of TiO_2 [6.41,131]. GONZALEZ-ELIPE et al. [6.136] reported that the extent of hydroxyl coverage of TiO_2 surfaces strongly influences radical forms of adsorbed oxygen, as well as the mechanism of photoadsorption. A similar conclusion about the influence of hydrate coverage of MgO on photoadsorption of O_2 was drawn by LISACHENKO and VILESOV [6.137].

In order to obtain a comprehensive understanding of the mechanism of adsorption and charging, one needs information on both the adsorption centers and the adsorbed molecules. WATANABE [6.138] conducted simultaneous EPR and NMR investigations of chemisorption of hydrogen on ZnO, and offered convincing proof of the presence of two forms of adsorbed hydrogen on the surface. The surface charging was monitored via the EPR signal coming from free conduction electrons.

Let us now turn to the electrophysical manifestations of donor-acceptor complexes created on the surface. As indicated earlier, these complexes modify the effective charges of EA and ED centers, as well as their energetic (ε_{tA} and ε_{tD}) and kinetic (c_n and c_p) parameters in the band scheme of the crystal. The model proposed here differs qualitatively from the idealized VOLKENSTEIN model [6.22], which fails to specify the mechanism by which donor states (Fig.6.14a, Schemes 1) and acceptor states (Fig.6.14a,Schemes 2) are created and is based on the a priori assumption that the values ε_{tA} and ε_{tD} of the emerging states depend on the structure of the adsorbed molecules and the properties of the adsorbent. An equally speculative model of the donor-acceptor complex was adopted by GREEN and MAXWELL [6.139], who assumed the parameters of the complex to be defined by the difference between the ionization energy and the electron affinity of the adsorbed molecule and the adsorptive center. The ambiguity of the parameters of the adsorptive bond when defined in this fashion was pointed out in [Ref.6.3,Chap.2].

According to the donor-acceptor mechanism of adsorption and surface charging proposed by us, the energetic parameters of the adsorptive complex are often determined not so much by the structure of the adsorbed molecule as by the immediate neighborhood of the ED (or EA) center and its chemical connections. For instance, the energetic parameters of slow adsorptive states for different molecules (H_2O and NH_3) on the surface of germanium have similar values, which nevertheless exhibit a strong dependence on the hydrate environment of adsorption centers (Sect.6.4). From the standpoint of their charge status, the neutral donor-acceptor complexes (Fig.6.14b,c) are similar to the neutral form of

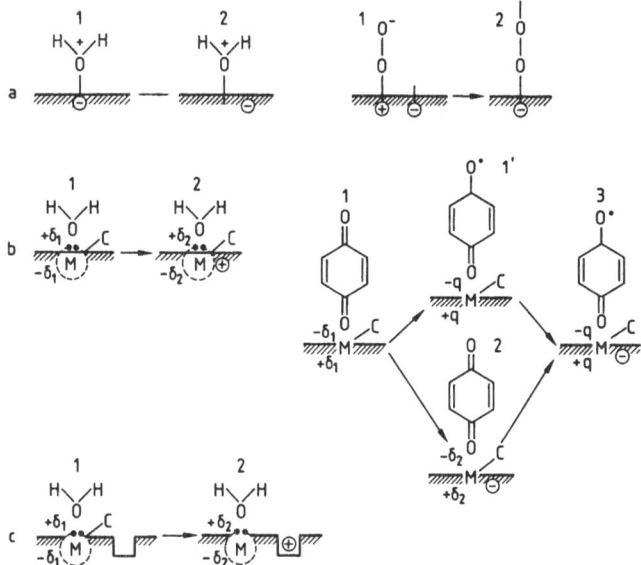

Fig.6.14a-c. Schemes of adsorption of donor-type and acceptor-type molecules according to [6.21] (**a**) and according to the donor-acceptor mechanism (**b,c**). (1,1') "weak" bonds, (2,3) "strong" bonds

adsorption (Fig.6.14a) as defined in [6.22]. However, from the standpoint of their chemical structure, Schemes a [6.22].and Schemes b and c [6.2,57] are different. Scheme a says nothing about the forces that fasten the adsorbed molecule to the surface (especially in the case of covalent surfaces). Schemes b and c correspond to donor-acceptor (coordinative) bonds, whose existence is confirmed by spectroscopic data.

The transfer of charge in the complex, be it partial or complete, is equivalent to the creation of effective dipole μ_{ef}. Numerous investigations of donor-acceptor complexes [6.140] indicate that the magnitude of μ_{ef} in such cases may amount to many Debye units. The captured charge carrier can be retained by Coulomb forces near chemically saturated but highly polarized bonds in the complex (Fig.6.14b). The electron can also be trapped by a defect adjacent to the adsorption site, with the defect's parameters being altered by the field of the complex (Fig.6.14c). The influence of adsorbed molecules on the effective charge of surrounding atoms was theoretically analyzed for crystals with partially ionic bonds in [6.141].

In order for the electron to be captured on an adsorptive state, it must lose its energy; otherwise it will be thrown back into semiconductor's allowed band by thermal fluctuations. The probabilities (and therefore the capture cross sections c_n and c_p) are determined by the efficiency of dissipation of the energy released in the act of capture of the charge carrier on the adsorptive state. This efficiency is rather low. As indicated by experiments (Sect. 6.4), the majority of adsorptive states in the case of Si, Ge, and certain binary semiconductors are slow states with low capture cross sections (10^{-25} to 10^{-28} cm^{-2}). An analysis of the background of this phenomenon would go beyond the scope of the band theory in its conventional form, so we shall postpone a detailed discussion until Chap.8.

Unfortunately, reliable information about the energy spectrum of surface states is not yet available for the overwhelming majority of semiconductor adsorbents and catalysts. For this reason, in order to keep our discussion on firm ground, we must turn again to the thoroughly investigated surfaces of germanium and silicon. For semiconductors with an oxide cover (germanium, silicon, intermetallic compounds), the adsorbed molecules diffuse through the defect-ridden oxide layer and anchor somewhere near the boundary between the semiconductor and the oxide. This is confirmed by experiments on the adsorption of differently sized molecules [6.37,39]. The interface, apparently owing to the entirely dissimilar structures of the semiconductor and the oxide, provides the most favorable conditions for the formation of stable donor-acceptor complexes. As demonstrated by one of the authors in [6.15], these interfaces are are most conveniently approached via the electron theory of disordered systems.

6.7 A Model of a Disordered Real Surface

Let us first discuss those systems that are most important from a practical point of view: semiconductors covered with an oxide film, or, to be more precise, $Si - SiO_2$ and $Ge - GeO_2$ which are the most thoroughly investigated and most extensively employed systems in solid-state technology. The properties of the interfaces between the insulator and the semiconductor naturally depend strongly on the genesis of these two phases. Of course, a distinction must be drawn between interfaces obtained naturally (real surfaces) and interfaces in MOS structures. However, both of these interfaces have a common feature: the peculiar junction of two solid phases which are entirely different in their chemical structure and parameters (the semiconductor and its oxide). This junction can give rise to "mismatching" dislocations and result in deformations of valence angles and bond lengths (this process leads to the decreasing of the interface energy) [6.142,143]. Fluctuations in the diffusional flows of reactants during oxidation and the presence of ambient water vapor (leading to hydration) [6.3] will promote the creation of a chemically inhomogeneous, amorphized boundary layer.

Consider the Si - SiO$_2$ interface. Owing to the high covalence of the bonds in SiO$_2$, one of the main causes of the interface's heterogeneity is the deformation of valence angles and bond lengths of the interfacial tetrahedra SiO$_x$ (x < 4) and hydrated tetrahedra SiO$_x$(H$_y$). This opinion, expressed by us in [6.15,143,144], was later confirmed by numerous spectroscopic investigations The presence of a layer with violated stoichiometry on the Si - SiO$_2$ interface which exhibits widely varying valence angles and bond lengths, is directly suggested by analysis of the fine structure of X-ray spectra (SSXA) [6.145,146], numerous data supplied by X-ray photoelectron spectroscopy (XPS) [6.147-149], ion scattering spectroscopy (ISS) [6.150-154], electrochemical methods [6.155]. and other techniques. (See also the references cited in [6.146-155] and the critical reviews [6.156,157]).The thickness of the interface, as assessed by different authors, varies from 5 to 0.3 nm. Such a large variation in the values of d$_{int}$ must be due to a number of factors: insufficient resolving power of analytic techniques; use of fluid etchants, which in themselves can cause changes in the chemical composition and structure of the residual oxide phase; stimulation of defects creation and dissociation of the surface groups by hard radiation (photodesorption and photodissociation); and nonuniform oxidation and heterogenety of the oxide film.

Many of these obstacles can be eliminated by using optical and radio spectroscopic techniques. The ellipsometric data reported in [6.158] give convincing proof that the thickness of the distorted oxygen-deficient layer (SiO$_x$, x < 2) is about 0.7 ± 0.2 nm. The violation of translational symmetry in the semiconductor layer adjacent to the interface is confirmed by data on the angular dependence of EPR signals on spin-dependent centers of surface recombination (see below).

All these experimental data cast doubt on the three-ply model of the interface proposed in [6.159,160]. According to that model, the interface includes a thin SiO$_2$ layer of perfect crystalline structure induced by the crystal field of silicon. The idea of an interfacial crystalline layer was also expressed in [6.161,162].

The literature does contain data on the creation of crystalline microinclusions in SiO$_2$ films at high temperatures [6.163]. However, the formation of a more or less continuous crystalline layer between silicon and amorphous SiO$_2$ at moderate (or even room) temperatures [6.164] does not seem plausible. Secondary ion mass spectroscopy (SIMS) alone cannot be assumed to furnish reliable data on such inclusions [6.164]; it must be supported by microdiffractional measurements.

Two opinions about the thickness of the interface d_{int} between Si and SiO_2 have been proposed in the literature:

1) The boundary is sharp and contains no more than a monolayer of SiO_x ($x < 2$) [6.148,157,165]. A theoretical attempt at conjugating the structures of Si and SiO_2 by varying the valence angles was made in [6.166,167]. The proposed models of an ordered [6.166] and stoichiometric interface resemble the three-ply model [6.160]; these models, however, hardly correspond to reality.

2) There is a smooth transition from Si via a nonstoichiometric layer of SiO_x which contains vacancy-type point defects to the layer of SiO_2 [6.143, 144,147,158,168]. As follows from [6.168], the deviations from stoichiometry affect the physical properties of SiO_x most strongly when x varies from 0 to 1.25. When $x > 1.5$ they differ little from those of the stoichiometric phase (SiO_2). For this reason the actual thickness d_{int} can be more greater than the value assessed on the basis of spectroscopic data.

The central question in the electrophysics of MOS structures is: which infractions of the interface structure can lead to the appearance of surface states in both Si and SiO_2? From the standpoint of the theory of disordered systems (Sect.3.2), this question brings us to the origins and statistics of fluctuative fields at the interface. Using simplified Pauling concepts about the structure of silicates, we demonstrated in [6.143,144] that the most general source of such fields is incomplete (SiO_x, $x < 4$), hydrated $SiO_x(Hy)$, or simply deformed SiO_4 tetrahedra. A change in the valence angle in siloxane Si - 0 - Si bonds [6.3,Fig.3.22] is known to lead to deformation of tetrahedra and to a change in the effective charge q_{ef} of the central Si atoms. Owing to the heterogeneity of the interface, the Coulomb fields here fluctuate, causing the "tails" of the density of states to split off the edges of the allowed bands, thus and supplying a good explanation for the quasi-continuous spectrum of fast surface states derived from the capture curves. As we showed in Sect.3.1, this kind of spectrum is typical of both real Si surfaces and MOS structures. The energy spectrum of fast states on interfaces is strikingly similar to the spectrum of states in a typical amorphous a-Si semiconductor [6.169].

These qualitative ideas were confirmed theoretically in [6.170]. The authors of that work demonstrated, within the framework of the Bethe lattice approximation that the appearance of tails of the density of states in the forbidden band of Si is connected with the deformation and rupture of valence bonds on the interface. Deformation of valence angles and vacancy-type defects are also responsible for the appearance of such tails in SiO_2 [6.171,172]. This conclusion is confirmed by investigations of photocharging of superslow states in the dielectric in $Si - SiO_2$ structures [6.173] (Sect.4.3). For thermally oxidized Si surfaces the energy range ΔW of these tails was 0.2 to 0.1 eV. MOTT [6.174] assumed that in a regular SiO_2 structure the value of ΔW is much smaller than kT. The fact that in a thermal oxide the value of ΔW is considerable indicates a high concentration of defects. Unfortunately, the theory has as yet failed to give proper treatment to the influence exerted on the energy spectrum by localized states of silicon-oxygen tetrahedra (including hydroxyl and hydride groups) which are always present near the $Si - SiO_2$ interface.

Our ideas regarding variations in the effective charges q_{ef} of silicon atoms due to variations in the angles of silocsane bonds [6.143,144] were directly confirmed in studies of the $Si - SiO_2$ interface using high-resolution XPS spectra [6.147,149]. In this technique the changes that take place in the SiO_4 tetrahedra are monitored through chemical shift of Si(2p) in the photoelectron spectrum. It was demonstrated [6.147] that the structure of the oxide in the interfacial layer (~ 1.6 nm thick) consists mainly of four-unit rings with $Si - 0 - Si$ angles of about $120°$. The authors of [6.147,149] assumed that the semiconductor is adjoined by a thin layer (close to a monolayer) composed of $Si_2O_3 : SiO : Si_2O$ in a ratio of $2 : 3 : 2$. The presence of paramagnetic Si^{3+} atoms on the interface is confirmed by EPR data [6.175,176]. Less indisputable are structures containing Si^{2+} and Si^+ atoms.

Many researchers have examined the possibility that a monoxide phase (SiO) exists within the boundary layer (e.g., [6.146,149,152] and references therein). However, the hybride bonds in SiO (GeO) are the π and σ bonds, as in a CO molecule, and thus the creation of a more or less stable three-dimensional monoxide layer is not possible [6.143]. A thin layer of SiO, whose existence is indicated by XPS [6.148] and SSXA [6.146], is apparently composed of two-dimensional clusters of adsorptive $\geqq Si = 0$ complexes formed during thermal oxidation. We demonstrated in [6.3] that such complexes are formed when an atomically clean silicon surface interacts with O_2. Note that SiO complexes were detected via SSXA spectra during oxidation of an atomically clean silicon surface. Another approach to explaining the nonstoichiometric oxide phase is the random bonding model, which considers the probability of occurrence of tetrahedral coordination of Si atoms, which would be based on X atoms of oxygen and $(4 - x)$ atoms of silicon [6.167,177,178]. This statistical-thermodynamical model was used to estimate the values of the shift of Si(2p) in photoelectron spectra, the density of states, the concentration of dangling $Si - 0$ bonds, and other parameters of the SiO_x structure [6.168].

It should be noted that at high oxidation rates the oxide phase exhibits microinclusions of silicon. Additional surface states may appear on the boundaries of silicon clusters and steps. The authors of [6.179] assumed these clusters and steps to be responsible for the roughness of the Si - SiO$_2$ interface which further complicates assessments of interface thickness d_{int}.

Let us now consider the Ge - GeO$_2$ interface. Because Ge - O bonds are longer than Si - O bonds, the GeO$_4$ tetrahedra in the oxide are less rigid. This is confirmed by figures on compressibility [6.180] and is probably connected with the fact that s states play a greater role in germanium-oxygen bonds than in siloxane bonds. Therefore variations in the length of a tetrahedron's edges play a significant role in the deformation of the GeO$_2$ structure. This implies greater topological disorder, which, according to the theory, must cause more extensive tails of the density of states to appear in the forbidden band of the oxide. This conclusion is supported by data on photocharging of Ge - GeO$_2$ structures (Sect.4.3). By comparing the tails (ΔW) for thermal oxide SiO$_2$ and GeO$_2$ (ΔW for the latter being mainly determined by deformation of the tetrahedra themselves), one learns that ΔW is greater for GeO$_2$ ($\sim 0.5 - 1$ eV).

All of the considerations developed above refer to oxidized structures of Si and Ge. For real surfaces of silicon and germanium the picture is somewhat different. The length of the tail in the case of Ge (as estimated from photocharging) differs little from that for thermal oxide ($\Delta W \sim 0.5 - 1$ eV) while for silicon ΔW is much greater ($\sim 1.5 - 1.8$ eV). This appears to be connected with the fact that a real surface is not completely oxidized: it contains uncompleted SiO$_3$ tetrahedra and, possibly, patches devoid of oxide phase ("patchy surface"). Surface Si atoms in such patches, after exposure to fluorine etchants, can form $\geqq$ Si - F or $\geqq$ Si = O groups which at room temperature persist on the surface for a long time. The possible existence of $\geqq$ Si = O groups is indicated by the above-mentioned SSXA data [6.146] and by our data on adsorption on atomically clean surfaces [6.3]. The existence of $\geqq$ Si - F groups is confirmed by mass spectroscopy [6.30]; the release of SiOF$_2$ molecules was detected during heat treatment of real Si surfaces. This also revealed centers which were extremely active in terms of irreversible adsorption of oxygen. Reversible negative charging of Si surfaces under alternating adsorption and desorption of oxygen also suggests incomplete oxidation of the silicon surface. This effect is not observed with real surfaces of germanium which is more readily oxidized.

We have not yet come close to constructing reliable structural and energetic models of the semiconductor-oxide heterojunction which would embrace all of the spectroscopic and electrophysical data. Nevertheless, we shall venture to present here, if only for the sake of perspicuity, a very approximate and qualitative diagram which may be considered to reflect the structure of real germanium and silicon surfaces (Fig.6.15), however open to criticism it might appear. To a certain extent this scheme does not contradict the above-discussed spectroscopic data and the results of our experiments on the effects of adsorption on the energy spectra of slow, fast, and recombination states (see Sects.3.1,3;6.2-6, and [6.47-49]).

Let us consider the real Ge surface and move from Ge towards GeO_2 on (Fig. 6.15). Fast surface states (FS) and recombination centers (RC), which readily exchange with the allowed bands of Ge, are found in Layer 1 which generally retains the structure of Ge, while the oxygen atoms here can be considered interstitial defects. The regularities in the curves of capture on fast states and in the rate of surface recombination S (Sect.6.2) imply that recombination and capture are each due to a separate quasi-continuous set of surface states. These can exhibit different capture cross sections and different spatial distributions within Layer 1 [6.49]. Reversible molecular adsorption of water vapor (Fig.3.6) and certain other polar molecules had no effect on Q_{sf} and S in the initial stage of occupation. This implies that defects in Layer 1 are

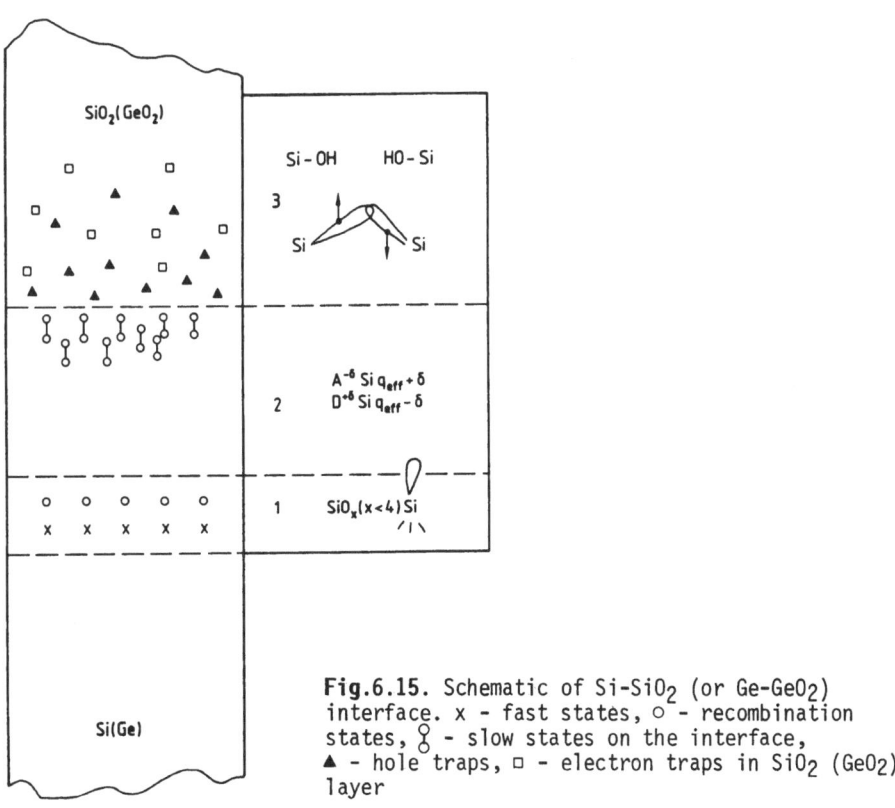

Fig.6.15. Schematic of Si-SiO_2 (or Ge-GeO_2) interface. x - fast states, $\circ$ - recombination states, $\begin{smallmatrix}\circ\\\circ\end{smallmatrix}$ - slow states on the interface, $\blacktriangle$ - hole traps, $\square$ - electron traps in SiO_2 (GeO_2) layer

155

inaccessible to environmental molecules. Only the smallest particles (protons created through dissociation of water molecules at high occupancies) can penetrate the amorphized layer (2) and modify, by virtue of their fields, the parameters of FSs and RCs [6.47]. Another cause of change in the parameters of FSs and RCs is associated with reconstruction of the interface due to irreversible dissociative chemical adsorption, which will be discussed later.

As yet we do not have any reliable experimental information about the nature of the defects which form the basis of fast states and recombination centers. Such information cannot be obtained from electrophysical measurements especially as the specific features of these states are masked by the cooperative properties of the quasi-continuous spectrum. Investigations of the spin-dependent channel of surface recombination offer some promise in this direction. The influence of spin polarization of electrons on the probability of recombination was first discovered with the aid of photoconduction measurements [6.181] and EPR [6.182]. Analysis of the parameters of the EPR signal from the paramagnetic centers responsible for this path of recombination indicates that they are the dangling bonds $\geqq Ge^{\cdot}$ and $\geqq Si^{\cdot}$ [6.182]. In the course of adsorption this channel was found to exhibit changes which were symbatic with the monotonic component of the rate of surface recombination S_m. Therefore at least part of the defects that contribute to S are free radicals (which are known to possess similar capture cross sections c_n and c_p). Analysis of the dependence of EPR signals on the orientation of the magnetic field reveals that Layer 1 (which contains spin centers) is partly amorphized [6.182]. This is evidence in favor of our identification of the spectrum of recombination centers as a spectrum characteristic of a disordered system (Sect.6.2). The participation of the spin centers Si^{3+} in the fast capture was recently direct confirmed in [6.183,184].

More reliable information is available on slow states of interface (SSI), found in the amorphized layer (2). These states are responsible for slow relaxation in the field effect. As demonstrated in Sect.3.3, these states exchange charges with the semiconductor mainly via the tunnel mechanism through an amorphized layer about 1 nm thick. Tunneling is not the limiting stage of slow capture. Low values of capture cross sections for SSIs (Chap.8) are connected with the peculiarities of electron-phonon interaction in adsorption complexes.

The majority of SSIs are adsorption states (Sects.3.3,6.4). Their basis is formed by coordination-unsaturated Ge (or Si) atoms centered within deformed incomplete, and hydrated germanium-oxygen (or silicon-oxygen) tetrahedra, whose existence in Layer 2 has been proved by spectroscopic methods. These atoms develop good contact with the environment, owing to the flaws and porosity of the oxide film (Fig.6.15). It is the donor-acceptor complexes $A^{-\delta}Ge^{q_{ef}+\delta}$ and $D^{+\delta}Ge^{q_{ef}-\delta}$, formed in the course of adsorption (Sect.6.6), that act as dipole neutral centers of slow capture. Coordination-unsaturated atoms are the primary centers of adsorption [6.3] . As was demonstrated in Sect.3.3, (see also Fig.3.6), charge starts to accumulate in SSIs in the earliest stages of adsorption.

Owing to the heterogeneity of Layer 2 and wide variations in the effective charges of atoms q_{ef} (see above), the magnitude of transferred charge δ in adsorptive complexes takes on random values [6.185]. Because the concentration of SSIs on real surfaces is rather high (Sect.6.4), the Coulomb fields of these complexes overlap, which is typical of a disordered system. Experimental results [6.186] indicate that the time constants of slow relaxation of germanium and silicon surfaces vary widely. We have already discussed (Sect.3.3) the distribution of SSIs over the height and width of barriers. From the standpoint of the physics of nonuniform surfaces it would be advisable to establish correlations between these distribution functions and the form of functional $P[v(r)]$, which describes the statistics of fluctuative fields on the surface (Sect.3.2).

Owing to the cooperative nature of random fields (Sect.3.2) and the diversity of the agents that influence the parameters of slow states, the energy spectrum of slow states does not reflect the individuality of adsorbed molecules. The effective levels of these states, derived from experiment (Table 6.1) are but convenient phenomenological characteristics of the quasi-continuous spectrum in the sense ascribed to them by the theory (Sect.3.2).

The effects of adsorption on the energy spectrum of SSIs are not limited to an influence on the amplitudes of random fields. The development of a donor-acceptor bond, as indicated above, alters the coordination number (c.n.) of the adsorption center. For example, with tetrahedral structures (SiO_2, GeO_2, etc.) the formation of a donor-acceptor bond elevates the coordination number of the atom from 4 to 5. This has two consequences.

6.7.1 Expansion of the Sample

An increase in the coordination number of an adsorption center due to the formation of a donor-acceptor bond must result in expansion of the specimen [6.15]. We pointed out in [Ref.6.3,Chap.2] that such an expansion is observed during adsorption of H_2O and NH_3 on porous SiO_2. The range of occupancies which corresponds to this expansion agrees well with the concentration of coordination-unsaturated Si atoms ($\sim 10^{13}$ cm^{-2}) as estimated by spectroscopic methods. Considerable deformation of thin films of $A^{II}B^{VI}$ compounds during adsorption of oxygen was reported by VOVSI and STRAKHOV [6.187].

LEVINE and FREEMAN [6.188] had indicated earlier that the displacement of atoms during adsorption and the resulting change in the Madelung constant may lead to the appearance of shallow levels in the forbidden band of the semiconductor. BYKOVA and LAZNEVA [6.189] observed the appearance of a set of shallow levels of sticking during adsorption of oxygen on CdS films, which they attributed to deformation of the surface structure of the film. Considerable variations in the surface energy spectrum are to be expected when coordination bonds are formed, and especially during dissociative adsorption. As demonstrated in Sects.3.1, 6.2 (Figs.3.7,6.6), the greatest variations in the spectrum of recombination centers which are responsible for the monotonic component S_m of the rate of surface recombination, occur during dissociative irreversible adsorption, accompanied by the rupture of $\geq$ Ge - O - Ge $\leq$ bonds and the creation of new surface groups (GeH, GeOH, etc.). This results in considerable variations of valence angles and bond lengths and reconstruction of the structure of germanium-oxygen (silicon-oxygen) tetrahedra adjacent to the adsorption center. These distortions are relayed to the lower-lying layers and can modify the structure of Layer 1 which contains recombination centers. The greatest structural distortions are observed during adsorption of water vapor (at high occupancies), accompanied by hydration of tetrahedra which may involve Layer 1. Simultaneously, a group of states may drop out of the quasi-continuous spectrum which would account for the appearance of the bell-shaped component S_b (Sect.3.1,Fig.3.7). A peculiar (and as yet unexplained) feature of fast states is their lower sensitivity towards reconstructions of this kind, as compared to recombination centers. It is possible that the polarization effects connected with reconstruction have a greater influence on the capture cross sections of defects responsible for recombination than on their location in the crystal's forbidden band.

Note that all of these experiments on surface recombination were conducted at low excitations of the semiconductor's electron system. In this case recombination involves comparatively deep levels [6.190]. At the same time the variations in the Madelung constant caused by any surface reconstruction will affect the shallow levels first [6.188]. Consequently, studies of recombination at high (laser) excitations when the spectrum of shallow traps assumes an important role in the process of recombination are of great interest.

6.7.2 Modification of the Surface Phonon Spectrum

It was demonstrated in [6.3] that the very creation of a surface leads to the appearance of localized surface oscillations which have become known as surfons. Adsorption which takes a coordinative path should modify the spectrum of surface oscillations. The first qualitative investigations [6.191] of adsorption of H_2O on SiO_2 indicated that vibration bands of Si - O bonds in silanol groups are sensitive to the presence of adsorbed molecules. The effects of adsorption on the IR spectrum of hydrated and fluorinated rutile are reported in [6.192].

As seen from Fig.6.16, the spectrum of the rutile monocrystal displays a strong band (740 cm^{-1}) which lies close to the high-frequency edge of the phonon spectrum of rutile and corresponds to vibrations of adjacent TiO_6 octahedra in the bulk. This band is poorly discernible in the spectra of disperse specimens where the bands $v_s = 940$ cm^{-1} (initial specimen) and $v_s = 880$ cm^{-1} (fluorinated specimen) become more intensive. The observed shift Δv, as com-

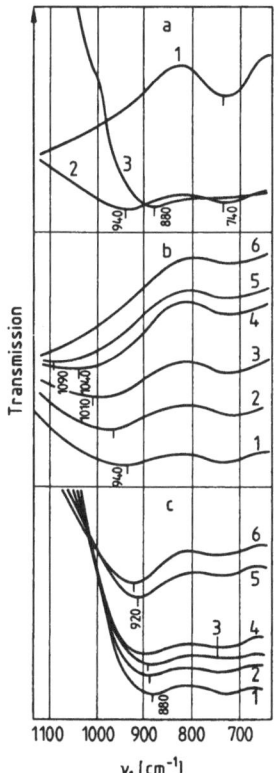

Fig.6.16a-c. IR spectra of rutile. (a) Single crystal (1), initial disperse specimen (2), fluorinated specimen (3). (b) Desorption of water from the initial disperse specimen at T = 300 (1), 400 (2), 500 (3), 600 (4), 700 (5), and 800 K (6). (c) Desorption of water from fluorinated specimen at T = 300 (1), 400 (2), 500 (3), 600 (4), 700 (5), and 800 K (6)

pared with a single crystal, as well as the drop in ν_s due to fluorination of the surface suggest that these bands are attributable to local vibrations in the subsurface region of the crystal.

From Fig.6.16b,c it is clear that desorption of water during vacuum heat treatment of the specimens results in a shift of ν_s towards higher frequencies which implies a lowering of the coordination number of the surface atoms. Similar shifts are detected in the IR spectra of titanium silicate glasses. Adsorption of water brins the spectrum back to its initial form in a reversible fashion. It is worth noting that the greatest variations in ν_s correspond to adsorption and desorption of coordination-bound water molecules [6.3]. This is yet another important argument in favor of the coordination mechanism of adsorption which is also in agreement with the EPR data from Ti^{3+} ions. A similar shift of the band of valence oscillations of the Ge - 0 bond (towards 880 cm^{-1}) was observed when coordination bonds were developed between water molecules and a surface of the tetrahedral modification of GeO_2 [6.192]. Adsorption was reported in [6.193] to have a strong influence on the phonon spectrum of the surface of oxides of transition metals.

The change in the phonon spectrum of the surface due to the creation of donor-acceptor complexes suggests the cooperative nature of the adsorption process. Since the kinetic parameters of the electron states depend implicitly on the value of the constant of electron-phonon interaction, the shift in the phonon spectrum during adsorption may influence the distribution of these parameters. The study of collective effects is the principal concern of surface science. Recent literature reflects a growing interest in surfons, and various techniques have been used to investigate them: electron spectroscopy [6.194], surface photoconductivity [6.195], disturbed complete internal reflection [6.196,197], and Raman backscattering spectroscopy [6.198]. Very promising in this connection are investigations of surface radiative recombination, extensively developed by ZUEV et al. [6.199,200]. By including the specific problems of adsorption in such investigations, we could increase our knowledge of the nature of adsorptive states and their kinetic parameters.

So far we have been considering disordered real surfaces of monatomic semiconductors. It is even more justified to use the theory of disordered systems in the case of compound semiconductors $A^{III}B^V$ and $A^{II}B^{VI}$, where the violations of stoichiometry on the interface are still more pronounced. With oxide semiconductors the subsurface layer of the crystal often contains an amorphized phase [Ref.6.3,Chap.3], and the same is true of atomically clean surfaces.

The effects of disorderliness of atomically clean surfaces on adsorptive and catalytic activity were analyzed by MARK et al. [6.201,202]. They assumed that topographic features such as steps (of atomic size) act as active centers of adsorption and catalysis on such surfaces. These features modify the value of the Madelung constant and lead to the appearance of electron states in the crystal's band scheme. Adsorption on active centers of this kind will further modify the value of the Madelung constant and the parameters of surface states. These topographic features can be produced during a catalytic reaction. The structure and electronic properties of vicinal atomically clean surfaces have been investigated in a large number of works (see, e.g., [6.203]). Particularly noteworthy are the theoretical calculations of hybride bonds on <111> Si faces carried out by RAJAN and FALICOV [6.204] in the strong bond approximation. They demonstrated that on steps there is an increase of charge on dangling $-Si\cdot$ bonds, which lowers the Fermi level and causes emptying of traps. Indeed, ROWE et al. [6.205], using UPS, discovered stepwise occupancy of surface states ($Si\cdot$) on <111> Si faces. These facts confirm the concept proposed by Mark and agree with our model of paramagnetic centers on atomically clean Si surfaces: at step locations one observes singly occupied electron states of Type A (Sect.3.4 and [Ref.6.3,Sect.3.1.4]).

Mark's model does not contradict our model of a disordered surface; the crucial point is the concentration of active centers. If the Coulomb fields of these centers overlap, their energy spectrum exhibits collective properties. The spectrum of the stepped surface given by MARK as an illustration [Ref.6.201, Fig.11] is equivalent to the spectrum of a disordered system (Figs.3.3,10).

The defects on the surface are mobile. Under certain conditions (e.g., by virtue of diffusion and interactions between defects), they may form associations (e.g., double vacancies) and, ultimately, clusters. Such associations and clusters can be brought about by exposure of the surface to radiation, or during crushing and redox treatment of adsorbents at elevated temperatures. Due to reconstruction of the surface phase, defects must associate more readily here than in the bulk, and aggregations of defects must certainly affect the adsorptive and catalytic properties of the surface, as well as the energy spectrum of surface states. Unfortunately, electrophysical methods that yield information on the collective properties of the semiconductor's electron system cannot be used to ascertain the spatial distribution of surface defects. Certain prospects in this direction are, however, opened up by radio spectroscopic techniques, in particular EPR.

By analyzing the dependence of the shape of the EPR line from paramagnetic Ti^{3+} ions (on a rutile surface) on their concentration, the authors of [6.206] demonstrated that reduction of rutile is accompanied by the formation of clusters of over-stoichiometric Ti^{3+} ions. The local concentration of ions n_{loc}

in the range of temperatures of reduction $T_{cal} \sim 700 - 800$ K is about 10^{14}cm^{-2}, which is an order of magnitude greater than the mean surface concentration of these ions ($n_s \sim 2 \cdot 10^{13}$ cm^{-2}). These clusters are about 10^{-7} cm in diameter, and the separation between them is about 10^{-6} cm. In this range of T_{cal} the increase in n_s is due mainly to an increase in the number of clusters while n_{loc} remains constant.

Aggregation of surface defects is certainly connected with the semiconductor's electron subsystem, for one thing because the defects have nonzero charge (as determined by their location in the forbidden band). The energy released in the acts of capture and recombination of free charge carriers can stimulate quasi-chemical reactions between defects (Chap.8). Another source of energy is acts of exothermal adsorption and catalysis. We discussed reconstruction of the surface of adsorbents during adsorption and catalysis in [6.3].

Aggregates can be built not only by defects, but also by adsorbed molecules and atoms. Clusters of adsorbed molecules have been reported to form as a result of the establishment of hydrogen bonds ([Ref.6.3,Chap.4] [Ref.6.66,Chap.4]). When not only attractive forces but also tangential forces of repulsion act between molecules, ordered surface structures can be formed such as those detected by ELEED [6.66,207]. The formation of clusters is often attributed in the literature to Coulomb interactions between oppositely charged forms of adsorption. The possibility that interaction between charged and neutral chemisorption can cause clusters to form was demonstrated in [6.208]. This interaction takes place via the transfer of an extra electron, trapped by the adsorptive state, to the neutral adsorption complex. Since the energy of an extra electron in a cluster is lower than in an isolated molecule such clusters can be formed even when capture of an electron on an isolated adsorptive state is not possible. Also conceivable is indirect interaction between adsorbed atoms via the phonon field of the crystal [6.209]. The linkage between active centers and the lattice in this case is established via common electron-phonon interactions.

An increase in the concentration of defects is accompanied by an increase in the configuration entropy. The inverse process, i.e., the formation of clusters and organization of the surface structure, takes place via interactions between particles in the surface phase. The study of self-organization on the surface is most closely associated with problems of synergetics [6.210]. A semiconductor surface can provide a convenient testing ground for nonequilibrium phase transitions and experiments on the organization of a molecular system at high levels of excitation of the semiconductor's electron subsystem, and for the elucidation of the role of fluctuations in these phenomena.

Finally, let us consider the fourth class of traps in the semiconductor-oxide system: superslow traps (SSD) in the insulating layer (Fig.6.15). These states exchange charge carriers with the bulk mainly via the over-the-barrier mechanism (Sect.3.3). This is one reason why the capture cross sections have quite small values (with respect to the semiconductor): $c_{n,p} \sim 10^{-27} - 10^{-31}$ cm^{-2} [6.185]. At the same time these traps can behave like fast traps with respect to the insulator ($c_{n,p} \sim 10^{-15} - 10^{-17}$ cm^{-2}). Superslow traps are detected in experiments on photocharging of insulator-dielectric structures and are responsible for the optical memory exhibited by these structures (Sect.4.3). On real germanium and silicon surfaces, superslow traps are found at a distance of about 2 nm from the semiconductor itself (5 - 10 nm on thermally oxidized surfaces).

Investigations of photocharging of Ge-GeO$_2$ and Si-SiO$_2$ reveal (see references in Sect.4.3) that at low energies of light quanta hν the charge is accumulated in electron traps, while light quanta with high energies stimulate charging of hole traps. Traps of both types are found in the most defect-ridden portion of the oxide film: Layer 3 in Fig.6.15. The three layers pass smoothly into one another: from the amorphized layer with violated stoichiometry (2), to the defect ridden (mostly with point defects) layer (3) to the more organized layer of GeO$_2$ (SiO$_2$).

The strong influence of dissociative adsorption of water on negative charging of SSDs which was discovered by Kozlov, Kashharov, and Kuznetsov (see references in Sect.4.3) brought them to the conclusion that it is complexes of binary hydroxyl groups (Fig.6.15) that are responsible for the capture of electrons. The centers of capture of holes are vacancy-type defects of GeO$_2$ (SiO$_2$), so called E' centers. In the initial nonparamagnetic state, these vacancy-type centers contain two unpaired electrons on overlapping sp^3 orbitals of two adjacent Ge(Si) atoms (Fig.6.15). After capturing a hole, the center becomes paramagnetic (Transitions 3,4 in Fig.4.3). Such centers have been thoroughly investigated in oxides of silicon and germanium and in IS structures during irradiation with high-energy quanta [6.211-213]. The effects of heat treatment of Ge-GeO$_2$ structures and photoinjection on the genesis and recharging of E' centers in oxide films were studied in [6.214,215]. The strong influence of adsorption and desorption on the parameters of SSDs which is emphasized in the above-cited publications, correlates well with data on the effects of these processes on the volt-ampere characteristics of IS structures (Fig.4.2).

In the literature on catalysis one frequently encounters the terms "local effects" and "collective effects", which are often contrasted with one another. In VOLKENSTEIN's electron theory [6.22], "local interactions" refer to the creation of neutral adsorption, while "collective effects" are understood as

all variations in the adsorptive and catalytic activity of the surface that
are connected with displacement of the semiconductor's Fermi level. With regard to real surfaces, as follows from the above discussion, collective effects
in the semiconductor-adsorbate system are connected not so much with the displacement of the Fermi level as with the change in the entire energy spectrum
of the surface as a consequence of the statistical nature of random fields on
it [6.15].

Reconstruction of the interface which is connected with the processes of
adsorption, gives rise to changes in the energy spectra of fast states and recombination centers. In this respect it resembles the phase transitions which
take place on atomically clean semiconductor surfaces and are accompanied by
abrupt changes in the electrophysical parameters of the surface (Fig.3.17).
An interesting task of surface science in the future will be study the mechanism
of such structural changes and elucidate the role of the electron subsystem in
these transformations.

7. Catalysis and Electronic Phenomena on Real Semiconductor Surfaces

In this chapter we discuss the linkage between the catalytic activity of semiconductor surface and excitation of the electron subsystem as well as the
importance of donor-acceptor interactions in catalysis. Captured charge carriers
are shown to modify the activity of electron-donor, electron-acceptor and
proton-donor sites. We also establish the connection between the catalytic
sites and the surface states. Finally, we analyze a tentative model for the
elementary act of dissociation of adsorbed molecules.

7.1 Ion Radicals on the Surface

Most researchers working with the electron theory of catalysis draw a connection between the reactivity of the semiconductor surface and the existence of
charged ion radical forms of adsorption. It is not possible to obtain information about the localization of an "extra" electron in the A-M (D-M) complex
within the framework of electrophysical measurements, but EPR is very promising
in this respect. In the ultimate case of CTC creating (Sect.6.6), the ESR
spectrum of ion radicals yields valuable data on the location of the wave-func-
tion maximum of an unpaired electron in a complex and the concentration of ion
radicals. Additional information concerning charge transfer complexes (CTC) can
also be obtained from electron and infrared spectra. Here we shall discuss the
concatenation of two phenomena: the creation of a CTC and variations in the surface charge.

Let us consider a model system of rutile-pBq. Quinone molecules can form
ion radicals by capturing an electron. In [7.1], the following were measured
simultaneously: adsorption, the IR and EPR spectra of adsorbed p-Bq molecules,
and variations in the surface potential of the rutile monocrystal during adsorption [with the aid of the contact potential difference (CPD) technique].
Adsorption of p-Bq was accompanied by the appearance of EPR signals ($g = 2.003$;
$\Delta H = 5$ G) from anion radicals $(p\text{-}Bq)^-$, whose concentration grew as adsorption
continued. Variations in surface potential Y_s indicated negative charging of
the surface (Fig.7.1). Typical IR spectra of adsorbed p-Bq are shown in Fig.
7.2; the bands $\nu_c = 1505$ cm^{-1} correspond to vibrations of the $\geq C = 0$ bond in the
p-Bq molecule, which has established a coordination bond with over-stoichiometric Ti^{3+} ions (Fig.6.14b, Scheme 1). The frequency shift $\Delta\nu \sim 165$ cm^{-1} (with
respect to the frequency of a free molecule) indicates the increase in the

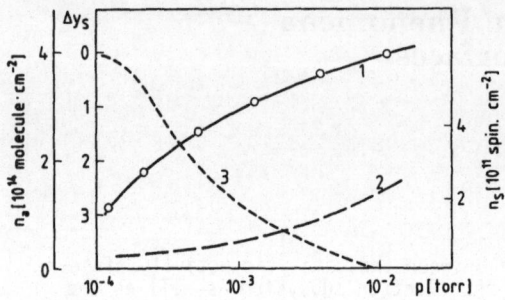

Fig.7.1. Adsorption (*1*) and concentration (*2*) of anion radicals on a dispersive rutile surface, and surface potential (*3*) of rutile single crystal versus pressure of p-Bq vapor at T = 300 K

effective negative charge of oxygen in the p-Bq molecule [7.2]. The band ν_{an} = 1470 cm^{-1} belongs to an anion radical (p-Bq)$^-$, which owes its existence to the transition of an electron from the adsorption center onto the orbitals of p-Bq, resulting in the rupture of a double $>C = 0$ (Fig.6.14b, Scheme 1'). With an increase in n_a the bands ν_c and ν_{an} become stronger, although ν_{an} remains much weaker than ν_c. This is in agreement with the data presented in Fig.7.1: the ratio between the number of anion radicals and the total number of adsorbed molecules is about 10^{-3}. All the remaining molecules are linked to the surface by weak van-der-Waals (ν = 1675 cm^{-1}) and hydrogen (ν'' = 1657 cm^{-1}) bonds (Fig. 7.2). The ratio of the number of coordination-bound molecules to the total is about 0.5.

According to the donor-acceptor mechanism, the majority of p-Bq molecules form complexes of (p-Bq)$^{-\delta}\cdot$Ti$^{3+\delta}$, where $\delta < q$. A complete transfer of charge δ = q) is observed only with a small fraction ($\sim 10^{-2} - 10^{-3}$) of the total number of molecules. The newly formed complexes of (p-Bq)$^{-\delta}\cdot$Ti$^{3+\delta}$ create a set of energy levels, some of which are occupied by electrons (Fig.6.14b, Scheme 3). The occupancy of levels increases with the elevation of the Fermi level. As seen from Fig.7.2b, the intensity of the ν_{an} band increases as the reduction of the rutile surface becomes greater (the Fermi level on the surface goes up). When the temperature of heat treatment is varied from 500 to 800 K, the concentration of spin centers n_S grows from $4\cdot10^{10}$ to $2.5\cdot10^{11}$ spin$\cdot$cm^{-2}. Oxidation of the specimen prior to adsorption (the Fermi level on the surface goes down) leads to a complete disappearance of the EPR signal and the ν_{an} band (Fig.7.2b). However, such experiments provide no basis for any quantitative conclusions, since the displacement of the Fermi level is certainly accompanied by a change in the energy spectrum of the surface. The transition from coordination bond to CTC is an activated one: the intensity of the ν_{an} band grows with an increase in the temperature at which adsorption takes place (Fig.7.2c). The same phenomenon is observed with EPR spectra. At room temperature the activation energy is about 0.1 eV.

Similar pecularities connected with the process of formation of ion radicals were reported by Zarif'yants, who took simultaneous measurements of adsorption, surface charging, and the formation of ion radicals for systems of PbS-tetracyanethylene and PbS-p-Bq using a piezoresonance balance. The kinetics of $n_S(t)$ was always considerably slower than that of $\Delta Q_s(t)$. The concentration of anion radicals constituted only about ten percent of the total number of charge appearing on the surface due to adsorption of the above-named acceptor-type molecules.

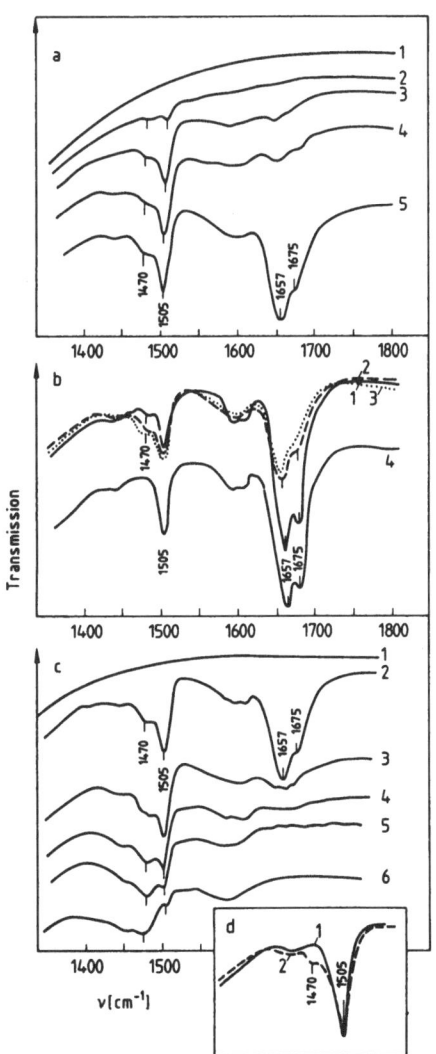

Fig.7.2a-d. IR spectra of disperse rutile. (**a**) Reduction at 700 K: (1) initial specimen, (2-5) specimens after adsorption of p-Bq (2) $n_a = 1.2 \cdot 10^{14}$, (3) $2 \cdot 10^{14}$, (4) $3 \cdot 10^{14}$, (5) $4 \cdot 10^{14}$ molecules/cm^{-2}. (**b**) After adsorption of p-Bq ($n_a = 4 \cdot 10^{14}$ molecules·cm^{-2}) on specimens reduced at (1) 500, (2) 700, (3) 800 K and (4) on an oxidized specimen. (**c**) Reduction at 800 K: (1) initial specimen, (2-6) specimens after adsorption of p-Bq at (2) 300, (3) 420, (4) 500, (5) 600, (6) 700 K. (**d**) Reduction at 450 K: (1) specimen after adsorption of p-Bq, (2) after exposure to light

So far we have been considering the adsorption of an acceptor-type molecule on an oxide surface in terms of local interactions. In the literature on the EPR of adsorbed oxygen molecules it has long been emphasized that there is a sharp discrepancy between the concentration of newly formed oxygen radicals n_{rad} and the concentration of disappearing paramagnetic adsorption centers n_s (e.g., Ti^{3+} in TiO_2 [7.3,4] and ED centers in ZnO [7.5]). As estimated in [7.3,4], the ratio n_s / n_{rad} is about 10 to 15. In order to clarify this contradiction, the authors of [7.6] took simultaneous measurements of EPR spectra at 77 K and variations in σ and surface potential Y_s (using CPD) on rutile monocrystals during adsorption of p-Bq (Fig.7.3). As is evident from Fig.7.3a, a slight increase in the concentration of ion radicals $(p-Bq)^-$ ($g = 2.003$) is accompanied by a sharp decrease in the concentration of paramagnetic Ti^{3+} ions ($g = 1.96$), which appear in clusters on the reduced rutile surface. At the initial stage of adsorption the ratio n_s / n_{rad} is about 30.

The discrepancy between n_{rad} and n_s can be attributed to the collective properties of the energy spectrum of the surface. The appearance of the first complexes of $(p-Bq)^{-\delta} \cdot Ti^{3+\delta}$ causes a shift of the entire energy spectrum (similar to adsorption of oxygen on PbS; see Fig.3.4), negative charging of

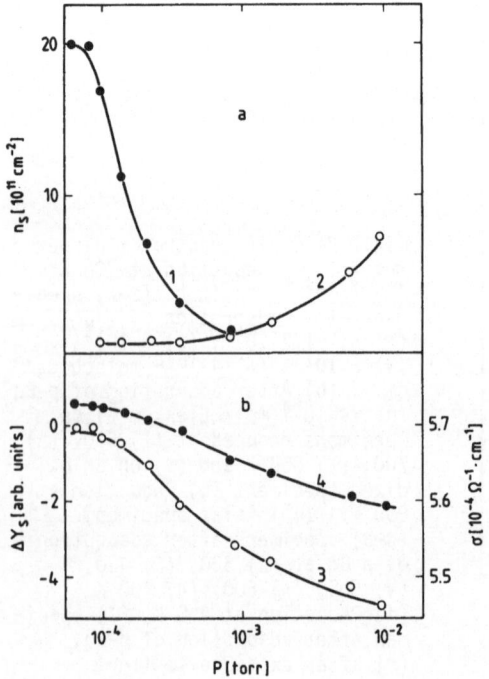

Fig.7.3. (a) Variations in EPR signal of rutile with $g = 1.96$ (*1*) and 2.003 (p-Bq) (*2*). (b) Variations in ΔY_s (*3*) and conductivity σ (*4*) during adsorption of p-Bq

the surface ($-\Delta Y_s$), and a decrease in σ_s (Fig.7.3b). Owing to downward displacement of the Fermi level, the remaining donor levels, in particular, the levels of Ti^{3+} ions, are emptied. The values of n_s and n_{rad} can be brought into qualitative agreement if one allows for a change in the occupancy of levels of Ti^{3+} ions with an increase in $|\Delta Y_s|$ according to (3.1). A similar phenomenon was observed for ZnO-O_2 in [7.7]. Those authors attributed the discrepancy between n_s and n_{rad} to the change in the concentration of electrons in the conduction band, assuming that it is conduction electrons that are responsible for the paramagnetic properties of the surface (n_s). The EPR signal from these centers was also observed in [7.8]. The temperature dependence $n_s(T)$ measured in [7.1] for rutile indicates that the value of n_s for TiO_2 (Fig.7.3) describes the concentration of localized spin centers.

A similar appearance of anion and cation radicals was observed during adsorption of p-Bq and diphenylamine, respectively, on real germanium surfaces [7.9,10]. The number of created spin centers n_{rad} turned out to differ only slightly from the total surface charge $\Delta Q_s = \Delta Q_{sf} + Q_{ss}$. It is noteworthy that with an increase in the temperature of adsorption, the change in surface charge is negligibly small, while the value of n_{rad} grows by nearly an order of magnitude. Once again, this indicates the activated nature of the transition from $A(D)^{-\delta} \cdot M^{qef+\delta}$ to CTC: $\delta \to q$. The function $n_{rad}(T)$ is different from $Q_s(T)$.

From this discussion of these preliminary investigations, it is clear that the mechanism by which ion radicals are created is by no means always equivalent to the mechanism of surface charging, as is assumed in VOLKENSTEIN's model [7.11]. In general, when adsorption follows the donor-acceptor mechanism, three activated processes seem to take place on the surface:

1) Creation of adsorption complex $A(D)^{\pm\delta} \cdot M^{qef\pm\delta}$ with activation energy E_a,
2) Capture of a charge carrier on the local level created by the complex in the forbidden band of the crystal. Here the activation energy E_{cap} depends on the parameters c_p and c_n of the newly formed state ε_t; these parameters in turn are determined by the structure of the whole complex, including its ligand surrounding.
3) Transition from donor-acceptor bond ($\delta < q$) to CTC ($\delta = q$). The activation energy of radical adsorption is E_{rad}. The transition from (2) to (3) may affect the values of capture cross sections c_n and c_p.

The preliminary experimental results indicate that the activation energy E_{rad} cannot always be neglected, as was done in VOLKENSTEIN's model [7.11], where the process of formation of ion radicals is described only by E_{cap}. It is to be hoped that further consistent investigations of temperature dependences $n_{rad}(T)$ and $Q_s(T)$ will provide the basis for estimating E_{cap} and E_{rad} and establishing the relationship between these quantities and the structural features of adsorbed molecules and the adsorption center.

There have been reports of electron transfer between adsorbed ion radicals [7.12,13], or between ions and oxygen (with creation of O_2^- radicals). However the lack of electrophysical data prevents us from ascertaining when this phenomenon is connected with a change in the surface potential and recharging of traps (Fig.7.3), and when it is due to direct exchange of charge between traps. The latter is quite possible for wide-band amorphous semiconductors, where the high density of fluctuative levels makes it possible for the transfer of charge to follow the "hopping" mechanism.

7.2 Catalytic Activity and Surface Electronics

According to VOLKENSTEIN's model [7.11], the catalytic activity of a semiconductor is wholly determined by the relationship between charged and neutral adsorption n^+, n^-, and n^0; i.e., it is assumed that in adsorption of valence-saturated particles the catalysis proceeds due to charged ion radicals whereas with valence-unsaturated particles it is neutral radicals that are responsible for catalysis (Chap.5). For the sake of clarity in the following we shall consider only chemisorption of valence-saturated molecules as the act preceding catalysis. For this case the very rough assumption is made in [7.11], that the concentration of reactive ion radicals n_{rad} equals the concentration of charged adsorption, i.e., $n_{rad} = n^\pm$ (Fig.6.14a). The whole theory which is supposed to establish correlations between catalytic activity (i.e., n_{rad}) and the Fermi level (and, consequently, $\Delta\sigma$ and $\Delta\varphi_t$), is based on this assumption[1].

We see that out of all the conceivable electron transitions between the bulk and the surface states, VOLKENSTEIN's theoretical model [7.11] considers only a very special case: transitions of electrons and holes directly onto the orbitals of the adsorbed molecule (transition 1-2, Fig.6.14a). As we demonstrated in [7.14,15], this is not the case for a great many valence-saturated molecules dealt with in adsorption and catalysis. Much more often the situation is one in which the electron, trapped in a local level of the chemisorbed molecule, does not become part of the electron ensemble of this molecule [7.14,15]. Precisely, this case is considered in the donor-acceptor mechanism of adsorption and charging (Fig.6.14b,c).

---

[1] In adsorption of valence-unsaturated particles such as H, it is assumed that the radical form is equivalent to the neutral (in terms of charge) form; $n_{rad} = n^0$. Then one uses (3.1) to establish the relationship between catalytic activity and $\Delta\sigma$ and $\Delta\varphi_t$.

The appearance of surface charge according to Models b,c (e.g., during adsorption of H_2O, State 2) cannot be interpreted solely as the appearance of
the ion radical form of adsorption ($n_{rad} \neq N_.$, where $N_.$ is the concentration
of charged surface states). The charge can be localized at polarized bonds in
the complex (Fig.6.14b) or on a defect (Fig.6.14c). The molecule then remains
in a valence-saturated state, while $\delta_2 > \delta_1$, and $\delta \to q$ upon transition to CTC
(transition 2-3). The symbol C denotes chemical groups connected with adsorption centers, e.g., OH groups always present on oxide surfaces [7.16]. The
value of δ depends on the nature of C. For wide-band semiconductors, when C
groups are sufficiently electronegative, δ can become so great as to allow
the transition 1-1' into the radical form without surface charging.

As regards the electrophysical manifestations, the differences between
Schemes a and b are purely quantitative. Variations in integral quantities
such as σ_s, $\Delta\varphi_t$ and Y_s during adsorption are determined by the change in the
total surface charge Q_s. It matters little where exactly the extra electron
(or hole) is localized on the surface. Much greater errors arise from neglect
of biograph states n_{sB}. As we have emphasized, $n^{\pm} \neq N^*$, since $N^* = n^{\pm} + n_{sB}$.
The picture is quite different when it comes to catalysis: here the reactivity
of adsorbed molecules depends directly on where the extra electron is localized. If $n_{rad} \ll n^{\pm}$, then any correlations between catalysis and variations
in $\Delta\sigma$ and $\Delta\varphi_t$ are out of the question, even within the framework of VOLKEN-
STEIN's model [7.11].

If we consider an extreme case: capture of a charge carrier on the orbitals
of an adsorbed particle, then it certainly will not always be valid to identify the ion radical form or neutral radical form of adsorption with the reactive particles responsible for catalysis. In order to estimate the reactivity of a molecule, we must know the symmetry of its wave functions. There
are many examples of very stable radicals that have zero reactivity. EPR data
indicate that in the case of the above-discussed anion radicals of $(p\text{-}Bq)^-$
type, the unpaired electron is smeared in the plane of the benzene ring. At
the same time, when p-Bq is adsorbed on a strong EA center (Al^{3+}), a cation
radical $(p\text{-}Bq)^+$ is formed, and the maximum electron density for the transferred
electron will be observed near the nucleus of Al^{3+}. In each case the reactivity
of ion radicals will be quite different. The same applies to radicals on real
solid-state surfaces. A defect having a free valence need not be the most active chemisorption center. A vivid example is unsaturated valences on an atomically clean germanium surface, which are indifferent to chemisorption [7.16].

From the above discussion it follows that the commonly accepted classification of adsorption into charged and neutral, strong and weak, radical and

valence-saturated forms [7.11] is rather cumbersome, and fails to reflect the actual state of adsorbed molecules and their reactivity. This classification is even less suitable for real defect-ridden surfaces. As emphasized in [7.11], here the "neutral" ("weak") form can be charged, the "strong" form can be neutral, and a "weak" bond can prove stronger than a "strong" one. Therefore, until we have a reliable theory of the elementary act of adsorption, it would be more practical to use the classification accepted in quantum chemistry, taking into account the character of the surface. The vast quantity of experimental data on the optical and radio spectroscopy of surface compounds which as yet has been overlooked in the electron theory, allows us to specify the nature of chemical bonds on the surface.

According to the donor-acceptor mechanism discussed in Sect.6.6, surface reactivity does not require the presence of radical forms of adsorption. Any donor-acceptor complex is reactive even before it has captured an electron or hole. Consider a model-type example of adsorption of water on an EA center. As was demonstrated in [7.16], such a molecule is highly deformed because the undivided pair of electron is drawn to the M atom. The extent of protonization can be controlled by chemically modifying the surface [Ref.7.16,Chap.4], e.g., by replacing OH groups with F (Fig.6.14b,c,Group C). Owing to the disorderliness of the surface and the highly varying effective charges of the adsorption centers, we are dealing with some kind of distribution of reactivity over adsorption centers.

What is changed then, if an electron or hole is trapped next to the surface complex, in other words, if the surface becomes charged? If a hole is trapped, the result will be a strengthening of the coordination bond, further protonization of the water molecule (Fig.6.14b,c), and, consequently, increased reactivity [7.14]. If an electron is captured, the bond will be weakened, and desorption of the molecule will become more probable. If the bonds are dative, the situation is reversed. We see that the reactivity depends on charging by virtue of deformation of molecules, which nevertheless remain electrically neutral in the usage of [7.11]. This case is entirely foreign to VOLKENSTEIN's model [7.11] which attributes the increase in catalytic activity solely to the increase in the concentration of ion-radical charged adsorption. From the above discussion it follows that surface charging affects the activity of EA and ED centers of catalysis. If an EA center is occupied by a water molecule (which according to [7.16] can act as a PD center), then capture of a hole will raise the activity of this center, which plays an important role in acid-base catalysis [7.14,17]. The works [7.14] were the first to consider the role of electron transitions in acid-base catalysis which is a step forward with respect to the existing theory of catalysis where the electron mechanism is considered only in redox reactions.

The idea that adsorbed molecules undergo additional deformation in the field created by captured charge carriers [7.14] was later confirmed by quantum-chemical calculations [7.18]. The theoretical analysis carried out in [7.18] revealed that the strong Coulomb fields of adsorption centers cause elongation of certain bonds in the adsorbed molecule, thereby increasing its dipole moment. This also heightens the ionicity of the bonds. These effects are most pronounced in nonpolar molecules. For instance, in a CO molecule, a 50 % decrease in the interatomic distance causes a 30 % reduction in the bond strength [7.18]. Also reduced is the activation energy for the process of dissociation of such molecules on the surface. In the excited state the molecule is more likely to undergo heterolytic dissociation, which is connected with separation of charges. However, this is true only for weak complexes where the transferred charge in the ground state is sufficiently small [7.19]. For strong complexes the magnitude of the charge transferred in the excited state is less than complete by the value of δ (where δ is the charge exchanged between donor and acceptor in the ground state).

Our experiments on photodissociation of H_2O adsorbed on Ge and Si (Chap.9) fit in well with the concepts of quantum chemistry. The experimental results on the effects of excess surface charge on the catalytic activity of the surface are discussed in [7.20].

The reactivity of molecules which enter the donor-acceptor bond is primarily determined by the following factors [7.21,22]:

1) The configuration and symmetry of orbitals of undivided electrons, and the extent to which the latter are drawn onto the orbitals of the adsorption center (δ).

2) The configuration and symmetry of hybride orbitals of the adsorption center which strongly depend on its immediate surroundings, i.e., on the chemistry of the surface.

3) The magnitude and sign of charge localized on or near the adsorption complex.

The first two of these factors can be labeled "chemical". They also determine the energetic and kinetic parameters of the electron adsorption states (Sects. 6.2-6). By virtue of the heterogeneity of the surface, these states assume a cooperative nature (Sect.6.7). The location of the Fermi level and, most importantly, the character of the whole energy spectrum, determine the third factor, which might be referred to as the electron factor. In the general case the activity of Lewis and Brönsted centers is characterized by the combination of the three factors.

The optimum value of adsorptive bond energy for the catalytic reaction must not be too high (80 - 169 kJ · mol^{-1}) [7.23]. This condition is well satisfied by donor-acceptor bonds. The strength of the donor-acceptor bond, and, consequently, the deformation of the adsorbed molecule, will strongly depend on the correspondence between irreducible representations of the symmetry transformation groups of A(D) and M in the complex.

The donor-acceptor mechanism implies the existence of two interconnected ways of exercising control over the catalytic and chemisorptive activity of the semiconductor [7.21,22]:

1) chemical modification of the surface as discussed in [7.16], and

2) manipulations with surface electronics. Let us briefly discuss this point. Provided the energy spectrum does not change, surface electronic processes can be controlled by changing the concentration of charge carriers in the SCR (field effect, injection of carriers in p-n junction, and illumination). The phenomena of electroadsorption and photoadsorption were discussed within the framework of VOLKENSTEIN's model [7.11] in Sect.5.3.

The effects of an electric field on desorption of donor-type H_2O and CH_3OH molecules from germanium surfaces were investigated in [7.24-26]. Desorption of H_2O [7.25,26] and CH_3OH [7.24] was observed when a negative potential was applied to germanium (in [7.24] desorption was monitored directly in a mass spectrometer). According to the schemes in Fig.6.14b,c, the application of a negative potential to surface traps weakens coordination bonds and stimulates desorption. Reversing the field ("+" on germanium) ought to strengthen the bond, which agrees with the results reported in [7.26]. The influence of field on adsorption of acceptor-type O_2 molecules was studied in [7.27] (germanium) and [7.28] (ZnO). The application of a negative potential to the semiconductor stimulated adsorption. Using mass spectroscopy, LYASHENKO et al. [7.29] detected desorption of H_2 and N_2 upon application of an electric field to germanium, although they did not discuss the nature of the bonds involved. In [7.30], dehydrogenation of ethanol on ZnO was reported to be stimulated by an electric field ("+" on ZnO). This fits the scheme in Fig.6.14b,c and corresponds to increased activity of EA (Lewis) centers. The application of a field of the same polarity stimulated decomposition of alcohol on TiO_2 [7.31]. According to Fig.6.14b, this corresponds to increased activity of Brönsted centers. Coordination-bound H_2O molecules persist on the surface of TiO_2 in the range of vaccumization temperatures investigated in [7.31]. The effects of negative charging of TiO_2 surfaces on oxidation of CO and decomposition of N_2O were analyzed in [7.32].

It should be noted that many investigations of electroadsorption have neglected the possibility of discharge occurrences in the gap of the field capacitor (specimen—field electrode), which can strongly affect the interpretation of experimental results. The existence of strong currents in a gap with a field strength of $10^{-5} - 10^{-6}$ V·cm^{-1} is directly indicated by the data reported in [7.28,30]. In cases in which judgement was based on measurements of the conductivity of polycrystalline [7.28,30] or monocrystalline [7.26] semiconductors, the results are highly ambiguous owing to the effect of charge accumulation on the surface. Experiments on electroadsorption [7.33] carried out in an ultrahigh vacuum gave negative results. The application of a transverse field ($E = 10^4 - 10^5$ V·cm^{-2}) led to the accumulation of positive charge on germanium surfaces; however, electrodesorption was not detected within an accuracy of 10^{10} particles·cm^{-2}. As will be shown in Chap.9, the accumulation of charge is due to heterolytic dissociation of adsorbed molecules.

Manipulations with surface electronics using quasi-equilibrium effects (such as the application of a transverse field) are of academic interest, but hardly promise any technological applications. The density of states on most real objects is so high (10^{13} cm^{-2} or higher) that one can hardly expect noticeable changes in the charge status of active centers of catalysis in the field effect. Besides, adsorptive states (as demonstrated in Sect.6.4) are slow states with abnormally small capture cross sections. The characteristic time of capture on these states is many orders of magnitude greater than the characteristic times of most catalytic reactions. More or less efficient charging of adsorptive centers can be expected only at high levels of injection of nonequilibrium charge carriers [7.21]. The number of acts of capture Δn in time Δt is known to be expressed as (back ejection neglected) [7.34]:

$$\Delta n_c = n c_n N v_t \, \Delta t \quad , \tag{7.1}$$

where n is the concentration of charge carriers, c_n is the capture cross section, N is the concentration of traps, and v_t is the thermal velocity of charge carriers. A substantial change in the charge of defect surface atoms q_{ef} can be achieved only at a high injection level, when the concentration of nonequilibrium charge carriers in the SCR [n in (7.1)] is equal to or higher than the concentration of equilibrium charge carriers (Sect.2.2).

With this purpose in mind, FEDEROV et al. and KISELEV et al. [7.35,36] studied catalytic decomposition of formic acid adsorbed on the surface of a germanium p-n junction at high injection levels. The reaction yield was analyzed using a radio-frequency mass spectrometer. The p-n junction was operated

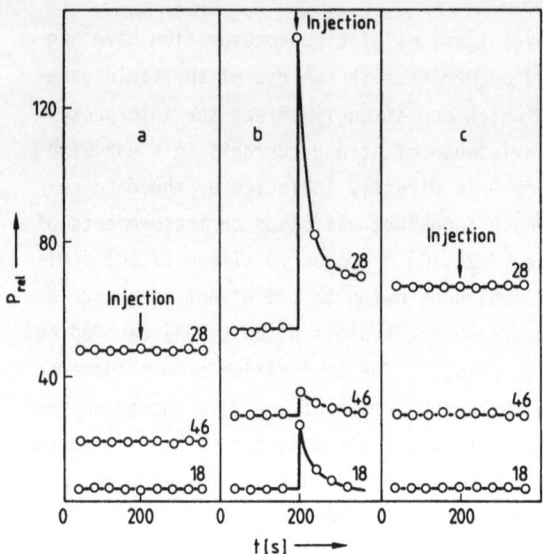

Fig.7.4a-c. Mass spectra of the products of catalytic decomposition of formic acid on the surface of a p-n junction upon pulsed application of forward current, where P_{rel} is the relative pressure of the components in arbitrary units

in pulsed fashion, so as to eliminate the effects of heating. As follows from Fig.7.4a, connecting the vacuumized p-n junction in the forward direction (injection of holes) failed to produce desorption in the relevant mass range. After adsorption of HCOOH, injection of holes caused the release of $CO(m=28)$, $H_2O(m=18)$, and $HCOOH(m=46)$ (Fig.7.4b). Injection raised the concentration of nonequilibrium charge carriers around the p-n junction to $10^{17} - 10^{18}$ cm^{-3}. All the adsorbed molecules of formic acid underwent decomposition after the first pulse of current; subsequent pulses failed to stimulate the release of reaction products (Fig.7.4c). Connecting the p-n junction in the reverse direction (injection of electrons) did not stimulate decomposition.

The effect of injection of holes on the decomposition of formic acid must be considered to be quite high, if one takes into account the minuteness of the area of the p-n junction. It is interesting to note that heating the germanium, as well as heating the p-n junction, promoted dehydrogenation by means of the reaction of dehydration: mass spectra indicated the release of CO_2 and H_2. This fits in well with the results reported in [7.37,38].

Experimental data [7.20] and theoretical results [7.39] indicate that it is protons that are responsible for the reaction of dehydration. At these temperatures the surface of germanium contains a large number of coordination-bound water molecules (Sect.6.2). According to Fig.6.14b,c injection of holes gives rise to additional deformation of O-H bonds, which improves the proton-donor activity of H_2O molecules (Brönsted centers). Thus we see that the mechanism of reaction is entirely different from the radical mechanism considered in [7.11].

Finally, a third way of controlling the electronics of the surface is photoinjection of nonequilibrium charge carriers. Photoadsorption has been the subject of an enormous number of investigations, the majority of which have dealt with it as a function of the intensity and spectral composition of light, the biography of the specimen, doping, temperature, etc. Valuable information has been obtained about the sign of the effect and the spectral dependence of the absolute quantum yield [7.5,40-43]. However, as we demonstrated in Sect. 5.3, many important features of photoadsorption receive no explanation within the framework of the model of light-generated "free radicals" [7.44]. A great number of recent experimental results indicate that the predominant role in photoadsorption is played by biographic local surface states [7.45-47]. Simultaneous investigations of EPR and photoadsorption provided the basis for elucidating the important role played by strongly bound charged forms of oxygen in the photoadsorptive activity of the surface [7.47]. A number of researchers have emphasized the noticeable effect of surface hydroxyl groups on the parameters of the surface defects that are responsible for photoadsorption [7.48, 49].

The effects of light on a molecular subsystem on a solid-gas interface are now quite well known. Far less profound is our knowledge of the relationship between photoadsorption and the photoelectric processes induced in the solid by illumination. Theory examines the connection between photoadsorption and the location of the Fermi level, as well as the bending of energy bands on the surface of dark specimens [7.44], but fails to reveal the concatenation between photoadsorption and the photoelectric parameters of the surface. Investigations of changes in conductivity and CPD caused by illumination [7.5,7] have been few in number, and most have been conducted on powders. Consequently, according to Sect.4.4 [7.50], they yield little information about the mechanism of electronic processes on the surface and the energy spectrum of the surface state responsible for photoadsorption. One of the first quantitative investigations of CPD as a function of photoadsorption on the surface of a single crystal (with simultaneous AES and SPS) was performed by LAGOWSKI and GATOS [7.51].

As we do not endeavor to give an exhaustive analysis of photoadsorptive phenomena, we shall restrict our discussion here to a few results in which the electron mechanism is most vividly manifested. First let us discuss the work of BYKOVA and LASNEVA [7.52], who investigated the effects that simultaneous action of light and a transverse electric field have on desorption of oxygen from a CdS film. As follows from Fig.7.5, photodesorption (monitored using a

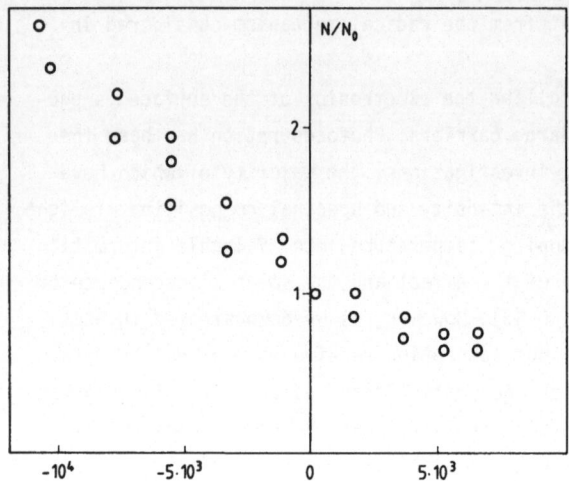

Fig.7.5. Relative variations in photodesorption of oxygen (N/N_0) from CdS film under the action of transverse electric field E [7.52]

mass spectrometer) increases after application of an electric field which draws holes to the surface. This is attributed to the capture of holes by negatively charged oxygen centers, which brings oxygen into a weakly bound state. Part of this oxygen is desorbed, while another part recaptures electrons; this is reflected in the kinetics of relaxation of photoconductivity and photo emf. Photodesorption of oxygen from the surface of a ZnO single crystal, as reflected by variations in surface conductivity, was investigated in [7.53]. It was explained there as being a result of recombination of photoinjected holes with electrons localized on chemisorbed oxygen.

Photoinjection of minority carriers can give rise not only to photoadsorption or photodesorption, but may also start a catalytic act. Very interesting in this connection are the data obtained by KOTEL'NIKOV and TERENIN [7.54] who observed photodecomposition of molecules of water and ammonia adsorbed on Al_2O_3. Illumination of the specimen (with preadsorbed water or ammonia) with light in the 230-300 nm spectral range resulted in the release of hydrogen which was detected by a mass spectrometer (Fig.7.6). It appears that decomposition of both H_2O and NH_3 exhibits three maxima of hydrogen release, corresponding to 248, 270, and 290 nm. The fact that these maxima are observed at the same wavelengths for two different molecules indicates that it is the light-stimulated electron processes in the solid that are responsible for the decomposition of molecules. The release of hydrogen cannot be attributed to dissociation of surface hydroxyl groups; vacuum heat treatment at 600 K com-

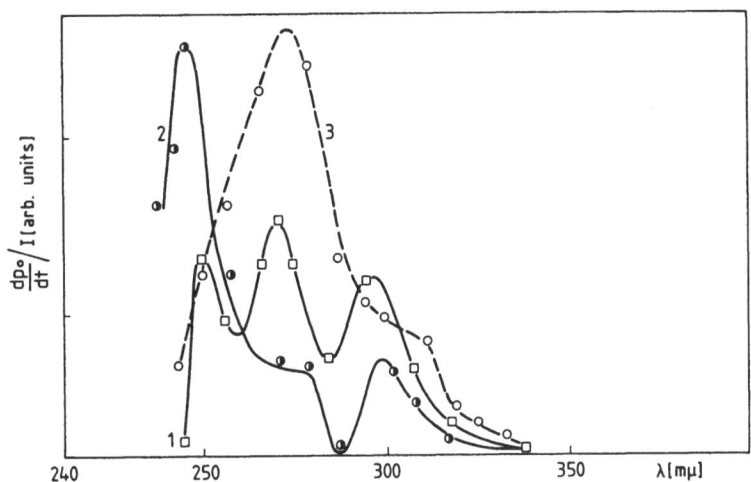

Fig.7.6. Ratio of the initial rate of hydrogen release dp_0/dt to light flux I as a function of the spectral composition of light [7.54]. (*1*) Photolysis of water adsorped on aluminum oxide, (*2*) the same for ammonia, (*3*) photoadsorption of oxygen

pletely prevented photodissociation. However, dehydration at these temperatures is largely due to the removal of coordination-bound water molecules [7.16], while the loss of hydroxyl groups is quite small. Photoadsorption of oxygen was observed on a dehydrated Al_2O_3 surface (Fig.7.6,Curve3), accompanied by an intensive EPR signal from oxygen ion radicals [7.54,55].

These results find a natural explanation in our model of the donor-acceptor mechanism of charging [7.21,56]. Molecules of H_2O and NH_3 are adsorbed according to the coordination mechanism. The role of active centers of chemisorption of these molecules can be played by tricoordinated Al atoms located next to oxygen vacancies (Fig.6.14c). An investigation of the energy spectrum of traps in Al_2O_3, performed with the VAC technique (Fig.4.2), revealed that H_2O and NH_3 molecules are adsorbed as donors. This results in the capture of a hole by an $H_2O^{+\delta} \cdot Al^{3-\delta}$ complex, which, provided δ is large enough, leads to dissociation of the molecule. Since high-dispersion Al_2O_3 is a dielectric having a rather high density of fluctuation levels (Sect.4.3), an important role in the catalytic act can be played by charge carriers "hopping" over these levels between acceptor-type and donor-type adsorptive states.

With regard to coordination bonds, the "electronic factor" is capture of holes; this conclusion is confirmed by experiments on photodissociation of water (Fig.7.6) and reactions on p-n junctions (Fig.7.4). The role of photoinjected holes in the decomposition of formic acid was also demonstrated in electrochemical experiments by MORRISON [7.57,58] and FREUND et al. [7.59]. We shall return to the mechanism of photocatalytic processes in Chap.9 in connection with photodissociation of water molecules adsorbed on semiconductors.

While in VOLKENSTEIN's electronic theory [7.11,44] the catalytic act depends completely on the capture of either an electron or a hole, our donor-acceptor mechanism does not impose such rigid restrictions. The incipient complex is reactive by itself, and capture of a charge carrier can only increase or reduce the deformation of intramolecular bonds in adsorbed molecules. The effect of light on the reactivity of adsorbed molecules can be connected not only with excitation of the electronic subsystem, but also with stimulation of local transitions within the $A(D) \cdot M$ complex. For instance, GRIZKOV et al. [7.60] reported a change of charge $(V^{5+} \rightarrow V^{4+})$ and decomposition of ammonia adsorbed on V when the system was illuminated in the SCR band. In this system of ions deposited on a dielectric the principal role belongs to local transitions. We see that the donor-acceptor mechanism is able to account for both local and collective effects of the interaction between light quanta and the semiconductor, in contrast to VOLKENSTEIN's model [7.44], which reduces the entire variety of effects of photoadsorption (photocatalysis) to variations in $n^{\pm}$.

With real catalysis that have a partly amorphized subsurface layer, where the density of surface states is extremely high, the electronic subsystem can only be expected to have a noticeable influence on catalytic activity under highly nonequilibrium conditions (at high levels of excitation) which would stimulate efficient recharging of all surface traps, see (7.1). Such conditions can be created by exposing the specimen to intense laser radiation and simultaneously applying a transverse electric field which draws nonequilibrium carriers towards the surface.

Such a situation is contrary to the model adopted in most of VOLKENSTEIN's works [7.11,61,62]. These works are based on contradictory assumptions: on the one hand, electroadsorption, photoadsorption, and catalysis are considered at almost complete thermodynamic equilibrium when the system in question can be described with a single Fermi level. On the other hand, these works totally ignore the abnormally small capture cross sections of adsorptive states, which can capture electrons with reasonable probability only at high levels of excitation of the system.

For wide-band semiconductors which include many important catalysts, hard radiation appears very promising. By virtue of shock ionization, it can sharply increase the quantum yield of nonequilibrium charge carriers. Traps in thin dielectric films and MIS structures can be recharged with the aid of contact injection (either metallic or semiconductor). We discussed monopolar injection in Sect.4.3 in connection with investigations of the energy spectrum. Of especial interest is the study of double injection in MIS structures. In this case the volt-ampere characteristics (VAC), (Fig.4.2) display sections with negative resistance (N-shaped or S-shaped VAC), and the system may start to generate electric oscillations. This case is well known to physicists but has never been investigated in connection with catalytic reactions. By employing a MIS structure with a sufficiently porous insulator layer under conditions of double injection, one might be able to realize an oscillatory system which would generate products of a catalytic reaction due to cyclic recharging of adsorptive centers. Certain possible applications of such systems in catalysis and biocatalysis as well as in the construction of sensitive elements of gas analyzers, were discussed in [7.63].

All the experimental material discussed in this section provides convincing proof of the close relationship between the activity of EA, ED, and PD centers on the one hand, which are responsible for catalytic transformations, and surface charge on the other. This relationship, however, is much more intricate than is supposed in various electronic models of catalysis. The numerous controversies regarding the role of the electronic factor in catalysis have arisen large around the central question of where the "extra" electron responsible for charging is localized on the surface. In Volkenstein's model (Sect.5.3) the whole theory is based on a simplified calculation of the interaction between a univalent atom and an ionic unidimensional crystal (similar to the problem of an H_2 molecule). Without sufficient grounds, these results are further generalized for comparatively complex molecules. At the same time, a very rigid condition, $n_{rad} = n^{\pm}$ (or $n_{rad} = n^0$), is imposed. Only valence bonds (dashes in Fig.6.14a) are taken into account. Nonvalence bonds (which, as we have shown, can greatly enhance the reactivity of adsorbed molecules) are neglected. For this reason, neutral (in the terms of [7.11]) adsorption is in no way equivalent to the donor-acceptor (coordination) complexes considered in Sect.6.6.

It is noteworthy that at the earliest stage of development of the electron theory, BONCH-BRUEVICH [7.64] considered a more general quantum-mechanical model of the elementary act of adsorption and charging, which did not require ionization of the molecule (i.e., $n_{rad} \neq N$). According to that model, the electrons could move only in the vicinity of the adsorbed molecule. From our point of view, this model reflects the experimental results more correctly than the model adopted in [7.11].

In the usage accepted in the literature on catalysis, our EA, ED, and PD centers are equivalent to Lewis basic and Brönsted acid centers [7.20,65]. According to the donor-acceptor mechanism of adsorption and charging proposed by us in [7.14,17,21], adsorption on these centers leads to the appearance on the surface of electron surface states which act as efficient traps for the semiconductor's electrons and holes. Later, MORRISON [7.58,66] also suggested a similarity between acid surface centers and surface electron states. However, Morrison's model involved an idealized homogeneous surface devoid of biographic states (Sect.5.1), and he had derived his information on electron surface states from the temperature dependence of the conductivity of powdered catalysts which is not correct. As we pointed out in Sect.4.4, these measurements only give information on the temperature dependence of the height of potential barriers between powder particles.

In our model it is defect acceptor-type atoms of the surface (e.g., coordination-unsaturated metal atoms) and acid hydrogen atoms that are responsible for acid centers and surface states. The basic centers can be donor-type surface atoms (e.g., coordination-unsaturated oxygen O^{2-} ions, or hydroxyl OH^- groups) [7.67-69]. Analysis of the electron-donor properties of various oxides (CaO, MgO, ZnO, Al_2O_3, etc.) using the EPR spectra of adsorbed nitrobenzene radicals reveals a correlation between the number of basic centers and donor-type surface states.

The donor-acceptor mechanism proposed by us for the charging and catalytic decomposition of an adsorbed molecule is but one of a number of possible mechanisms for the elementary precatalytic act; it is perhaps more common than others. This model makes it possible to explain many experimental correlations between surface charging and the reactivity of adsorbed molecules. However, in order to obtain reliable quantitative estimates, quantum-chemical calculations of model donor-acceptor complexes are necessary.

A number of quantum-mechanical calculations of the energetic parameters of the simplest defects in the bulk and on the surface of heteropolar and covalent crystals have already been carried out in the cluster approximation (see references in [7.16,18]). For semiconductor catalysts these calculations must include model-type computations of the interaction between captured charge carrier and complex. Since such calculations are rather approximate, a phenomenological theory of catalysis would require at least general correlations between surface charging and the corresponding parameters of the complex, similar to the reactivity indices now widely employed in catalysis and adsorption [7.70]. Only then will we have a more or less reliable basis for considering correlations between catalytic and electronic processes on the surface.

8. The Phonon and Shock Mechanisms of Charge-Carrier Capture in Adsorption and Catalysis

The analysis of experimental data which is carried out in this chapter indicates that the capture of charge carriers by the adsorptive surface states is a necessary but not a sufficient condition for the dissociation of adsorbed molecules. An important role in this process belongs to the excitation of vibrating modes of adsorbed molecules by the energy released in capture and recombination of charge carriers by surface states. We also consider the paths of dissipation of energy released in the process of adsorption and capture of charge carriers.

8.1 Phonon Excitations in Photoadsorption and Photocatalysis

The electronic theory developed in [8.1-6] explains all adsorptive and catalytic phenomena on the surface in terms of variations in the charge status of adsorbed atoms and molecules, which are caused by the capture or recombination of the semiconductor's free charge carriers on adsorptive surface states. Let us discuss photoadsorption and photocatalysis where the concatenation between adsorption and the electron processes on the surface is most vividly manifested. According to theories of photoadsorption and photocatalysis [8.7-12], the change in the adsorptive (catalytic) activity of the surface which occurs when the semiconductor is illuminated by light quanta $h\nu \geqslant E_g$ is associated with the appearance of nonequilibrium charge carriers and consequently with a change in the concentrations of charged and neutral adsorption ($n^{\pm}$ and n^0 in the usage of [8.1]). The "memory effect" in photoadsorption (residual phenomena after removal of illumination) is attributed to the capture of nonequilibrium charge carriers on structural defects of the semiconductor surface.

This approach, based entirely on the charge status, fails to give a consistent explanation of all of the experimental data on photoadsorption and photocatalysis. For instance, in many cases the spectral range of photoadsorption and photocatalysis, as well as that of photomemory, does not coincide with the intrinsic absorption band of the crystal, and there is no correlation between the magnitude and sign of the effect of photoadsorption and the electronic processes (dark conductivity and photoconductivity) [8.13,14].

In [8.13] an interesting correlation was discovered between photoadsorptive activity (PA) and the width of the forbidden band of the crystal, E_g: PA has a tendency to increase with an increase in E_g (Fig.8.1). Similar regularities were earlier observed for radiolysis of adsorbed substances (methanol and cyclohexane) [8.15,16]. The radiation yield G (number of adsorbed molecules undergoing decomposition per 100 eV of energy absorbed in the bulk of the semiconductor or dielectric) increases with an increase in E_g.

The electron theory [8.1] does not consider the dependence of adsorptive and catalytic phenomena on the width of the forbidden band; it analyzes dark chemisorption and catalysis under conditions of complete electronic equilibrium on the surface, which are actually never realized (Sect.5.2). When dealing with problems of photoadsorption and photocatalysis, the theory connects variations in adsorptive (catalytic) activity to the corresponding values for dark adsorption (catalysis), and the resulting mathematical expression does not include the value of E_g [8.9,17]. However, if one considers the steady-state change in the surface concentration of nonequilibrium charge carriers under the constant action of some agent (irradiation, transverse field, steady-state catalytic reaction, etc.) then both dark catalytic activity k_0 and PA will depend on E_g.

In the idealized model of chemisorption in [8.1], k_0 and PA are determined primarily by the surface concentrations of nonequilibrium electrons (n_s) and holes (p_s). For a nondegenerate semiconductor, according to (2.5),

$$n_s = n_i \exp[\beta(\Psi_s - \varphi_n)] = c_1 \exp(-E_g / 2kT) \exp\left[\frac{\Psi_s - \varphi_n}{kT}\right] \quad ,$$

$$\tag{8.1}$$

$$p_s = p_i \exp[\beta(\varphi_p - \Psi_s)] = c_2 \exp(-E_g / 2kT) \exp\left[\frac{\varphi_p - \Psi_s}{kT}\right] \quad ,$$

where c_1 and c_2 are coefficients that depend on temperature and on the effective masses of the charge carriers, which, in turn, are dependent on E_g and on the structure of the crystal's energy bands. For semiconductors with sufficiently wide bands ($E_g > 1$ eV) the band curvature Ψ_s is usually rather small, owing to self-compensation of acceptor-type and donor-type states [8.18] (see below). From (8.1) it is evident that n_s and p_s (and, therefore, k_0 and PA) will, according to the electron theory, decrease with an increase in E_g, and increase with an increase in the injection level which depends on the se-

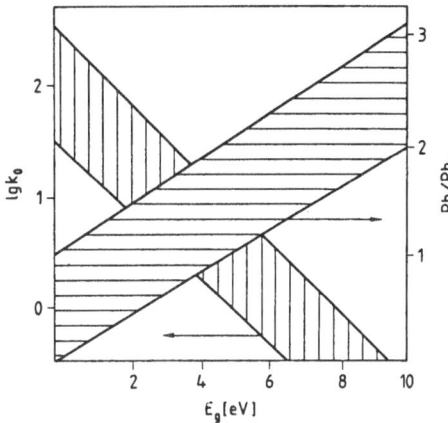

Fig.8.1. Correlative dependence of catalytic activity k_0 in the dark reaction of oxidation of CO [8.19] and photosorptive activity PA (PA_N is the noise level) towards O_2, H_2, and CH_4 [8.13] upon the width of the forbidden band E_g, of oxide specimens, Ph_N —background noise

paration between the Fermi quasi levels of electrons and holes ($\varphi_n - \varphi_p$). The activation energy E_a will increase with an increase in E_g, in agreement with conclusions made in [8.4]. This point was discussed in Chap.5 (5.5,6)[1].

The relationship between k_0 and E_g that follows from the purely electronic mechanism fits the experimental correlations between k_0 and E_g obtained for a large number of dark redox catalytic reactions [8.19] (Fig.8.1). However, the situation is quite different with regard to photoadsorptive activity. The increase in PA with a decrease in E_g, which is predicted by the electron model (8.1), is contrary to the experimental data [8.13] (Fig.8.1). It must therefore be assumed that together with the mechanism of electron capture there is an alternative mechanism of photoadsorption (photocatalysis). In [8.20-22] we proposed a new donor-acceptor electron vibrational mechanism of adsorption and catalysis on semiconductor surfaces which takes into account both the variations in the charge status of adsorption centers and the energetics of charge-carrier capture on adsorptive states. Let us discuss this in greater detail.

Let nonequilibrium free electrons or holes appear in the allowed bands of a semiconductor under the action of external excitation (adsorption, irradiation, transverse electric field, contact injection, etc.). The relaxation of the system into equilibrium or a stationary state is accompanied either by capture or by recombination of nonequilibrium charge carriers on defects. Which of the two mechanisms plays the predominant role is determined by the parameters of the defects.

---

[1] Note that if photoadsorption is treated as a steady-state process [8.12], the criterion for the effect's sign will include E_g, which implicitly enters the expressions for Fermi quasi levels similarly to (8.1).

The energy released in these processes can be dissipated in the following ways:

1) emission of photons,   2) transfer of energy to lattice phonons,
3) Auger effect[2] [8.23].

Let us for the present restrict our discussion to nonradiative transitions and focus our attention on the phonon mechanism of capture of, for example, an electron from the conduction band (reference point of energy) into a sufficiently deep level ε_a. The energy released in this process can be partly transferred to lattice phonons (the energy of Debye phonons is about $2 - 7 \cdot 10^{-2}$ eV), and partly used to create electrically active defects. For example, an interstitial impurity atom, inactive in electron processes, can pass over (owing to the excess energy) into an empty lattice site and become electrically active. Excitation of the electron subsystem of the solid can stimulate the creation of F centers and Schottky and Frenkel defects.

The participation of nonequilibrium charge carriers in defect formation in semiconductors and dielectrics was suggested as early as the 1950s. For instance, the rupture of deformed bonds in fused quartz was attributed to the capture of charge carriers in [8.24], and these concepts were successfully applied to MIS structures in connection with the problem of radiation stability [8.25,26]. This mechanism of defect formation was also assumed to be responsible for the increase in the density of fast surface states during injection of charge carriers from silicon to oxide [8.26-28]. It was demonstrated that the application of strong electric fields to MIS structures [8.29-31] and exposure to hard (though subthreshold) radiation [8.27,31-33] leads to alterations in the properties of MIS structures and to an increase in the density of surface states. These processes often result in degradation of MIS structures, as reflected by their voltage-current characteristics [8.30]. Transitions of intrinsic and impurity defects of a crystal into an electrically active state due to capture of charge carriers are the cause of self-compensation of conductivity of wide-band semiconductors [8.34-36].

As early as the 1930s FRENKEL [8.36b] demonstrated that the probability that a defect will be created by virtue of capture of an electron is $\exp(\varepsilon_t / kT)$ times greater (where ε_t is the depth of the trap) than the probability that it will be created owing to purely thermal fluctuations. The expenditure of energy for the creation of a new defect (ε_0) is compensated by the energy released in the act of capture. The greater the depth ε_t of the defect in the forbidden

---

[2] By Auger processes we mean here the use of the energy of recombination for the excitation of a third charge carrier (either an electron or a hole) from an adjacent center.

band, the more pronounced the effect of defect formation will be. The effect will be greatest when ε_t is close to E_g. Note that the main channel of nonradiative dissipation of energy is the multiphonon mechanism. It becomes less efficient with an increase in ε_t, which also is beneficial for defect formation. Because localized phonon modes associated with the defect are present in the crystal and owing to considerable localization of the electron's wave function on the defect, such a situation must give rise to strong electron-phonon interactions. Therefore the greater part of the energy released in the act of capture must be expected to go over into the vibrational energy of the defect [8.20], which may alter the normal coordinate of the defect on the configuration diagram. This process will be stimualted by polarization of the lattice, caused by the capture of the charge carrier. The creation of defects leads to an increase in the entropy of the crystal and to a decrease in its free energy, so that in terms of energy this process is very advantageous [8.35].

The statistical and quantum-mechanical theories of defect formation involving captured charge carriers were developed in [8.35,37-39]. A model for the generation of vacancies on dislocations utilizing the energy of recombination of electron-hole pairs was considered in [8.40]. VINEZKY [8.35] estimated the value of the temperature threshold (T_{thresh}) of defect formation in crystals having high ionicity of bonds. Owing to interaction between electron and lattice phonons in such systems, one must take into account the energy of creation of polaron states ε_p. According to [8.35], the value of T_{thresh} is

$$kT_{thresh} > \frac{\varepsilon_0 - \varepsilon_t - \varepsilon_p}{\ln(N_s^2 / RV)} \quad , \tag{8.2}$$

where R is the density of states in the band (usually $\sim 10^{19}$ cm^{-3}), N_s is the number of lattice sites ($\sim 10^{23}$ cm^{-3}), and V is the volume of the crystal. To illustrate, the authors estimated the value of T_{thresh} at $\varepsilon_0 = 2.5$ eV, $\varepsilon_p = 0.2$ eV, $\varepsilon_t = 2.0$ eV, and $V = 1$ cm^3. It turned out that defect formation starts at rather low temperatures ($T_{thresh} \sim 150$ K). A considerable drop in T_{thresh} is to be expected for structures with a disordered lattice, amorphous bodies, crystalline dislocations [8.41], and crystalline surfaces [8.20].

ZUEV et al. [8.18] were the first to use these concepts to explain charge compensation on semiconductor surfaces and consequent stabilization of the Fermi level on the surface. Owing to distortion of the coordination sphere of surface atoms, rehybridization of bonds, and the presence of various chemical surface groups, the value of ε_0 will be lowered, and therefore the process of defect formation can start at a lower temperature, $T_{thresh}^s < T_{thresh}^v$ (s stands for surface, v for volume). According to the estimates made in [8.18], defect formation on a silicon surface ($E_g = 1.1$ eV) can start at as low as $T_{thresh} \sim 300$ K. For semiconductors with a wider band, defect formation at such temperatures can be quite pronounced, as confirmed by investigations of surface charge compensation. BARDEEN [8.42], in his studies of rectifying junctions, was one of the first to direct his attention to compensation of charge on real surfaces. In VOLKENSTEIN's model [8.1] such surfaces are termed "quasi-insulated": the difference between positive (p_s^*) and negative (n_s^*) surface charges is by modulo much smaller than their sum: $|p_s^* - n_s^*| \ll p_s^* + n_s^*$. Self-compensation of surface charge due to defect formation can be one reason for the creation of a "quasi-insulated" surface.

The concept of self-compensation of electrically active defects that has been developed in the literature can be applied to adsorptive phenomena, in particular photoadsorption and photocatalysis, as we demonstrated in [8.20]. The energy of adsorptive bonds ε_{ads} is usually lower than the energy of lattice bonds: $\varepsilon_{ads} < \varepsilon_0$. Therefore adsorption effects which utilize the energy of electron transitions can be quite considerable at room temperature.

Let us first consider photodesorption. After capture of a photoexcited charge carrier on the level of the defect to which an adsorbed molecule is bound, part of the energy of the electron transition can be expended on desorption of the molecule. It is not necessary that the photodesorption be accompanied by transition of the charged form of chemisorption into the neutral form, as is assumed in [8.9]. The adsorbed molecule can remain in the neutral form both before and after illumination, while the photoexcited charge carrier is retained near the adsorbed molecule; the mechanism of retention is similar to that discussed in Sect.7.2.

GERSTEIN et al. [8.43] proposed two more possible mechanisms of photodesorption when a quantum of light interacts with an adsorbed molecule:

1) excitation of a plasmon and desorption of the molecule in the ground state, and

2) excitation of the molecule itself and desorption in the excited state.

The second mechanism was apparently observed in experiments on oxidation of silicon [8.44]. As opposed to the mechanism proposed by GERSTEIN et al. [8.43], our mechanism does not require direct interaction between the quantum of radiation and the adsorbed molecule. Unfortunately, the qualitative nature of the existing theory of chemisorption does not allow estimates to be made of the quantum yield for various mechanisms of photodesorption.

The same considerations can be applied to photodissociation of adsorbed molecules (precursory act of photocatalysis). To illustrate, let us consider dissociation of a donor-type molecule, which is coordination-bound with surface centers. According to Sect.6.6, the bond is created owing to the pulling over of a pair of undivided electrons onto the orbitals of the atom acting as adsorption center. The localization of a photogenerated hole near such a center increases the drawing of the electron cloud and stimulates deformation of the molecule. Part of the energy released in the capture of the charge carrier can be expended on resonant excitation of individual bonds of the molecule ($h\nu_c$), which stimulates its dissociation.

A vivid illustration of the above discussed phenomenon is furnished in [8.45,46]. The authors of those works observed dissociation of H_2O molecules adsorbed on real Si and Ge surfaces, stimulated by pulsed illumination of the semiconductor with low-energy light quanta ($h\nu \sim 1.8 - 3$ eV). As seen in Fig.8.2, the quantum yield of the dissociation product $(H_2)\eta_{H_2}$ from the silicon surface increases with an increase in the energy of light quanta $h\nu$, while the quantum yield of electron-hole pairs $\eta_{n,p}$, generated in the semiconductor by light, remains the same [n_s, $p_s = $ const in (8.1)]. The surface charge Q_s also remains constant; this situation is quite contrary to the simplified electron models of photoadsorption and catalysis in [8.1,17]. We demonstrated in [8.45,46] that a change in the charge status of centers is a necessary, though not sufficient, condition for adsorption and catalysis. The regularities in these phenomena of photodissociation will be discussed in greater detail in Chap.9.

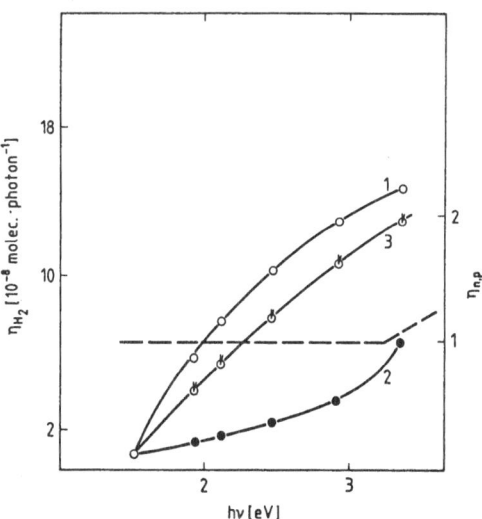

Fig.8.2. Spectral dependence of mean spectral yield of hydrogen η_{H_2} from silicon single-crystal surface illuminated in pulsed fashion. Temperature of vacuum heat treatment of specimens= $T = 300$ K (1), 620 K (2); rehydrated specimens (3). Dashed line: quantum yield $\eta_{n,p}$ of electron-hole pairs

189

Finally, let us consider photoadsorption. The release of energy during exothermal electron transitions may give rise to the following processes:

(1) generation of new intrinsic surface defects,

(2) transition of some adsorption-inactive surface states into adsorption-active ones, and

(3) quasi-chemical reactions between surface defects, resulting in the creation of more active adsorption centers.

These processes are similar to the bulk processes considered in [8.34]. In all these cases illumination will cause an increase in the concentration of adsorption centers, thus promoting photoadsorption. If illumination gives rise to sufficiently deep traps, these may later act as photomemory centers. In contrast to the photomemory centers in VOLKENSTEIN's model [8.9] (where they are assumed to be represented by innate surface defects), in our case these centers are generated by light or other nonequilibrium agents in the semiconductor's subsurface region.

So far our energy source has been exothermal capture of charge carriers on adsorption surface centers. If the capture cross sections for electrons and holes have similar values, these states can act as recombination centers. At high injection levels, Shockley-Reed nonradiative multiphonon recombination can take place on these centers (Sect.2.4). Precisely this case was realized in experiments on photodissociation of water in [8.45,46]. The model of photodissociation proposed in those works is more akin to the kinetic models of Garrett [8.3] and LEE [8.4] than to the quasi-equilibrium VOLKENSTEIN model [8.17].

The creation of energy-nonequilibrium states on the surface can be due not only to the capture or recombination of charged particles but also to localization of excitons provided the excitant light belongs to the exciton band. Annihilation of an exciton and the energy released in the act of capture of an electron and a hole on a trap can stimulate the creation of a new defect. Also possible is the inverse process when annihilation of a defect (e.g., a Frenkel defect: the passing of an interstitial atom into a lattice vacancy) produces an exciton.

The deeper the levels of the traps responsible for capture and recombination of charge carriers the greater is the influence of energy-nonequilibrium states which owe their existence to exothermal electron transitions. The contribution of exothermal electron transitions to photoadsorption and catalysis will increase with an increase in the width of the forbidden band E_g of the adsorbent which fits the correlations in Fig.8.1. If the incipient traps are deep enough, and the probability of thermal ejection of charge carriers is low, then photoadsorption on such centers will be irreversible; this is sometimes observed in experiment. Since photoadsorptive (catalytic) activity depends on both the concentration of nonequilibrium charge carriers (8.1) and the energy factor these two processes may in some cases compensate each other, which eventually will mask the dependence of PA on E_g (Fig.8.1).

The creation of energy-nonequilibrium states on real defect-ridden surfaces can be a result not only of the passing of electrons onto adsorptive surface states but also of transitions to biographic states in the subsurface layer. The energy released in such transitions can be transported by phonons to adsorbed molecules over distances of several lattice constants. This process is easily imagined if one recalls the clusterwise distribution of defects. As indicated by EPR data [8.47], the concentration of defects within a cluster can be much higher than their average concentration on the surface. The creation of clusters has been most thoroughly investigated for transition-metal ions deposited on a dielectric (Al_2O_3, MgO). The transfer of energy between defects in a cluster with the aid of phonons is quite probable.

In electron transitions on biographic states, the correlations between photoadsorptive effects and the electrophysical parameters of the surface can be highly disguised. This may be one reason for discrepancies between experimental results and the predictions of the electron theory of chemisorption [8.9]. In particular, the lack of correlations between photoadsorption and photoconductivity mentioned in [8.13] can be attributed to self-compensation of photoconductivity which is due to the formation of defects.

The processes of defect formation stimulated by exothermal electron transitions on the surface can also take place during the dark nonequilibrium processes of adsorption and catalysis. In its initial stage, adsorption causes variations in the occupancies of both adsorptive and biographic surface states (Sect.6.5). The energy released in the capture of nonequilibrium charge carriers can be expended on excitation of adsorbed molecules and on the creation of defects. We discussed the interaction between charge carriers in metals and oscillations of adsorbed molecules in [8.48]. The same approach can be employed in considering the interaction between thermally excited charge carriers in semiconductors and adsorbed molecules. Such an attempt was made by SCHEVE and SCHULTZ [8.49] who discovered certain correlations between the activation energy of catalytic reactions and that of conductivity. It should be noted, however, that the latter quantity is rather ambiguous in the case of polycrystalline specimens, since it strongly depends on the shape and height of the potential barriers between powder particles.

The efficiency of phonon excitations in surface phenomena can be quite high even at room temperature. Statistical estimates for the bulk of the crystal indicate [8.35] that in the presence of sufficiently deep donor-type states ($|\varepsilon_0 - \varepsilon_c| < 0.5$ eV) at T = 300 K the absorption of about 10^{15} quanta of light can give rise to $10^{15} - 10^{17}$ defects per cubic centimeter, or $10^{10} - 10^{11}$ defects $\cdot$ cm^{-2} when reduced to the subsurface layer. The actual density of surface defects will be even higher since $\varepsilon_0^s < \varepsilon_0^v$. Recall that the photoadsorptive capacity is $10^{10} - 10^{11}$ molecules $\cdot$ cm^{-2} [8.13] while the concentration of biographic states on real surfaces near the band edges is no less than 10^{13} or 10^{14} cm^{-2} (Sect.3.3). If the energy spectrum of biographic states does not change in the course of illumination then the semiconductor surface acts as a kind of photosensitizer during photoadsorption and photocatalysis.

As we emphasized in Sects.4.2 the energy spectrum of electron states on real semiconductor surfaces is most adequately described within the framework of the electron theory of disordered systems. The peculiarities of a disordered surface should affect the processes of photoadsorption and photocatalysis. Absorption of light at the edge of the fundamental band increases exponentially according to Uhrbach's rule. A rigorous analysis of the applicability of this rule was carried out in [8.50]. Photoexcitation of the semiconductor in this case can take place via electron transitions between the valence band and the "tail" of acceptor-type levels that have split off from the bottom of the conduction band (Fig.8.3). Relaxation of the electron subsystem via backward exothermal transitions to "tails" of donor-type states will result in the above mentioned stimulation of photoadsorption and photocatalysis. It is true that the dissipation of energy of nonequilibrium carriers will be higher than in the case of a classical ordered system [8.51], owing to energy transfer to phonons. But on the other hand, the energy of defect formation ε_0^s in a disordered system can have a lower value.

Vigorous electron transitions between "tails" of allowed states in the forbidden band of the semiconductor may be one cause of the shift of photoadsorptive activity towards greater wavelengths, $h\nu < E_g$. However, it is not the only one. It should be kept in mind that the electron transitions in the subsurface region of the semiconductor often take place in strong electric fields ($\sim 10^4 - 10^5$ V $\cdot$ cm^{-1}) created by the charge in biographic and adsorptive surface states. The appearance of an electric field (which often is approximated by a uniform field) causes bending of the crystal's energy bands (Sect.2.2). The presence of an electric field raises the probability that an electron (or hole) will fall into the forbidden band. This probability is described by an exponentially decreasing function stretching from the classical limits of bands

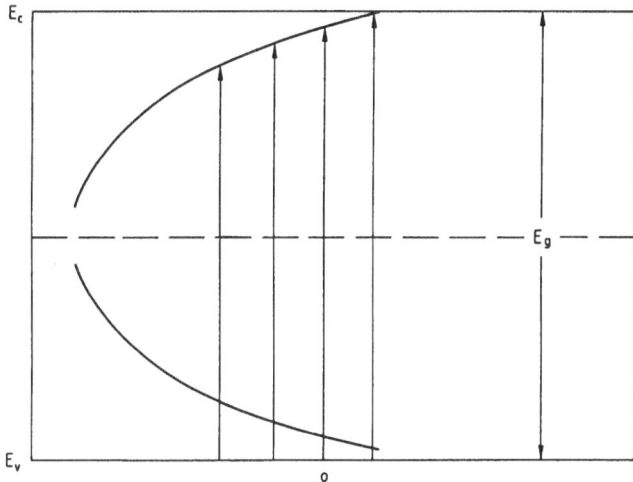

Fig.8.3. Optical excitation of disordered semiconductor surface. E_c, conduction band; E_v, valence band; ρ, density of states

into the forbidden band. This leads to the appearance of so-called extra-band levels of electrons (and holes). Optical transitions of electrons onto these levels (tunneling with involvement of a photon) take place at $h\nu < E_g$. Here we are dealing with the Franz-Keldysh effect which manifests itself in the exponential drop in the extinction coefficient beyond the limit of intrinsic absorption of the crystal [8.23]. The shift in the limit of intrinsic absorption in metallic iodides during adsorption of acceptor-type molecules observed in [8.52,53] was attributed to the Franz-Keldysh effect (Sect.6.3).

The shift in the intrinsic absorption band due to the Franz-Keldysh effect is rather small. For instance, at a field strength of about 10^5 V $\cdot$ cm^{-1} the change in the effective optical width of the forbidden band is about 10^{-2} eV. Much greater effects can be expected in the range of extrinsic extinction when optical electron transitions take place from deep levels in the forbidden band [8.54]. This case is most interesting from the standpoint of the above-discussed problems of photoadsorption and photocatalysis. Interactions between electrons and lattice phonons play a considerable role in optical transitions between deep impurity levels and allowed bands. Detailed calculations of impurity extinction coefficient, with electron-phonon interactions taken into account, were carried out in [8.55]. The estimates made there indicate that in the presence of electric fields having a strength of about 10^5 V $\cdot$ cm^{-1} (which is usually the case on the surface), the energy of optical transition between a deep level (ε_t) and the allowed band can be considerably lowered: $\varepsilon_t + B - h\nu > 0.2$ eV, where B is the parameter of electron-phonon interaction; e.g., for deep centers of germanium $B \leqslant 0.1$ eV. We see that the above-mentioned effects of photo-desorption and photoadsorption which are associated with dissipation of the energy of electron transitions will exhibit a shift towards greater wavelengths with respect to the edge of the extrinsic absorption band.

If electron transitions stimulated by light involve adsorptive surface states, the shift in the intrinsic absorption band $(E_g - h\nu)$ will be determined by the interaction of a trapped electron not only with lattice phonons but also with vibrational modes of the adsorbed molecule. In other words, the shift will depend on the nature of the adsorbed molecules. This phenomenon has not yet been thoroughly investigated.

From the above discussion it follows that phonon excitation caused by electron transitions on a disordered semiconductor surface can strongly influence both the dark and light stages of adsorption and catalysis. These processes can be termed "the phonon factor" for short. As we have already mentioned, this factor may be responsible for certain discrepancies between the purely electronic treatment and the experimental results, e.g., an increase in photo-adsorptive activity with an increase in E_g, and irreversibility of photoadsorption. The role of the phonon factor in correlations between dark and photo-catalytic activity and the width of the forbidden band is directly opposite to that of the pure electron factor. There is a mutual relationship between phonon and electron processes on the surface. Phonon excitations, caused both by adsorption itself and by electron transitions that accompany adsorption, constitute the primary concern of the theory of surface phenomena in semiconductors and insulators.

8.2 The Participation of Vibrational Modes of Adsorbed Molecules in the Capture and Recombination of Free Charge Carriers

Let us consider in greater detail the phonon mechanism of capture of free charge carriers on surface states. The localization of a charge carrier at a trap requires the removal of energy released in the act of capture (ε_t). If the traps are deep enough, the value of the released energy is greater than half the width of the forbidden band $(\varepsilon_t \geqslant E_g / 2)$. For instance, ε_t for germanium and silicon is $0.3 - 0.5$ eV, which is an order of magnitude greater than the energy of a phonon [8.56-58]. The probability of a multiphonon energy dissipation process[3] involving a large number of phonons is usually low, but not infinitesimally small [8.56]. The greater the number of phonons necessary to dissipate the released energy, the smaller the probability of such transitions is. In other words, this probability decreases with an increase in E_g.

The theory of nonradiative transitions on deep traps usually considers three situations [8.56-58]:

(1) the charges of trap and charge carrier have opposite signs,

(2) the charges of trap and charge carrier have the same sign,

(3) the trap is neutral.

In the first case, in the presence of Coulomb forces of attraction, the most probable mechanism of energy dissipation is the cascade mechanism proposed by LAX [8.60]. According to Lax, the existence of attractive forces gives rise to a cascade of excited levels of an impurity atom, which acts as the center of capture. An electron from the allowed band passes consecutively from one excited level to another, each time emitting only one phonon. The greater the number of excited levels created in the act of capture, the more effective this process is.

In the second case, Coulomb forces of repulsion act upon the charge carrier at long distances while at short distances it comes under the influence of short-range attractive forces. To be captured, the electron must overcome a potential barrier. Because it is possible to tunnel through this barrier, the capture cross sections in some cases turn out to be close to the corresponding values for neutral centers [8.57]. It is hardly possible for a large number of excited levels to be created in the case of negative or neutral capture centers. In the latter case LAX [8.60] and SMITH and LANDSBERG [8.61] assumed the possibility of polarization interaction between the center and the incident carrier. However, calculations carried out by BONCH-BRUEVICH and GLASKO [8.62] indicate that such interactions give rise to only a few excited states which renders the cascade process of energy dissipation impossible. More probable in such a situation are multiphonon processes or radiative recombination. A review of works giving a theoretical analysis of nonradiative capture on neutral and negative centers can be found in [8.56].

The Lax cascade mechanism was employed by RZHANOV [8.63] and NOVOTOTSKY [8.64] to explain the quite high values (10^{-11} - 10^{-12} cm^{-2}) of the capture cross sections of discrete recombination centers, which those authors assumed to be responsible for the bell-shaped component S_b of the rate of surface recombination (Sect.3.1). As a model of the center they took a defect with a di-

---

[3] Multiphonon transitions in disordered semiconductors were analyzed in [8.51]. Estimates for the probability of multiphonon nonradiative transitions in the SCR were made in [8.59].

pole molecule attached to it. A theoretical analysis of nonradiative capture on such a neutral dipole center was carried out in [8.65,66]. According to the estimates made there, under certain conditions the dipole molecule attached to the defect can generate a sufficiently great cascade of levels to effectuate capture of charge carriers. However, the efficiency of capture on a dipole center (within the framework of the cascade mechanism) can hardly be expected to match that on an oppositely charged Coulomb center [8.58]. Unfortunately, no quantitative estimates of the relative contributions of the cascade and the multiphonon mechanisms were made in [8.65,66].

There are a large number of theoretical estimates of the capture cross sections for dipole centers in the bulk [8.56]. The efficiency of capture on such centers depends considerably on the separation ℓ of charges in the dipole. At low values of ℓ, bound states are absent. When ℓ is sufficiently great, the two poles act as separate charges. STONEHAM [8.56] rightly stated that our understanding of capture on neutral centers is far from complete. Theoretical estimates of $c_{n,p}$ for such centers are strongly dependent on the choice of the model for the defect, and even more so in the case of surface centers. The location of the dipole on the interface is assessed rather speculatively. In [8.67] the dipole is assumed to be represented by a coordination complex. However, the value of ℓ in such complexes is quite high (Sect.6.6). According to estimates made in [8.68], this must result in very low values for the capture cross sections of a recombination center on the semiconductor-vacuum interface. A more realistic model of a neutral center on a semiconductor-insulator interface was proposed by VLASENKO and SURIS [8.69]. They proposed an elegant method of determining the charge status of surface traps based on the entirely dissimilar dependence on electric field exhibited by the capture cross sections of charged and neutral traps.

The experimental results of Matveev and Grinev mentioned in Sects.3.1 and 6.2 do not support the dipole model of surface recombination centers. The donor-acceptor dipole complexes created during adsorption act as slow states, having quite small capture cross sections. This point will be discussed later in more detail. The regularities in surface recombination are more naturally explained in terms of electron transitions via the quasi-continuous spectrum of surface states. This being so, all theoretical estimates made for isolated traps seem rather unrealistic.

Another model of discrete recombination centers on semiconductor-electrolyte interfaces was proposed by GROMOV et al. [8.70]. They assumed the contact electric fields on a heterogeneous semiconductor-insulator interface to be responsible for the transformation of fast states into recombination centers.

This model is actually based on the well-known FRENKEL effect [8.36]: the influence of strong electric fields on the parameters of deep traps (in particular, on capture cross sections).

These electric fields on the semiconductor-insulator interface can be generated by charged slow adsorptive states. However, there was no noticeable influence of these fields on the rate of surface recombination S on the semiconductor-gas interface in [8.71,72]. As follows from Fig.3.7, dissociative adsorption of water vapor and molecular adsorption of pyridine resulted in almost equal charging of Ge surfaces, while the effect on S was quite different. The magnitude and distribution of charge on the surface being the same, an increase in S was observed only during adsorption of H_2O, accompanied by reconstruction of the interface.

The classical phonon mechanism is not the only conceivable mechanism for dissipation of the energy of electron transitions on semiconductor surfaces. KISELEV et al. [8.21,22,73,74] discovered another mechanism: energy dissipation on vibrational modes of adsorbed molecules. As will be demonstrated, this mechanism explains certain regularities in the slow relaxation of charge in adsorptive states, and, most importantly, reveals why these states have abnormally low capture cross sections.

In Sect.3.3 we showed that relaxation of charge in slow adsorptive traps in the field effect is well approximated by Koc's formula (3.7). The characteristic time of slow relaxation, τ_s, according to (3.8), exhibits an exponential dependence on T in a rather wide temperature range:

$$\tau_s^{-1} = \tau_0^{-1} \exp(-\Delta E_\tau / kT) \quad , \tag{8.3}$$

where τ_0^{-1} is the frequency factor, and ΔE_τ is the activation energy of the effective time of slow relaxation. The value of ΔE_τ is close to the value of the mean thermal activation energy of charge carriers exchanged between slow states and the bulk.

Detailed studies of temperature-dependent slow relaxation reveal that adsorption of donor-type (H_2O, D_2O, CH_3OH, NH_3) and acceptor-type (p-Bq) molecules has a strong effect on the characteristics of relaxation [8.73,74]. The experimental results indicate that both parameters in (8.3) (τ_0^{-1} and ΔE_τ) depend on the occupation of the surface by adsorbed molecules and on the structure of those molecules (Table 8.1).

From Table 8.1 it follows that the removal of weakly bound molecules from the surface by prolonged vacuumization at 300 K results in a lowering of activation energy ΔE_τ and an abrupt decrease in the frequency factor by several orders of magnitude. As seen from Fig.8.4 $\lg \tau_0^{-1}$ is a linear function of ΔE_τ.

Table 8.1. Dependence of parameters of slow relaxation τ, τ_0^{-1}, and ΔE_τ on the structure of adsorbed molecules

Adsorbed molecules	$h\nu$ [eV]	$p/p_s(\Theta)$	In adsorbate vapors			After prolonged vacuumization at 300 K		
			τ [s]	τ_0^{-1} [s^{-1}]	ΔE_τ [eV]	τ [s]	τ_0^{-1} [s^{-1}]	ΔE_τ [eV]
H_2O	0.41 – 0.45	0.15(0.9)	40	$1.6 \cdot 10^{10}$	0.70 ± 0.05	170	$2.8 \cdot 10^4$	0.40 ± 0.10
H_2O	0.41 – 0.45	0.37(1.2)	35	$1.8 \cdot 10^{10}$	0.70 ± 0.05	170	$2.8 \cdot 10^4$	0.40 ± 0.10
H_2O	0.41 – 0.45	0.55(1.6)	6.4	10^{15}	1.00 ± 0.07	170	$2.8 \cdot 10^4$	0.40 ± 0.10
D_2O	0.31 – 0.34	0.15(0.9)	120	$7.4 \cdot 10^5$	0.47 ± 0.05			
D_2O	0.31 – 0.34	0.55(1.6)	33	$7.0 \cdot 10^9$	0.65 ± 0.05			
CH_3OH	0.38 – 0.46	0.37(0.6)	43	$4.0 \cdot 10^5$	0.43 ± 0.05	900		
CH_3OH	0.38 – 0.46	0.55(0.7)	26	$9.0 \cdot 10^6$	0.49 ± 0.05	900		
C_6H_{14}	0.37 – 0.38	1	55	10^8	0.58 ± 0.05	$1.2 \cdot 10^3$	$1.1 \cdot 10^2$	0.30 ± 0.05
CCl_4	0.087 – 0.093	1	41	$7.1 \cdot 10^4$	0.38 ± 0.05	170	$4.2 \cdot 10^3$	0.34 ± 0.05
NH_3	0.37 – 0.42	10^{-2}(0.1)	36	$3.2 \cdot 10^8$	0.60	120	$1.3 \cdot 10^2$	0.25
p–Bq	0.20 – 0.21	1	6	$8.4 \cdot 10^5$	0.40	66	5	0.15

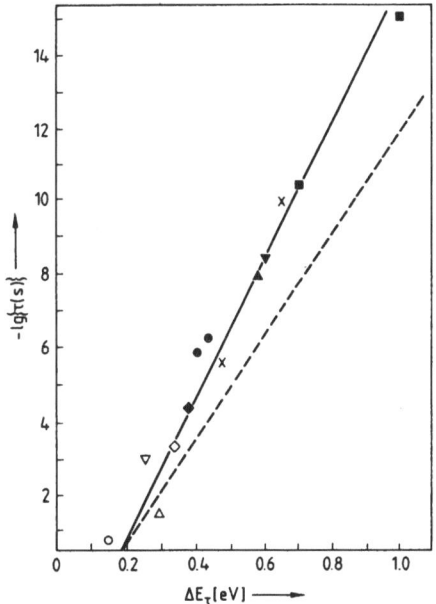

Fig.8.4. Experimental dependence of frequency factor τ_0^{-1} on activation energy ΔE_τ of slow relaxation (Constable's law) in adsorbate vapor (solid symbols) and after evacuation of adsorbate (open symbols): H_2O (■), D_2O (X), CH_3OH (⊙), C_6H_{14} (▲,△), CCl_4 (◆,◇), NH_3 (▼,▽), p-Bq (●,○). *Dashed line* corresponds to $\tau_0^{-1} = \exp(\Delta E_\tau/kT)$

After vacuumization the surface retains molecules firmly bound by donor-acceptor bonds. These molecules are responsible for slow states and surface charging (Sect.6.4). Subsequent adsorption and desorption result in reversible variations in τ_0^{-1} and ΔE_τ. The results of these experiments are basically the following:

(1) Van der Waals adsorption of nonpolar hexane and CCl_4 molecules (which does not give rise to slow states and surface charging) produced reversible variations in τ_0^{-1} and ΔE_τ (Table 8.1).

(2) Measurements of the temperature dependence of slow relaxation under isosteric conditions ($\theta = const$) indicate that at the same occupancies θ, the values of τ_0^{-1} and ΔE_τ differ for different molecules. While the electron structure of H_2O and D_2O is the same, the values of τ_0^{-1} and ΔE_τ are higher for H_2O than for D_2O (Table 8.1).

The reversible variations in τ_0^{-1} and ΔE_τ observed during adsorption in these experiments (when the concentration and the energy spectrum of slow adsorptive states remained constant), the isotopic effect, and the acceleration of relaxation by nonpolar molecules cannot be explained within the framework of any existing model of the electron theory of adsorption and catalysis. These features also do not agree with the ion model of slow relaxation which takes as the limiting stage either adsorption or dissociation of adsorbed molecules (Sect.3.3). In fact, the acceleration of relaxation by nondissociative C_6H_{14} and CCl_4 molecules directly disproves the latter mechanism. In [8.73,74] it was demonstrated that the only possible cause for acceleration of relaxation during adsorption is variations in the kinetic parameters of adsorptive states. These variations are associated with additional dissipation of energy released in the act of capture, upon excitation of vibrational modes of adsorbed molecules.

First let us consider a situation in which slow adsorptive states do not interact with adsorbed molecules (Table 8.1: specimens vacuumized at 300 - 350 K). A distincitve feature of adsorptive states is the strong connection between a charge carrier localized on such a state, and vibrational modes of the adsorptive complex. As demonstrated in Sect.6.6, the capture of a charge carrier on an $A^{-\delta}_+ \cdot M^{+\delta}_-$ complex results in an increase in δ, an increase in the polarization of the complex, and a change in the force constants of the molecular bonds. All this strengthens the connection between the charge carrier and the complex. In terms of configuration diagrams (Fig.8.5) the transition of the charge carrier from a free state (Parabola I) into a localized state (Parabola II), in the case of adsorptive states, is accompanied by considerable reconstruction of the neighborhood of the trap (displacement of configuration coordinate $Q_I - Q_{II}$). The greater the displacement of normal coordinates, the lower are the values of the integrals of overlapping of vibrational wave functions (dashed line in Fig.8.5), which define the probability of transition from

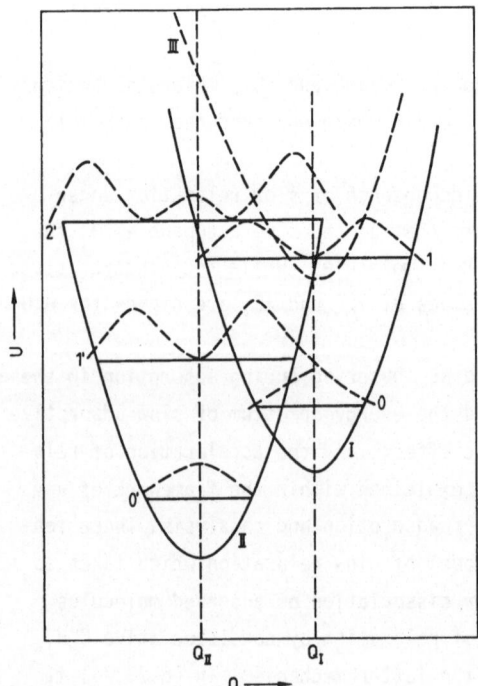

Fig.8.5. Total energy U of system of electron and lattice versus configuration coordinate Q

I to II. In an isolated adsorptive complex, the energy released in the electron transition excites the most vigorous vibrational modes of the complex, the so-called perceptive modes [8.56].

Let us consider, for the sake of definiteness, a real germanium surface where $=\!\!>\!\text{Ge}:0\!<^H_H$ complexes act as slow states (Fig.6.14). In this case the perceptive modes are represented by vibrations of $0-H$ bonds ($h\nu_{OH} \sim 0.45$ eV). There are several reasons for the low probability of capture on such a complex (small values of $c_{n,p}$). The high polarizability of $0-H$ bonds in the complex leads to large displacements ΔQ, and to low values of the overlapping integral of vibrational wave functions, and consequently to a low value of preexponential factor τ_0^{-1} in (8.3). At moderate temperatures owing to a rather large separation between sublevels of $0-H$ bond vibrations, it is mainly the zero level I_0 that is occupied. Therefore the nonactivated transition $I_0 \rightarrow II_0$ is the most probable. In this case the activation energy ΔE_τ in (8.3) is determined by the energy of excitation E_{cs} of the electron into the conduction band on the surface of germanium, i.e., $\Delta E_\tau = E_{cs} - F$, where F is the Fermi level. The experimental values of ΔE_τ (Table 8.1) do not diverge strongly from this difference.

Another reason for the low probability of capture is that a rather large amount of energy (~ 0.45 eV), accumulated by the perceptive mode, must be dissipated. This energy can be exchanged with no less than 10 lattice phonons. The local oscillations of the $0-H$ bond in the complex are weakly linked to the phonons in the semiconductor. The transfer of energy is possible only by virtue of the anharmonicity of the bonds [8.75]. Furthermore, as demonstrated in [8.48], with a disordered real surface the energy spectrum of surface phonons is separated from that of phonons in the bulk by an energy gap. As a result of all this, the lifetime of local oscillations in the complex can exceed the time of vibrational relaxation in condensed media by many orders of magnitude [8.76]. Indeed, the frequency factor in this case (Table 8.1) is many orders of magnitude lower than the usual values of τ_{0b}^{-1} for electron transitions in the bulk of germanium ($\tau_0^{-1} \approx 10^{12} - 10^{13}$ s^{-1}).

Owing to the low efficiency of the path of energy transfer from complex to lattice, another path of energy dissipation becomes possible. Under certain conditions the energy liberated in the act of capture can be great enough to destroy the adsorptive complex. In that case the adsorptive state itself vanishes. As we mentioned earlier, dissociation of H_2O molecules during recombination of charge carriers on adsorptive states was observed in [8.45,46]. The efficiency of this process increased as the energy of recombining electron-hole pairs became greater (Fig.8.2).

Our electron-vibrational model of charging is based on the assumption that there is strong electron-phonon interaction between the captured charge carrier and the local modes of the adsorptive complex, considered as a surface defect. We pointed out in [Ref.8.48,Chap.3] that the spectrum of surface phonons on a disordered surface is separated from phonons in the bulk by an energy gap. It is this gap that is responsible for slow relaxation of energy stored in the complex and, therefore, for the smallness of capture cross sections $c_{n,p}$. The assumption of strong electron-phonon interactions is confirmed by analysis of the shape of the band of surface luminescence of GaAs [8.77]. The electron-phonon interactions with local modes can give rise to surface polarons. A detailed analysis of electron behavior in surface phonon fields was carried out in [8.78].

An alternative explanation is furnished by the electron theory of adsorption. The authors of [8.79] attributed the small capture cross sections of adsorptive states to the existence of ion radical adsorption (Sect.7.1). According to their model, the electron, localized on the molecule's orbitals, is weakly linked with the phonon field of the crystal. In other words, they considered the case of weak electron-phonon interactions and neglected interaction between the electron and molecular vibrational modes. However, slow relaxation is also observed in cases when the creation of ion radicals is not possible (adsorption of H_2O, NH_3, CH_3OH). In contrast, in the presence of ion radicals (adsorption of p-Bq, Sect.7.1), τ_0^{-1} and ΔE_τ assume lower values. As we emphasized in Sect.5.3, the model of a slowly relaxing ion radical does not fit in well with the electron stage of catalysis where ion radicals (as assumed in [8.1]) play the crucial role.

Let us now turn to a situation in which weakly attached molecules are present on the surface (relaxation in an atmosphere of adsorbate vapor). In this case the vibrational energy, stored in the adsorptive state in the act of capture of a charge carrier, can be transferred to adjacent adsorbed molecules. The dissipation of energy here will require a considerably smaller number of lattice phonons. Since the integral of overlapping of vibrational wave functions increases with an increase in the number of levels (Fig.8.5), transitions between higher-energy levels now become more probable, e.g., $I_1 - II_1$, or $I_1 - II_2$, in Fig.8.5. The activation energy ΔE_τ is then augmented by a value close to the energy of the quantum transferred to the ensemble of adsorbed molecules: $\Delta E_\tau = E_{cs} - F + h\nu_v$. The value of $h\nu_v$ corresponds to the energy of the uppermost mode of molecules adsorbed around the complex; this is confirmed by experiment. From Table 8.1 it can be derived that at $p/p_s \geqslant 0.4$ the experimental value of ΔE_τ lies close to $E_{cs} - F + h\nu_v$ [for germanium $\Delta E_\tau \sim (0.2 - 0.3) + h\nu_v$].

Owing to the increased efficiency of the process of dissipation of energy stored in the adsorptive complex, in the presence of weakly bound molecules the preexponential factor τ_0^{-1} is increased more than $\exp(-\Delta E_\tau / kT)$ times.

Measurements of slow relaxation at high relative pressures of H_2O and D_2O vapors indicate that the value of ΔE_τ increases considerably while τ_0^{-1} grows faster than $\exp(- E\Delta / kT)$ (Table 8.1, Fig.8.4). Apparently, in this range of occupancies (when polymolecular clusters are formed on the surface; see [Ref. 8.48,Chap.3]) the energy exchange proceeds very actively. The time of vibrational relaxation in such clusters is nearly the same as in condensed media ($\sim 10^{12}$ s) [Ref.8.48,Chap.5]. This time is lower than the relaxation time of the energy of a free electron in the conduction band of germanium. In this case it is possible for the system to pass from State III (Fig.8.5) which corresponds to an electron located above the bottom of the conduction band, into a vibrational-excited state ($II_{2'}$).

The assumption about the excitation of vibrational modes of adsorbed molecules in the capture of charge carriers by adsorptive states is directly proved by the isotopic effect discovered in [8.74]. The parameters of relaxation (Table 8.1) vary in accordance with changes in the vibrational modes of adsorbed H_2O and D_2O molecules: $h\nu_v = 0.45$ eV for H_2O; $h\nu_v = 0.3$ eV for D_2O.

The proposed electron-vibrational mechanism of the elementary act of capture on adsorptive states seems to be sufficiently general. For instance, it would explain the results reported in [8.80], where the activation energy of slow relaxation in the presence of water adsorbed on the surface of ZnO was close to 0.6 eV. After adsorption of erythrosin on ZnO (the value of $h\nu_v$ for erythrosin is close to that for p-Bq), the value of ΔE_τ was reduced to 0.35 eV, which is close to the corresponding figure for germanium. This mechanism is also valid for MIS structures. The authors of [8.81] observed a considerable increase in the efficiency of capture of electrons injected from a semiconductor on dielectric traps in the presence of adsorbed molecules. This may result in unstable performance of MIS electronic devices.

It is quite probable that these processes play an important role in a wide variety of surface phenomena. For one thing, they can give rise to the well-known "compensation effect" in catalysis which consists in symbatic variations in the activation energy of various catalytic reactions and in the corresponding values of the frequency factor. A linear dependence between $\lg \tau_0^{-1}$ and ΔE_τ, similar to that depicted in Fig.8.4, has been reported more than once in experiments dealing with catalysis on semiconductors (the so-called empiric Constable law). Various theoretical interpretations of this law and the pertinent data are reviewed in detail in [8.82].

8.3 Dissipation of the Energy of Adsorptive Interactions

So far we have been discussing the transfer of vibrational energy from a center of capture to an adsorbed molecule. The inverse process is equally important. As was pointed out in [8.83], energy exchange and relaxation on the surface play a significant role in catalysis, especially in the fast stages of reactions. These questions are discussed in detail in [Ref.8.48,Chap.5]; see also [8.84]. Energy exchange between the adsorbed molecule and the solid is closely related to radical luminescence [8.85], adsorboluminescence [8.86] and exoelectronic emission [8.87]. Let us see what role electron processes on the surface and phonon excitation play in these phenomena.

Luminescence of the solid stimulated by recombination of radicals on the semiconductor was considered by VOLKENSTEIN et al. [8.85] from the standpoint of highly idealized concepts of the electron theory of chemisorption (Sect.5.1). The same position is assumed by those authors in their explanation of adsorboluminescence, which was discovered in 1966 by RUFOV et al. [8.88]. Adsorboluminescence is emission of light during adsorption of gases and was attributed by its discoverers to electron transitions in the solid, stimulated at the expense of adsorption energy. Later this phenomenon was observed in various oxides and on silicon during adsorption of oxygen, CO, NO, and some other gases (see the review in [8.86]). The kinetics and mechanism of luminescence during adsorption of O_2 on MgO was most thoroughly investigated by Rufov and his colleagues. They proposed a mechanism that fits in nicely with the kinetic data presented in [8.89,90]. The mechanism takes into account the electron acceptor properties of oxygen and is based on the assumption that some of the oxygen molecules diffuse over the adsorbent surface:

$$O_2 + S\dot{V}_L \xrightarrow{k_a} (O_2 \ldots S)V_L \xrightarrow[\overleftarrow{k_a''}]{\overrightarrow{k_a'}} \begin{array}{l} O_2S \\ (\dot{O}_2S)V_L + h\nu \end{array} \quad ,$$

where S is the surface of the adsorbent (MgO), $\dot{V}_L$ is a donor-type center with an electron, $O_2 \ldots S$ is an adsorbed O_2 molecule wandering over the surface O_2S is a firmly bound chemisorbed O_2 molecule, and k_a, k_a', k_a'' are reaction rate constants.

The emission of light is due to the transition of the electron from $\dot{V}_L$ to the adsorbed molecule which has fallen within the sphere of action of the donor-type center. The probability of transition w depends on the distance r between $\dot{V}_L$ and O_2 as

$$w(r / r_0) = w_0 \, e^{-r/r_0} \quad , \tag{8.4}$$

where r_0 is the radius of the donor electron ψ-function.

The intensity of adsorboluminescence is described by the expression

$$I(t) = r_0 C_D k_a' \left(\frac{1 - \exp(-k_a't)}{k_a't + \exp(-k_a't) - 1} \right) \left\{ 1 - \exp\left[-\varkappa_a't - \varkappa \exp(-k_a't) \right] \right\} \quad , \tag{8.5}$$

where $\varkappa = 2\pi w_0 k_a r_0 zN \cdot [\exp(r / r_0) / k_a'^2]$, C_D is the density of V_L centers, z is the lattice constant, and N is the number of surface lattice sites per square centimeter. Equation (8.5) accurately describes the kinetics of adsorbolumin-escence in the entire range of investigated temperatures and pressures. The calculation for adsorption of O_2 on MgO indicates that the reaction rate con-stants k_a and k_a' are exponentially dependent on temperature (the activation energies being $E_a = 3600$ cal $\cdot$ mol^{-1} and $E_a' = 7200$ cal $\cdot$ mol^{-1} respectively).

At $\varkappa(k_a't - 1) \gg 1$,

$$I(t) \simeq \frac{r_0 C_D}{t} \quad . \tag{8.6}$$

The concentration of donor-type centers, calculated using (8.6) at $r_0 = 0.5$ Å, turned out to be about $5 \cdot 10^{11}$ cm^{-3}. This implies that at the observed quantum yield (10^5 cm^{-2}) the electrons come from donors situated no deeper than 20 Å below the surface.

The diffusional model fails to explain all the peculiarities of adsorbolu-minescence, for example, the spectral dependence. Relatively few investigations have been conducted on powdered and polycrystalline adsorbents and luminophors. Owing to barrier effects (Sect.4.4), electrophysical measurements yield almost no information about the mechanism of electron processes on real surfaces. For this reason we shall not include them in our subsequent discussion. The use of single crystals, and investigations of surface electronic phenomena using the field-effect technique (Sect.2.3) have brought considerable progress in this direction.

In 1961 PEKA and KARCHANIN [8.91] discovered a new phenomenon: the lumines-cent field effect. The application of permanent and alternating transverse electric fields to crystals affected the intensity of luminescence [8.92].

Both recombination luminescence (Cds) [8.93] and the luminescence associated with annihilation of excitons on the surface (Cu$_2$O) [8.91] were investigated. Adsorption had two effects on luminescence [8.94]. The first was that owing to bending of bands, there was a change in the occupancy of the surface states which take part in luminescence. This mechanism is similar to the action of an electric field. Secondly, adsorption led to the appearance of new centers of surface recombination. In experiments on electroluminescence of single crystals of ZnO, ZnS, and SiC [8.95,96], the composition of the environment was shown to influence the intensity of luminescence. The main concern in these investigations was with recharging of surface traps and variations in the concentration of recombination centers during adsorption.

A theoretical approach to all of these luminescence effects was attempted in [8.85]. However, this approach involved the extrapolation of a simplified theory, developed for an idealized surface under the assumption of electron equilibrium in the system of semiconductor and adsorbed phase onto the necessarily nonequilibrium processes of cathode luminescence and adsorboluminescence on real surfaces. It therefore threw little new light on these phenomena. Our criticisms in Chap.5 regarding the limitations inherent in this theory apply here as well.

Another aspect of this problem has been even less deeply explored: the energetics of cathode luminescence and adsorboluminescence. The electron theory [8.1,85] fails to supply any reliable estimates for the probabilities of non-radiative and radiative transitions in the solid during adsorption, or to provide a theoretical consideration of the mechanism by which the energy of these transitions is dissipated. Model-type theoretical calculations of adsorption-stimulated phonon excitations in the solid were carried out in [8.97-99].

The estimates made for a unidimensional model of crystal [8.99] indicate that rapid thermal relaxation of the lattice results in phonon excitations having quite low levels. Energy exchange with phonons localized at surface impurity atoms and defects is more efficient.

Under certain conditions information on surface phonons (surfons) can be derived from luminescence spectra. When an atom is being excited by light, the light loses some of its energy due to scattering on phonons; this is the so-called Frank-Condon shift. As a result, the long-wave end of the luminescence spectrum exhibits additional lines or phonon reiterations. In the works of ZUEV et al. [8.100,101] (see also Sect.6.7) such phonon reiterations were observed in luminescence spectra of GaAs surfaces. As seen from Fig.8.6, the energy of surfons E_{ph}^{S} is 33 ± 0.5 meV, which is lower than the energy of bulk LO phonons at 4.2 K (37 meV). A similar approach to the problems of ad-

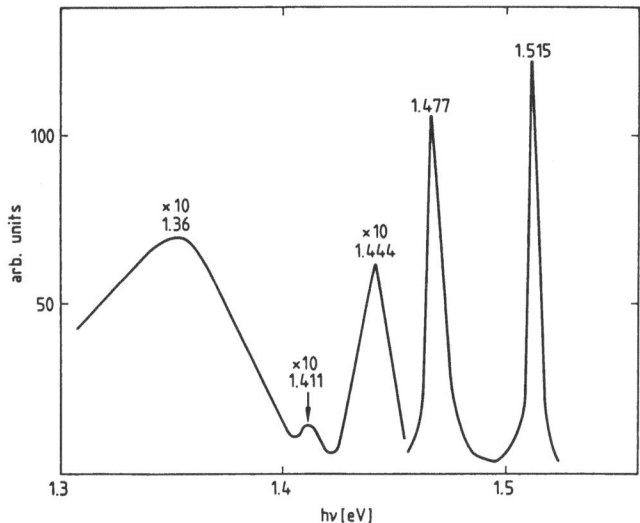

Fig.8.6. Luminescence spectrum of GaAs at 4.2 K, excited by laser light [8.101]. Peak at 1.477 eV: surfons. Peaks at 1.444 and 1.411 eV: phonon reiterations. Peak at 1.515 eV: surface excitons. Peak at 1.36 eV: out of the bulk (reproduced by permission of the authors)

sorption will bring considerable progress in the study of the phonon excitations associated with adsorption and the excitation of adsorbed molecules.

The effects of adsorption on radiative recombination are just beginning to be investigated. NAGAI et al. [8.102] observed reversible variations in the intensity of photoluminescence of InP during adsorption of Ar, N_2, H_2, O_2, and H_2O. However, the purity of the gases they used is doubtful, since at room temperature the physical adsorption of Ar, N_2, and H_2 is negligibly small and is known to induce no noticeable variations in the electrophysical parameters of the surface. The strong effect of adsorption of O_2 on the photoluminescence of InP was later attributed either to a competitive process of nonradiative recombination [8.103], or to bending of energy bands [8.104]. It should be noted that the rate of surface recombination S depends in turn on the surface potential Y_s (Sect.3.1). Simultaneous investigations of radiative and nonradiative recombination as a function of Y_s offer great promise in this connection. The picosecond technique proposed in [8.105] is also promising.

The same questions arise with regard to the mechanism of exoelectronic emission (low-temperature emission of electrons from a solid-state surface, stimulated by various physico-chemical and mechanical agents [8.87,106]). KRYLOVA [8.107] discovered the possibility that emission might be stimulated by adsorption (desorption) and catalysis, and showed correlations between these phenomena. The energy released due to recombination of active particles (radicals, charged particles) on the surface is not entirely dissipated on phonons, but is partially transferred to emission centers, thus leading to thermal ejection of an electron from the solid [8.87,107]. TOLPYGO et al. [8.108] assumed Auger processes to be responsible for exoelectronic emission. However, the efficiency of Auger processes, as will be shown in the next section, is sufficiently high only when the levels of excitation of the semiconductor's electron subsystem are high.

Still open is the question of where the levels responsible for exoelectronic emission are located in the energy spectrum. In [8.109] they are assumed to be represented by quasi-resonance levels in the conduction band. The possibility that discrete states exhibiting a potential barrier might appear in the conduction band was theoretically demonstrated by KELDYSH [8.110].

The mechanism of energy exchange between adsorbed molecules and the solid is closely associated with another important problem in semiconductor surface photoelectronics: spectrally selective sensitization of photoeffect by adsorbed organic molecules [8.111]. A solution to this problem is of vital importance in electrophotography and the development of artificial photosynthesis systems (Sect.9.2). Spectral sensitization consists essentially in a shift of the active spectral range of various photoelectric processes on the surface and in the subsurface region of the solid towards longer wavelengths owing to adsorption of dyes.

Sensitization of the photoconductivity of wide band semiconductors such as CdS [8.112], ZnO [8.111,113], and Se [8.114] has been most thoroughly investigated. FLYNN et al. [8.115] discovered pronounced spectral sensitization in a system of SiO_2 and organic dye. Silicon dioxide, being a dielectric, exhibits noticeable photoconductivity in the UV spectral region. After adsorption of a dye, the maximum photoconductivity of SiO_2 nearly coincided with the maximum absorption of dye (~ 500 nm). IONOV and AKIMOV [8.116] reported the possibility of sensitization of semiconductors having comparatively narrow bands (Ge and Si).

Two possible mechanisms of sensitization have been proposed [8.111,113]. The first is the purely electron mechanism, according to which a dye molecule ionized by light, captures an electron from the local level of the semiconductor. The molecule's electron is assumed to be knocked by the light into the conduction band of the semiconductor. In the second mechanism, proposed by MOTT [8.117], energy from a dye molecule excited by light is assumed to migrate to the local level, with subsequent ejection of an electron localized on this level into the conduction band. AKIMOV et al. [8.111] suggested that the energy transferred from dye molecules to the semiconductor can be accepted by centers not only on the surface but also in the bulk. The energy, as indicated in [8.111], can be transferred via the induction resonance mechanism to distances of up to 150 - 200 Å. In reality all these (mostly two-quantum) processes are often complicated by cooperative phenomena. Illumination of a system of adsorbed molecules and a semiconductor can give rise to excited states: excitons and polaritons.

Direct proof of the predominant role of energy transfer in the mechanism of photosensitization was obtained in [8.118]. The authors of that works investigated photocharging (Sect.4.3) of dielectric-semiconductor ($Ge - GeO_2$) structures with dye molecules adsorbed on the outer surface of the dielectric. Illumination of the outer surface with light in the dye's absorption band caused efficient emptying of negatively charged slow traps in the dielectric. In the absence of dye, the effect of light-induced emptying of traps (at the same energies of light quanta) was zero.

8.4 On the Possible Role of Auger Effects in Adsorption and Catalysis

At high levels of excitation of the semiconductor's electron subsystem, nonradiative shock mechanisms of capture and recombination of charge carriers begin to play an important role [8.57]. As opposed to the phonon mechanism (the Shockley-Reed mechanism of recombination) when the energy released in the acts of capture and recombination is transferred to the lattice, in the so-called Auger process this energy is transferred to a third charge carrier. Figure 8.7 gives some examples of Auger transitions in the bulk of a semiconductor [8.23]. The capture of a hole in an n-type semiconductor (i.e., the passing of an electron from level E_t into the valence band, Fig.8.7a) releases energy $(E_t - E_v)$, which is transferred to one of the majority carriers, an electron (of which there are many). The electron is excited into a higher level in the conduction band, E_1' ("hot" electron). This energy can be expended on the ejection of another electron in a p-type semiconductor to level E_t to level E_2'. The passing of an electron in a p-type semiconductor to level E_t (Fig.8.7b) stimulates the passing of a hole to a deeper level E'' ("hot" hole). Auger transitions between donor-type and acceptor-type levels are illustrated in Fig.8.7c, and those between bands in Fig.8.7d. Auger effects have also been observed in transitions to exciton levels [8.119].

Interband Auger recombination (Fig.8.7d) has been most thoroughly explored [8.120]. A statistical theory of surface Auger recombination via discrete centers was developed by ZUEV et al. [8.121], and these problems are also discus-

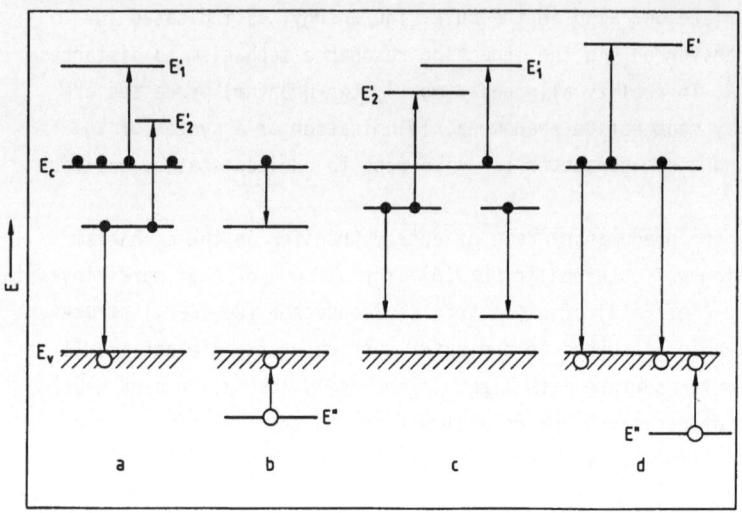

Fig.8.7a-d. Auger transitions in a semiconductor

sed in detail in [8.77]. Whereas Shockley-Reed phonon recombination via a defect is a monomolecular process, Auger recombination via a discrete center is bimolecular. At high levels of excitation when the concentration of nonequilibrium charge carriers Δn, $\Delta p \gg n_i [\delta \gg 1$; see (2.4)], monomolecular and bimolecular recombination may occur side by side with radiative bond-to-bond recombination and bulk trimolecular band-to-band Auger recombination. At extreme levels of injection, a considerable role is taken on by plasmonic recombination, when the minimal energy of incipient plasmons is close to the energy of ionization of the deep recombination center. The parameters that characterize each of these types of recombination, and their dependence on the intensity of incident light I were calculated in [8.121]. In order to determine which of the recombination paths plays the predominant role, one must compare the results of theoretical calculations with the data on various photoelectric effects (light extinction, luminescence, photoconductivity, etc.), measured for absorption of light on the surface and in the bulk [8.121].

To illustrate, Fig.8.8 shows the results of investigations in [8.121] on the concentration of nonequilibrium charge carriers Δn as a function of the intensity of light I incident on single crystals of silicon and germanium. Surface absorption of light was ensured by the use of a ruby laser; with irradiation by neodymium glass laser the absorption took place mainly in the bulk. A theoretical analysis of the data presented in Fig.8.8 reveals the following [8.121]:

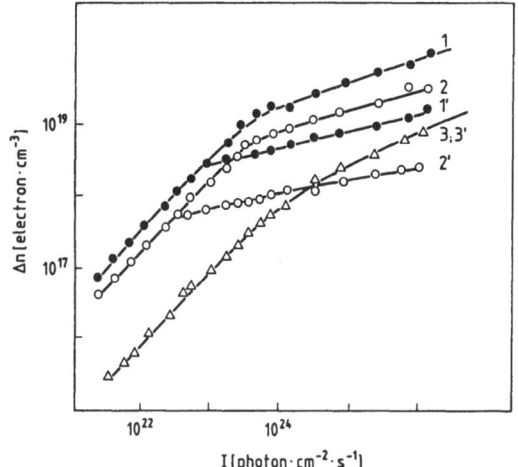

Fig.8.8. Light-intensity dependence of the concentration of nonequilibrium charge carriers [8.71,121]. (Reproduced by permission of the authors)

(1) With the initial etched and oxidized surfaces of Si (Curve 1) and Ge (Curve 2), bulk bimolecular recombination dominated over surface recombination.

(2) After a considerable increase in the density of recombination centers, induced by bombardment of the surface with He$^+$ ions, surface bimolecular Auger recombination started to prevail over recombination in the bulk (Curves 1' and 2'). Naturally, surface treatment did not affect the function $\Delta n(I)$ in the case of absorption of light in the bulk (neodymium glass laser, Curves 3 and 3')

We see that surface treatment (and possibly adsorption as well) can stimulate Auger recombination on the surface. According to estimates made in [8.121], surface Auger recombination in pure specimens of Si and Ge dominates over recombination in the bulk when the concentration of discrete centers of surface recombination N_r is greater than $5 \cdot 10^{12}$ cm^{-2}. In Shockley-Reed recombination, surface recombination dominates over bulk recombination at $N_r > 10^{10}$ cm^{-2}. With an increase in injection level δ, the rate of Shockley-Reed surface recombination S grows, and at high values of light intensity I, it comes to saturation. At high levels of excitation (e.g., laser; see Fig.8.8), the effective rate of Auger recombination S_A can prevail over S even in semiconductor in which at low injection levels monomolecular Shockley-Reed recombination predominates. The probability of Auger processes is much higher in wide-band, highly doped semiconductors [8.122], which are the ones most widely employed in experiments on adsorption and catalysis.

Hot electrons and holes generated in Auger processes (Fig.8.7) either expend their energy on the emission of phonons, or are captured on levels of defects having sufficiently large capture cross sections. An electrically inactive defect, having captured at hot charge carrier and taken over its energy, can pass into an electrically active state (Sect.8.1). When the Auger process takes place on the surface or in the subsurface region of the crystal, capture of an electron on a surface state can turn the state into an adsorption-active one. It is also possible for the energy of the Auger electron to be greater than the crystal's work function; then some of the Auger electrons can leave the crystal.

All these processes become probable if the levels of excitation are sufficiently high. Auger processes may play some role in photoadsorption and photocatalysis (Sect.7.2). For instance, irradiation of CdSe with a ruby laser led to desorption of charged particles [8.123]. Such processes might well be stimulated by Auger electrons. Capture of a hot Auger hole on a negatively charged adsorptive surface state of oxygen with release of energy leads to partial neutralization of oxygen ions and desorption. Capture of an Auger electron on an acceptor-type biographic surface state makes it an active center of adsorption. Auger processes can be responsible for partial recharging of surface states, as assumed in the dynamic mechanism of photoadsorption proposed by BASOV et al. [8.13].

A consistent theory of photoelectric phenomena in semiconductors at high injection levels was evolved by ZUEV et al. [8.77]. As we mentioned in Sect.2.4, the surface photoelectromotive force Y_{ph} serves as a convenient electrophysical characteristic of the surface. The value of Y_s is a function not only of the injection level and the light intensity I but also of the initial (dark) bending of bands Y_{s0}. The values of Y_{ph} as a function of I (as calculated theoretically in [8.77]) is plotted in Fig.8.9. At extreme levels of injection the surface photoelectromotive force tends to saturation ($Y_{ph} \approx Y_{s0}$) because nonequilibrium carriers completely shield all external fields, including those of surface states recharged due to illumination. This situation is encountered with irradiation by laser light. It is at these levels of injection that Auger processes of recombination begin to play a consider role. The theory constructed in [8.77] gives a satisfactory description of experimental data on Y_{ph} for both silicon and germanium.

Auger processes may also play an important role in radical photocatalysis. As demonstrated in Sect.7.1, the transition of a molecule from a state of strong donor-acceptor bond into the radical form (CTC) is sometimes associated with the surmounting of potential barrier E_{rad}. If the energy of the Auger electron $E_0 > E_{rad}$, then illumination of the surface will produce radical forms of adsorption. As we have noted before, the creation of radical forms of adsorption is not equivalent to the creation of charged forms of adsorption in VOLKENSTEIN's model [8.1,9]. Experimental results confirm the creation of ra-

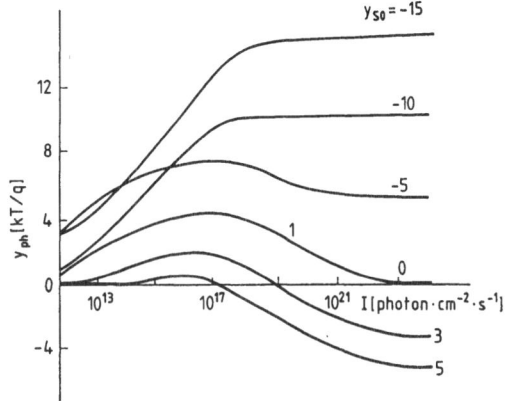

Fig.8.9. Theoretical dependence of surface photovoltage Y_{ph} on the intensity I of absorbed light [8.77]. Y_{s0} is the initial surface potential. (1) For flat bands [8.77]. (Reproduced by permission of the authors)

dical forms of adsorption upon illumination of a semiconductor with intense light [8.124,125].

The efficiency of Auger processes on the surface can be quite high, provided the density of states of both biographic and adsorptive origin is sufficiently high. At excitation levels such that Δn, $\Delta p > 10^{18}$ cm^{-3} [8.121], Auger processes start to dominate over purely phonon processes of capture and recombination of charge carriers. The probability of Auger processes at lower temperatures can be greatly increased by phonon excitations [8.126], which can be stimulated by adsorption and catalysis on the surface.

9. Proton Processes on the Surfaces of Semiconductors and Insulators

It might seem strange that a book on electronic phenomena should include a chapter dealing with the behavior of protons on the surface. However, according to our donor-acceptor mechanism of adsorption and charging of semiconductor surfaces, there is a close interconnection between the molecular and the electron subsystems of the ensemble composed of the solid body and the adsorbed phase. Excitation of the electron subsystem stimulates adsorption and dissociation of adsorbed molecules. The ions generated by virtue of heterolytic dissociation can, under certain conditions, have a considerable effect on surface conductivity and bring about changes in the parameters of surface states.

In our discussion of the physical nature of the pertinent processes we shall restrict ourselves to adsorption of the simplest molecules, primarily water. Our elucidation of the mechanism of dissociation of water molecules and of proton conductivity will also prove useful in a number of related branches of surface science.

9.1 Proton Conduction of Adsorbed Water

Adsorption of water is known to raise the surface conductivity of high-resistance materials, thus deteriorating their dielectric properties. SHOCKLEY and KOOPER [9.1], FEDOROVICH and DUMISH [9.2], HOFSTEIN [9.3], and other researchers have observed a strong influence of adsorbed water on ionic currents in the MIS structures which form the basis of modern microelectronic technology. Slow ionic processes in an insulator modify its parameters and cause unstable performance of MIS semiconductor devices. Direct measurements using tritium-labeled alcohol [9.3] have proved the existence of protonic conduction via a porous SiO_2 layer in MIS structures. The possibility of protonic processes in MIS structures is also indicated in [9.4-6][1].

---

[1] A large role in the ionic conduction of MIS structures is often played by "extrinsic" conduction of foreign ions (mostly Na^+) implanted during technology processes. This is discussed in detail in [9.7,8].

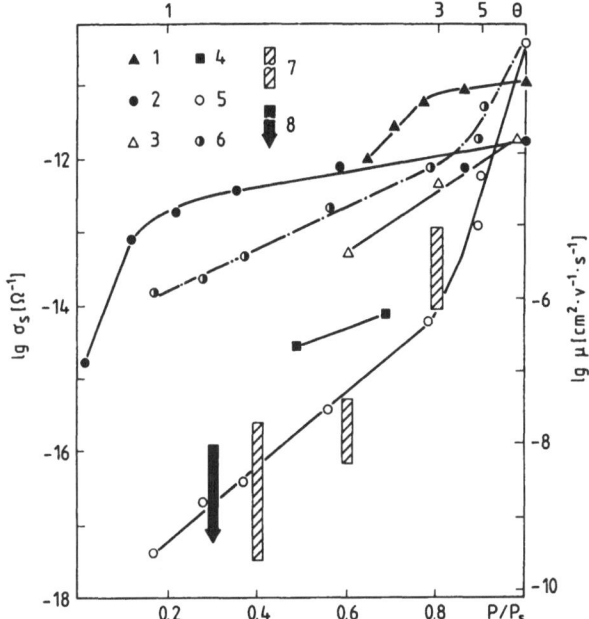

Fig.9.1. Dependence of the specific surface conductivity of quartz (1-6) and surface mobility (7,8) on relative pressure p/p_s and coverage with ab- sorbed water molecules (in terms of the number of monolayers θ), as reported by different authors: (1) [9.11], (2) [9.12], (3) [9.10], (4) [9.9], (5) [9.13], (6) [9.21], (7) [9.1], (8) [9.2]

The literature contains numerous data on the conductivity of films of ad- sorbed water. The high conductivity of water adsorbed on glasses and alumino- silicates can be attributed to the high density of foreign ions on the surface. The high conductivity of water adsorbed on pure oxide surfaces is much less easily explained. Water adsorbed on quartz has been most thoroughly investiga- ted [9.9-12]. The results are presented in Fig.9.1. The values of surface con- ductivity σ_{pr} differ even for equal values of p/p_s which must be attributed to unequal surface cleansing. The lowest values of σ_{pr} were obtained in [9.13], where the measurements were taken on the surface of α-quartz subjected to a special cleansing procedure. The initial value $\sigma_{pr} = 5 \cdot 10^{-18} \, \Omega^{-1}$ ($p/p_s = 0.2$) is several orders of magnitude lower than the values usually cited in the li- terature. This gives reason to conclude that we are dealing here with a predo- minantly protonic mechanism of conduction. This is also indicated by the close- ness of the values for the activation energy of electroconduction, calculated from temperature dependence $\sigma_{pr}(1/T)$ [9.13] and those for the bulk of pure water $\sigma_v(1/T)$.

SAFFER, FOLMAN, and ANDERSON, PARKS [9.14,15] noted that deuterization of SiO_2 surfaces alters the electroconductivity. The ratio $\sigma_{H_2O} / \sigma_{D_2O}$ is about

215

1.3 and differs little from the corresponding ratio for solutions of salts in H_2O and D_2O. The isotopic effect also confirms the assumption that the transport of charge occurs via the protonic mechanism.

According to current concepts based on IR spectroscopy [9.16] and quantum chemical calculations [9.17], a proton in pure water forms links with two molecules of water, thus forming an $H_5O_2^+$ ion:

$$\left[\begin{array}{c} H \\ \diagdown \\ {}^{\diagup}O - H \dots O^{\diagup} \\ H \end{array} \begin{array}{c} H \\ \diagdown \\ \diagdown \\ H \end{array} \right]^+ .$$

When current flows through pure water, the proton tunnels from the $H_5O_2^+$ ion to molecules of water, and moves along the chain of H bonds via the relay mechanism [9.18]. This mechanism also usually applies to conduction in a film of water adsorbed on an SiO_2 surface [9.9-12]. However, the sources of protons in the film and their density are still controversial. TSCHAPEK et al. [9.19] assumed that the dissociation constant for adsorbed water k_d^s is much greater than the corresponding value for the bulk ($k_d^v = 10^{-9}$). ANDERSON and PARKS [9.15] assumed that at pre-monolayer ($\theta < 1$) occupancies (according to the BET equation) the protons are supplied by hydroxyl groups and associated water molecules (linked by H bonds) while at $\theta > 1$ the sources are only molecules of water. The transport of charge is assumed to be performed by H_3O^+ ions.

ANDERSON and PARKS [9.15] based their mechanism of conduction on the assumption that adsorption of water proceeds on solitary Si - OH groups which is confirmed neither by spectroscopic nor by adsorption data [Ref.9.20,Chap.4]. According to these data, the Si-OH groups exhibit very low proton-donor activity. PAK et al [9.21] reported an antibatic correlation between the conductivity and acidity of OH groups in their experiments with adsorption on a quartz surface covered with a layer of TiO_2 (constructed using the molecular deposition technique). A much higher protonic acidity is displayed by water molecules that are coordination bound with hydrated Si atoms.

In the [9.2,22], which deal with slow protonic processes in silicon MIS structures, it is assumed that the SiO_2 film is a colloid structure in which the sources of protons can be water molecules or the molecules of polysilicon acids $(H_2Si_nO_m)^2$. The dissociation constant k_d for these molecules is assumed to be equal to the corresponding values for solutions or liquid colloids. The authors of [9.23] considered the possibility that the dissociation constant

---

[2] NICCOLIAN and SINHA [9.4] also considered SiO_2 films as electrolytes.

can assume greater values in strong electric fields applied to the IS structure. Usually, however, the intensity of such fields is about 10^5 V $\cdot$ cm^{-1}
(10^6 V $\cdot$ cm^{-1} in MIS structures). According to estimates made by BENNETT [9.24]
the electric field can be expected to have a strong effect on the parameters
of the chemisorptive bond only at intensities of 10^7 or 10^8 V $\cdot$ cm^{-1}, which are
achieved only in electrolytes or in a field emission microscope. The colloid
dielectric model seems unrealistic for experiments carried out in a vacuum on
thermally dehydrated surfaces of both SiO_2 crystals and MIS structures [9.25].

From the standpoint of our donor-acceptor mechanism of adsorption (Sect.6.6)
it would be natural to assume that the proton sources are highly deformed coordination-bound molecules on the surface of SiO_2 as well as many other oxides.
NMR data [Ref.9.20,Chap.3] indicate that with increased occupancy, the water
molecules build clusters around these firmly bound centers, and intensive proton exchange takes place within the clusters. Fusion of such clusters creates
channels of protonic conduction. Owing to nonuniformity of the surface, such
channels can appear even before a continuous polymolecular film is formed.

The majority of researchers [9.9-12,15] assume that the increase in protonic
conduction which accompanies adsorption is associated with an increasing concentration of protons. However, protonic conductivity is defined as $\sigma_{pr} = qn_{pr}\mu_{pr}$,
where q is the charge of the proton, and μ_{pr} is the proton's mobility. The
possibility of a change in μ_{pr} as occupation proceeds was not taken into account; direct measurements of μ_{pr} suggest that it increases with an increase
in θ [9.1,2]. PAK et al. [9.21] also observed an increase in μ_{pr} during adsorption of H_2O.

Detailed investigations of σ_{pr} as a function of θ were carried out by
KURZAEV and KOZLOV [9.13] (Fig.9.1). They were the first to relate the measured
values of σ_{pr} to the density of adsorbed molecules n_a, and not to the pressure
of the water vapor. Adsorption was monitored with the aid of a piezoresonance
balance. The monolayer according to B.E.T. ($\theta = 1$) is equivalent to $n_a = 8 \cdot 10^{14}$
molecules $\cdot$ cm^{-2}. The rather natural assumption was made that at $p/p_s = 1$ the
surface mobility of protons μ_{pr} is close to the corresponding value for liquid
water ($3.5 \cdot 10^{-3}$ cm$^2 \cdot$ V$^{-1} \cdot$ s^{-1}). Then the observed values of σ_{pr} imply (Fig.
9.1) that $n_{pr} = 10^{11}$ cm^{-2} which is five orders of magnitude greater than the
density of protons in the equivalent quantity of liquid water. This means that
the dissociation constant for adsorbed water $k_d^s \approx 10^{-4}$ is five orders greater
than k_d^v, exactly as assumed in [9.19].

If one considers the sources of protons to be coordination-bound water molecules whose concentration n_c (according to NMR data) is about 10^{13} cm^{-2}
[Ref.9.20,Chap.3], then the extent of dissociation of water molecules of this

kind is about 10^{-2} which agrees with the high proton-donor activity of these molecules. Since $n_c \gg n_{pr}$, the value of n_{pr} can be assumed to vary little in the whole region $\theta \geqslant 1$, while variations in σ_{pr} (Fig.9.1) are entirely determine by mobility as a function of occupancy $[\mu_{pr} = \mu_{pr}(\theta)]$. The calculated values of μ_{pr} fit in well with the experimental results (Fig.9.1).

Attempts have been made to estimate the mobility of protons in adsorbed water on the basis of data on nuclear relaxation [9.26]. These data, however, are rather ambiguous, owing to the strong effects of uncontrollable impurities on relaxation times. The NMR technique, used in [9.27], helped to demonstrate that the coefficient of self-diffusion of adsorbed water varies by a factor of 10 when the value of θ increases from 1 to 15. In the same range of occupancies the value of μ_{pr} varies by six orders of magnitude. The absence of a correlation between the coefficient of self-diffusion (which characterizes the movement of the water molecule as a whole) and the value of μ_{pr} is consistent with the relay mechanism of proton transport. The limiting stage of this process is represented by rotations of water molecules.

It appears that the same mechanism of protonic conduction is valid for IS and MIS structures. The mobility of protons in an insulator film is perhaps smaller than on the surface of a single crystal. HOFSTEIN [9.3] used measurements of transition currents in MIS structures to obtain the value $\mu_{pr} = 5 \cdot 10^{-11}$ $cm^2 \cdot V^{-1} \cdot s^{-1}$, which corresponds to a dry quartz surface (Fig.9.1).

As demonstrated in Sect.7.2, the donor-acceptor mechanism of adsorption implies that there are two possible ways of controlling protonization of coordination-bound water molecules:

1) by chemically modifying the surface; and

2) by controlling the density of holes in the subsurface region of the crystal.

Replacing surface hydroxyl groups with more active acceptors of electrons (e.g., fluorine atoms) increases the effective positive charge of metallic atoms. As a result, the undivided pair of electrons of an H_2O molecule is more effectively drawn onto vacant orbitals of the adsorption center. On the one hand, this strengthens the adsorptive bond; on the other hand, it weakens the O-H bonds in the water molecule and results in protonization, e.g.,

$$M^{-\delta} \cdot O \begin{smallmatrix} \nearrow H^{+\delta} \\ \\ \searrow H \end{smallmatrix} \quad .$$

Such a molecule is an active proton-donor (Brönsted) center. According to NMR data, the interproton distance in coordination-bound water molecules is increased from 1.54 to 1.7 or 1.8 Å after fluorination of the surface of SiO_2 or Al_2O_3. These results are discussed in detail in [Ref.9.20,Chap.4]. We see that fluorination ought to increase the extent of dissociation of coordination-bound water molecules and result in an increase in n_{pr} [9.28]. However, replacement of OH groups by fluorine atoms leads to hydrophobization of the surface [9.20]. For this reason the growth of clusters of water molecules and their fusion (giving rise to channels of conduction) occurs at much greater relative vapor pressures p/p_s than is the case with hydrated hydrophilic surfaces.

Fluorine is a major constituent of etchants used for treatment of germanium and silicon. Implantation of fluorine atoms into the oxide layer of these semiconductors can stimulate ionic processes. The introduction of F into an electrolyte in the process of anodizing silicon accelerates oxidation and increases the positive charge of the oxide. The positively charged traps which result from the implantation of fluorine atoms are found in the immediate proximity of the $Si-SiO_2$ interface. An exhaustive review of these investigations, together with vast original experimental material, can be found in [9.29]. The effect of F groups on photocharging of $Si-SiO_2$ structures in a vacuum was explored in [9.30]. The removal of F groups (defluorination) was accompanied by an increase in positive light-induced charge in SiO_2. On real surfaces, defluorination (flushing with an ammonia solution or water) caused a change in the density of fast states and the rate of surface recombination [9.31-33]. The introduction of F groups into amorphous silicon greatly improved the efficiency of solar cells [9.34,35].

Of especial interest in the technology of MIS electronic devices is the role of chlorine in the shaping of the $Si-SiO_2$ interface. It has been demonstrated [9.36-39] that oxidation of silicon in the presence of chlorine compounds (usually HCl) leads to a considerable decrease in the density of surface states on the $Si-SiO_2$ interface (to 10^{10} $eV^{-1} \cdot cm^{-2}$) and to an improvement in the recombination characteristics of the interface.

The literature still contains conflicting opinions regarding the effects of halogens on the electrophysical properties of MIS structures. KRIEGLER et al. [9.36] and ROHATGI et al. [9.39] assumed that halogens, interacting with metallic impurities in an insulator, neutralize them and put them into an electrically inactive state. CHEN and HILE [9.40] assumed halogens to form volatile compounds that are removed in subsequent vacuum heat treatments. Later we shall show that protonic processes stimulated by coordination-bound water can participate in all these occurrences. It should be noted that in interactions between SiO_2 and vapors of HF, the proton-donor centers can be coordination-bound HF molecules [9.28].

Let us now discuss the electron factor. Because an undivided pair of electrons is pulled over from an H_2O molecule to the orbitals of the electron acceptor, the positive charge on the latter will be reduced. This conclusion, implied by our donor-acceptor mechanism of adsorption (Sect.6.6), is confirmed by analysis of X-ray photoelectron spectra of SiO_2 [Ref.9.20, Chap.4]. Desorption of co-ordination-bound water increased the positive charge of Si. As we mentioned in Sect.6.6, the resulting complex $\geqslant Si : O\begin{smallmatrix} H \\ H \end{smallmatrix}$ acts as a center for the capture of holes. Localization of a hole will increase the pulling of the unshared pair of electrons of H_2O and lead to further protonization of the H_2O molecule. According to the experiments described in Sect.7.2, under certain conditions the injection of holes causes dissociation of coordination-bound molecules.

We see that the donor-acceptor mechanism of adsorption and charging implies that it is possible to control protonic conduction by injecting holes into the SCR. This can be achieved by illuminating the crystal. It is from this standpoint that we explained the release of hydrogen observed upon illumination of Al_2O_3 [9.41], Ge, and Si in Sect.7.2.

All these phenomena can take place in IS and MIS structures. Field injection or photoinjection of holes into the dielectric layer will stimulate protonic processes in the structure, thus changing the inherent charge of the oxide and the parameters of surface states on the insulator-semiconductor interface. The injection of charge carriers into IS and MIS structures has been explored in a large number of works. However, most of them do not pay due attention to the mechanism of ionic processes.

NICCOLIAN et al. [9.4,42], and later many other researchers, assumed that silanol groups are responsible for the capture of an electron: $\geqslant Si - OH + e \rightarrow \geqslant Si - O^- + H^{\cdot}$. However, there is no confirmation for this reaction, nor is it clear what the further fate of this very reactive hydrogen atom is. Investigations of photocharging of $Ge-GeO_2$ and $Si-SiO_2$ structures [9.30,43,44] (Fig. 4.3) indeed indicate that negative charging is associated with hydroxyl groups. However, the process of charging is completely reversible. At temperatures that are not too high, the traps are emptied while retaining their nature owing to thermal ejection of electrons. This can hardly be reconciled with a mechanism that assumes dissociation of OH groups. Experiments on fluorinated Si and Ge surfaces [9.28] also do not confirm Niccolian's scheme. The replacement of OH groups with fluorine groups lowered the activation energy of thermal desorption of hydrogen and enhanced the photodesorption of H_2 while the surface charge remained the same. The release of H_2 was not accompanied by the appearance of an EPR signal from $\geqslant Si^{\cdot}$.

REVESZ [9.5] considered one possible source of protons to be hydride Si-H groups, and assumed these groups to be the centers of adsorption of H_2O molecules. This, however, contradicts data on adsorption. A hydrogenated surface of SiO_2 is hydrophobic [9.20], so it is hardly reasonable to expect protonic conduction on it. Dissociative processes involving holes and coordination-bound H_2O molecules are probable. Capture of holes injected from Ge or Si into the oxide takes place on vacancy-type defects situated near the Ge-GeO$_2$ or Si-SiO$_2$ interface, as indicated by EPR data [9.43, Ref.9.20,Sect.3.2.3, Fig.3.23b]. If H_2O molecules form coordination bonds with atoms of Ge (or Si) belonging to such defects, then capture of holes can stimulate their photodissociation According to the arguments developed above, halogenization of the surface promotes this process. These questions will be discussed in greater detail in the next section.

Another possible source of protons is the metal-semiconductor interface. The metal most commonly employed in silicon-based MIS structures is aluminum. When aluminum is deposited on a hot SiO_2 substrate, diffusion of Al atoms causes the formation of an interfacial layer of aluminosilicate, a typical catalyst. The creation of such an interfacial structure has been established in experiment [9.45]. The active centers in aluminosilicates are Al atoms in triadic and tetradic configuration [Ref.9.20,Sect.3.2.3]. Water molecules adsorbed on such centers undergo complete dissociation and give rise to protons. Similar effects can be expected to take place on the boundary between SiO_2 and transition metals (Ti, V, Mo, etc.), as well as on interfaces between SiO_2 and transition-metal oxides.

9.2 Photostimulated Dissociation of Adsorbed Molecules

One of the simplest model reactions in the study of surface photocatalytic transformations is light-stimulated decomposition of water molecules adsorbed on a solid surface. This reaction is not only interesting from the standpoint of photoadsorption and photocatalysis; it is also closely connected with one of the most urgent problems of modern science: the conversion of solar energy into electric and chemical energy using semiconductors. There are several aspects to this problem. The first is photosynthesis. The first light-stimulated stage of photosynthesis (decomposition of water) is the least understood today. Secondly, there is the problem of constructing electrochemical semiconductor converters of solar energy. Thirdly, there is deterioration of photocells, photoresistors, and photodiodes which must operate in the presence of environmental water vapor.

Solar batteries which directly convert solar energy into electricity, are successfully operating in space. The efforts of leading scientific laboratories are currently aimed at the construction of low-cost solar energy converters to be used on earth [9.46]. In order to bring down the cost of such devices, scientists are exploring the possibility of substituting expensive monocrystalline cells with Schottky junctions (Si, GaAs) [9.47], and ultimately with amorphous and polycrystalline silicon films [9.48,49]. Cheap film elements have great economic advantages over monocrystalline elements. Investigations in this direction have focused mainly on obtaining structures with electrophysical parameters that will ensure the most efficient conversion of solar energy into electricity. The effects of surface occurrences on the parameters of solar converters have been almost completely neglected, despite the fact that their importance is greatly enhanced when dealing with films.

Another direction in the development of solar energy converters is connected with the use of electrochemical cells [9.50,51]. These kinds of converters are especially interesting insofar as the production of electricity here is accompanied by decomposition of water into hydrogen (the problem of hydrogen fuel) and oxygen (the problem of clean air). The operation of a photoelectrochemical cell can be understood from Fig.9.2 [9.50], light generates electron-hole pairs in the semiconductor. The holes in the electric field migrate to the semiconductor-electrolyte interface and take part in the dissociation of water molecules[3], which is commonly described by the reaction

$$H_2O + 2p^+ \rightarrow 1/2\ O_2 + 2H^+ \quad , \quad 2H^+ + 2e^- \rightarrow H_2 \quad .$$

One of the major obstacles in the development of photoelectrochemical cells involves the stability of the semiconductor electrode. Wide-band semiconductors have been used: TiO_2 (rutile) [9.52], SnO_2 [9.53], $SrTiO_3$ [9.54] and the like, which operate in the UV spectral range (3 - 4 eV). The efficiency of conversion of solar energy into electricity was 10 - 20 %. The maximum of the spectrum of solar radiation at ground level corresponds to about 2.4 eV. At such energies of light quanta the efficiency of conversion was quite low. In order to improve it, the wide-band semiconductors were sensitized with organic dyes [9.55] (Sect.8.3). For instance, when TiO_2 and SnO_2 were sensitized with rhodamine, the maximum shifted to 2.2 - 2.3 eV.

---

[3] Note that the band scheme of the semiconductor presented in Fig.9.2 is valid for a very low level of injection of nonequilibrium charge carriers. At the high levels of illumination usually employed, one must consider Fermi quasi levels for electrons and holes.

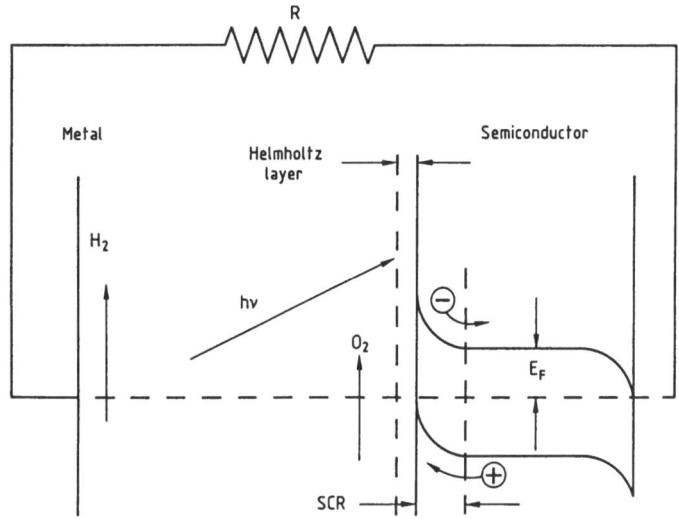

Fig.9.2. Electrolyte photocell. E_F is the Fermi level [9.50]

In connection with photoelectrochemical converters we must mention experiments on the possible use of chlorophyll [9.56-58] and its analogues (phthalocyanines) [9.58,59], supported on semiconductor and metallic substrates. The efficiency of such cells is 14 %. Instability of parameters and the aging common to all organic semiconductors prevent practical use of such converters. Nevertheless, investigations of such systems are very valuable for modeling and understanding the light stage of photosynthesis.

Although considerable progress has been made in the study of photoelectric converters, the mechanism of the elementary act of dissociation of water molecules is still unclear. The dissociation itself is usually discerned by means of variations in the current through the electrolyte, caused by illumination of the semiconductor electrode, while the mechanism of dissociation is deduced on the basis of overall reactions. The study of dissociation of adsorbed molecules is complicated by the presence of ions in the electrolyte and by complex chemical reactions that take place in the vicinity of electrodes. Experiments on photodissociation of adsorbed water molecules in a high vacuum appear to be helpful here; the reaction yield can be directly assessed with the aid of mass spectroscopy.

One quantum photodissociation of a free water molecule occurs in vacuum ultraviolet range, in the spectral region of absorption bands of water (6.7 - 10 eV). In the range from 6 to 7 eV the reaction $H_2O + h \rightarrow H^+ + OH^-$ is most probable, while dissociation into O and H_2 is less probable [9.60,61].

Table 9.1. Relative values A of maximal desorption rate of H_2 and O_2 durign photolysis of adsorbed water [9.64]

Catalyst	T_{cal} [a]	A	Catalyst	T_{cal} [a]	A	Catalyst	T_{cal} [a]	A
BeO	550	100	ThO_2	550	8	TiO_2	500	1[b]
γ-Al_2O_3	600	30	HfO_2	450	6	ZnO	500	1[b]
ZrO_2	550	10	SiO_2	500	3	GeO_2	500	1[b]
La_2O_3	450	9	MgO	700	1[b]			

[a] T_{cal} is the temperature of vacuum heat treatment of the catalyst in °C

[b] Only H_2 is released upon illumination

Dissociative excited triplet states of water molecules are found below the edge of optical absorption. For instance, the triplet state $^3B_2$ corresponds to an energy minimum of 5.11 eV (dissociation into H and OH); the state $^3A_2$ corresponds to 5.03 eV (dissociation into O and H_2).

More than 40 years ago TERENIN [9.62] suggested the possibility of a lowered dissociation threshold for adsorbed molecules. Reactions of decomposition can take different paths in the adsorbed phase than the same reactions would in the gaseous phase. These ideas are still being successfully developed by Terenin's disciples [9.63,64]. Their main concern has been with the search for adsorbents / catalysts active in photolysis of water.

This point is illustrated in Table 9.1, which presents data on the relative maximum rate of desorption A of the products of photolysis of water into H_2 and O_2 for certain especially active oxide semiconductors and insulators. All specimens were irradiated in a continuous mode with a high-pressure SVD-120A mercury lamp. The maximum quantum yield was 0.27 molecules $\cdot$ photon^{-1} for γ-Al_2O_3 ($\lambda \simeq 200$ nm), $2.2 \cdot 10^{-2}$ for BeO ($\lambda = 206$ nm), and 0.1 for HfO_2 ($\lambda = 185$ nm). The spectral dependence of the absolute quantum yield (with respect to hydrogen) of photolysis of water on disperse α-Al_2O_3 is presented in Fig.9.3. One sees that n increases rapidly with an increase in the energy of light quanta $h\nu$. Partial dehydration of the surface reduces the value of η in the region of lower-energy light quanta. The shape of curve $\eta(h\nu)$ is in qualitative agreement with variations in the coefficient of adsorption $K(h\nu)$. The rapid increase in η with $h\nu$ is attributed to different levels of excitation of Al_2O_3.

The results of this series of investigations lead to the following important conclusions:

1) Active photolysis of water occurs only in the far ultraviolet spectral range ($\lambda < 200$ nm, or $h\nu < 6$ eV); thus there is no considerable energetic advantage over photodissociation of molecules in a gas.

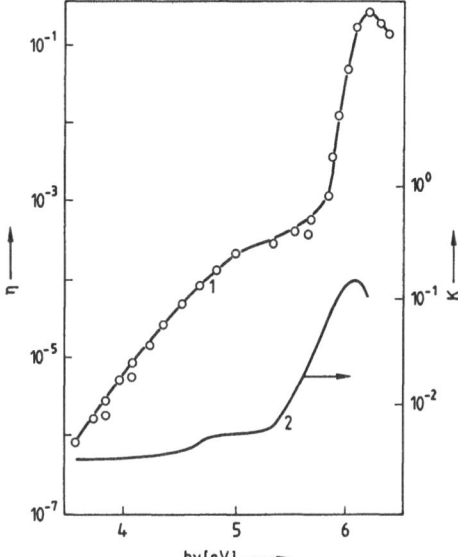

Fig.9.3. Spectral dependence of absolute quantum yield of hydrogen η in photolysis of water adsorbed on disperse α-Al$_2$O$_3$. (1) Adsorbent subjected to hydration in water vapor at 650 K for 6 hours. (2) Extinction coefficient K (in cm^{-1}) of α-Al$_2$O$_3$ single crystal [9.64]

2) Photolysis involves molecules that are firmly bound to the surface. The nature of the bonds is not discussed in [9.63,64].

3) The rate of photolysis depends on the status of the hydroxyl cover the oxide.

4) The rate of photolysis is a linear function of the light intensity which indicates that the process bears one-quantum character.

The spectral limit of photolysis of water on a dielectric (amorphous SiO$_2$) can be shifted towards visible and near UV spectral ranges (2 - 4 eV) by sensitizing the surface with adsorbed molecules of flavins [9.65] and aromatic substances (naphthalene) [9.66]. Then the reaction yield is a quadratic function of the number of light quanta which points to the two-quantum nature of the process. In the case of flavins (Fl), we are presumably dealing with photo-transfer of a proton from an H$_2$O molecule and the creation of an intermediate product, FlH$^+$ (first step), followed by phototransfer of an electron to FlH$^+$ (second step). In the case of naphthalene and other aromatic compounds, the shift in the spectral threshold is attributed to two-quantum excitation of an organic molecule into a high-excitation triplet state, with subsequent transfer of energy to an H$_2$O molecule, which then passes into a triplet dissociative state (E > 5 eV). Of all possible mechanisms of photosensitization, preference is given to resonant transfer of energy.

Let us now consider photolysis of water on real semiconductor surfaces. BASOV et al. [9.63,64] assumed that photolysis is a result of electron exci-

tation of the solid, caused by the effects of sufficiently hard radiation
($h\nu > 6$ eV). It is not clear what is meant by "electron excitation" in these
works.

In our earlier discussion of the experiments of Terenin and Kotel'nikov
on photolysis of H_2O and NH_3 on γ-Al_2O_3, we explained dissociation of molecules from the standpoint of the donor-acceptor mechanism of adsorption and
surface charging via the capture of light-generated holes on adsorptive states. The same ideas were used in our explanation of the dependence of photodissociation on the extent of surface hydration [Ref.9.20,Chap.3]. However, this
explanation remains speculative until direct electrophysical measurements become available. From Table 9.1 it is evident that dielectric and wide-band
semiconductors are most active in photolysis of water.

In [9.28,67-69] experiments were carried out on photolysis of water on single
crystals of classical semiconductors such as Ge, Si, and GaAs. The use of
single crystals allowed mass-spectroscopic analysis of the reaction products
to be supplemented with investigations of variations in the electrophysical
parameters of the surface caused by photolysis of adsorbed molecules. In order
to increase the level of injection of nonequilibrium charge carriers in the
semiconductor (and, consequently, to raise the reaction yield), the specimens
were illuminated in a pulsed mode with photoflash IFK-120. The pulse duration
was about 2 ms, and in this time the surface received about 10^{18} quanta per
square centimeter (with allowance for reflection). The ac field-effect technique (Sect.2.3) was used to measure the surface potential and plot the curves
of capture on fast states both before and after the flash. The photoconductivity
of the specimen was measured simultaneously with the flash. This effect has
been most thoroughly investigated for germanium and silicon for which we have
the most comprehensive information on the electrophysical properties of the
surface (Chaps.2,5). Let us discuss the main results of these experiments.

Investigations were carried out on <111> faces of high-resistance Ge and
Si, treated respectively in H_2O_2 and CP_4 (HNO_3, HF). Some of the Ge specimens
were treated in gaseous HF. The surface of Si was already well fluorinated
after treatment in fluorine-containing etchant. Pulsed illumination of a Ge
specimen covered with a hydrated GeO_2 film (peroxide etchant) with unfiltered
light caused H_2 to be released in a very small quantity (Fig.9.4, Curve 1). The
signal from hydrogen (n_{H_2}) was much greater in the case of fluorinated Ge specimens (Fig.9.4a). No signal from oxygen could be detected by mass spectrometer.
In the case of Si specimens, the release of H_2 was accompanied by liberation of
O_2 and H_2O_2. A considerable release of H_2 was observed at $n_{qu} \geqslant 5 \cdot 10^{20}$
quanta $\cdot$ cm^{-2} $\cdot$ s^{-1}. As follows from Fig.9.4a, dehydration of the surface caused

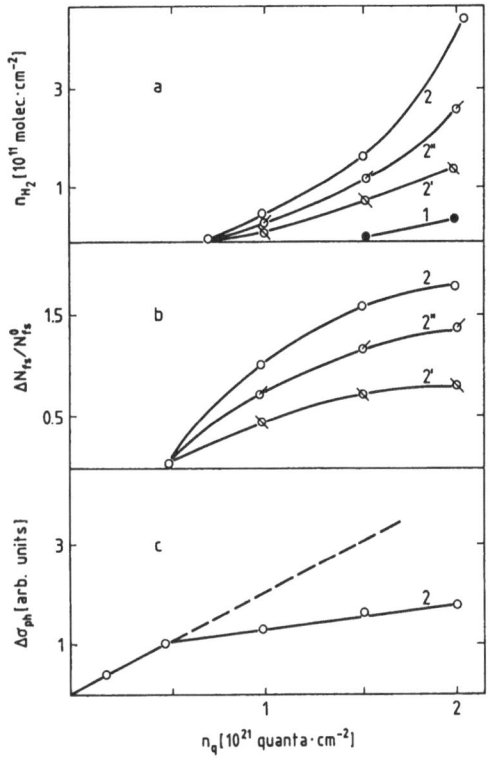

Fig.9.4. (a) The quantity n_{H2} of hydrogen released in pulsed illumination of hydrated (*1*) and fluorinated (*2*) surfaces of a germanium single crystal as a function of the number of incident light quanta n_q, the specimens having been vacuum heat treated at 300 K (*2*), 500 K (*2'*) and rehydrated in saturated water vapor (*2"*). (b) Relative variations in the density of acceptor-type fast surface states N_{sf}/N^0_{sf} upon illumination of fluorinated germanium surfaces [designations the same as in (a)]. (c) Steady-state photoconductivity versus n_q

a decrease in n_{H2}, while rehydration restored the effect to some extent. Subsequent adsorption and desorption of water vapor brought about reversible changes in n_{H2}.

The data on the field effect indicate that with fluorinated Ge specimens, the release of hydrogen is accompanied by a considerable increase in the relative density N_{sf}/N^0_{sf} of acceptor-type fast states in the upper half of the forbidden band of germanium (N^0_{sf} is the density of fast states prior to illumination, Fig.9.4b). After the release of H_2, the surface of Ge acquired additional negative charge ΔQ_s whose value was about an order of magnitude lower than n_{H2}. In the case of Si, the variation in N_{sf} and Q_s lie within the limits of experimental error.

We observed in Sect.8.1 that the mean quantum yield η_{H2} of photodissociation of water molecules on the surface of silicon in the explored spectral range increases with an increase in the energy of light quanta $h\nu$, while the quantum yield of electron-hole pairs $n_{n,p}$ ($n_{n,p} = 1$ [9.70]) and surface charge Q_s remain constant (Fig.8.2). This experiment does not fit the model of the elementary act of catalysis in VOLKENSTEIN's electron theory of adsorption [9.71,72], which attributes all variations in the adsorptive and catalytic activity of the surface solely to variations in the charge status of adsorption centers due to illumination. A similar increase in n_{H2} as $h\nu$ increases was observed in experiments with Ge [9.67]. Small variations in surface charge in this case are associated with the increased density of fast states due to illumination which is also overlooked in [9.71,72]. The release of H_2 in [9.67,68] was observed starting with $h\nu \geqslant 1.5\ E_g$ (where E_g is the width of the forbidden band of germanium or silicon), and not with $h\nu = E_g$, as would follow from VOLKENSTEIN's theory [9.71]. An increase in photoadsorptive activity with an increase in $h\nu$ (at $h\nu > E_g$) was also reported for other adsorptive systems (e.g., CdS-CO_2 [9.73] and TiO_2-CH_4 [9.74]), without any reference to the underlying causes. The observed regularities can be fully explained within the framework of the electron-vibrational model of adsorption and capture evolved in Chap.8. The number of vibration-excited molecules increases with an increase in $h\nu$. It must be remembered that in the case of a heterogeneous real surface we are dealing with a rather wide spread in bond energies and frequencies of valence oscillations of the adsorbed molecules, as is directly indicated by the wide absorption bands of valence oscillations in the IR spectrum of adsorbed molecules [9.20]. According to the electron-vibrational model, this fact is reflected in the distribution of activation energies of the capture of charge carriers on slow adsorptive states ΔE_τ. It was demonstrated in [9.75] that the distribution of ΔE_τ over the surface is close to Gaussian and its halfwidth is close to the halfwidth of the absorption bands of absorbed molecules.

Measurements of steady-state photoconductivity (dc field mode) lead to the following conclusions [9.76] (see Fig.9.4c):

1) The photoconductivity σ_{ph} of nonfluorinated Ge specimens dehydrated at 470 K is a linear function of the number of quanta n_{qu} received by the surface.

2) The curve $\sigma_{ph}(n_{qu})$ for fluorinated Ge specimens has a break at $n_{qu} = n' \approx 5 \cdot 10^{20}$ quanta $\cdot$ cm^{-2} $\cdot$ s^{-1}.

3) The breakpoint on $\sigma_{ph}(n_{qu})$ corresponds to the starting point for the release of H_2 from the Ge surface (as registered by mass spectrometer).

4) Dehydration of the surface of Ge specimens at 470 K lowers the slant d_{ph}/dn_{qu} at $n_{qu} \approx n'$ and the value of n_{H2} while rehydration tends to restore the initial values. Similar effects were also observed with Si specimens.

Since the mobility of charge carriers in the investigated range of n_{qu} does not change, the function $\sigma_{ph}(n_{qu})$ is equivalent to the function describing the concentration of light-generated nonequilibrium charge carriers $\Delta n(n_{qu})$ (Fig. 9.4c). The decrease in the slant of lines $\Delta n(n_{qu})$ at $n_{qu} = n'$ indicates a decrease in the lifetime of nonequilibrium carriers, or the actuation of an ad-

ditional channel of surface recombination. This additional channel was believed
in [9.76] to be associated with the participation of the products of heterolytic dissociation of adsorbed molecules in the process of recombination, and
was appropriately labeled "the chemical channel of recombination".

According to our donor-acceptor mechanism of charging (Sect.6.6), the most
natural centers of chemical recombination and sources of H_2 are the coordination-
bound H_2O molecules on the semiconductor-oxide interface. These molecules form
the basis of slow adsorptive states (Sect.6.4). The probability of dissociation
of such molecules is

$$w = w_0 \exp(-E_a / kT) \quad .$$

Fluorination of the surface (as demonstrated in Sect.6.6 and [9.28]) leads to
deformation of coordination-bound molecules and to a reduction of activation
energy E_a by ΔE_1. Qualitative estimates based on the results of experiments
on thermal desorption [9.28] indicate that $\Delta E_1 \sim 1.5 - 2$ eV. Apparently, a large
number of these deformed molecules act as sources of hydrogen. The AES data
for fluorinated Ge specimens [9.68] indicate that the density of F groups on
the surface of these specimens differs little from n_{H_2} (Fig.9.4a).

The molecules of H_2O create donor-type states which become positively charged
(Sect.6.4). As follows from Sect.7.2, the Coulomb field of positive charge
leads to further drawing of an unshared pair of electrons from an H_2O molecule
to an adsorptive center, to lengthening of O-H bonds, and to increased ionicity
of bonds in the adsorption complex. The latter further lowers the energy of
heterolytic dissociation of a molecule by the value ΔE_2.

The adsorptive states are slow states with low capture cross sections. They
are occupied only at a high level of injection, $\delta = \Delta n/n$, see (7.1). In the
experiments in question, the value of δ reached 10^3. Figure 9.4a shows that
the hydrogen yield n_{H_2} greatly increases with an increase in n_{qu} (or, which
is the same, δ).

The following possible schemes for the processes taking place upon pulsed
illumination of semiconductors were proposed in [9.28,67-69]:

1) $\geq Si : O\begin{smallmatrix} H \\ \diagup \\ \diagdown \\ H \end{smallmatrix} + n + p + h\nu_v \rightarrow \geq Si + H^+ + OH^- + p + n \rightarrow \geq Si + H^\cdot + OH^\cdot$,

2) $H^\cdot + H^\cdot \rightarrow H_2\uparrow$,

3) $OH^\cdot + OH^\cdot \rightarrow H_2O_2\uparrow$,

4) $H_2O_2 \rightarrow H_2O + O\uparrow$.

Here $h\nu_v$ is a quantum of vibrational excitation of a molecule. If $h\nu_v \geqslant E_a - \Delta E_1$ $- \Delta E_2$, then the molecule undergoes heterolytic dissociation.

In the case of silicon some of the OH radicals recombine to form H_2O_2, which is detected in mass spectra. Some of the H_2O_2 molecules dissociate into H_2 and O_2. The experimentally observed relation $H_2 : H_2O_2 : O_2$ in mass spectra differs little from the theoretically predicted yield of the proposed chain of reactions. In the case of germanium only H_2 is released. Some of the very reactive OH radicals modify the Ge-GeO$_2$ interface, as indicated by the growth in concentration of fast states (Fig.9.4b). The chemical channel of recombination is associated with interactions between charge carriers on the one hand and adsorptive states and products of dissociation of adsorbed molecules on the other. At high levels of excitation of the electronic subsystem, slow traps can act as centers of delayed recombination [9.77]. The said states exchange charge carriers with the semiconductor's bands mainly via the tunnel mechanism. The quantum yield of H_2 remained almost unchanged when the ambient temperature varied from 300 to 77 K [9.67].

The proposed mechanism of photolysis is sufficiently general. Its basic distinction from all existing electron models of photocatalysis and photodissociation is that vibrational excitations of adsorbed molecules are taken into account as a necessary stage of dissociation. The energy needed for excitation comes from acts of recombination of charge carriers on adsorptive states. Our donor-acceptor electron-vibrational model of adsorption and surface charging [9.20,28,67,68,77,78] is clearly confirmed by the isotopic effect in photolysis of H_2O and D_2O molecules [9.79]. To make the effect of photolysis more pronounced, a silicon specimen covered with a monolayer of Cr_2O_3 deposited by the molecular layering technique (see below) was used. Prior to vacuumization the specimen was subjected to prolonged deuterization in heavy water vapor (80% D_2O + 20% H_2O). As seen in Fig.9.5, the function $\eta_{D_2}(h\nu)$ is shifted towards lower values of $h\nu$ with respect to the corresponding functions for H_2 and HD. Here one can draw a parallel with isotopic shifting of the absorption bands of coordination-bound D_2O molecules with respect to those of H_2O molecules in the IR spectrum [9.20]. Because H_2O and D_2O molecules have the same electron structure, they create similar sets of donor-acceptor complexes whose effective levels in the forbidden band of the crystal, ε_t, are the same. The isotopic shifting of $\eta(h\nu)$ and the isotopic effect on the time of relaxation of charge in slow adsorptive states (Sect.8.2) are not explained in any of the earlier proposed electron mechanisms of adsorption and catalysis (Chap.5). The observed isotopic shifts follow directly from the electron-vibrational model of adsorption.

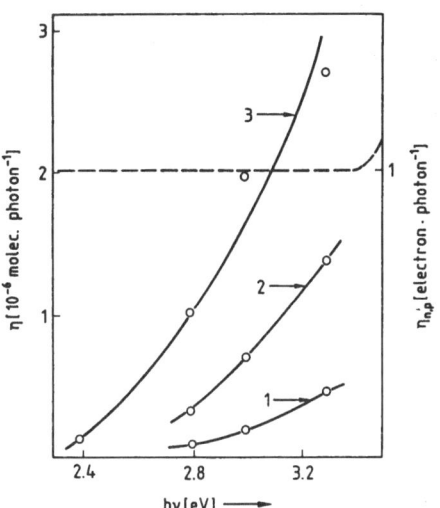

Fig.9.5. Spectral dependence of mean quantum yield $\eta(h\nu)$ of H_2 (1), HD (2), and D_2 (3) in pulsed illumination of a single crystal of silicon, doped on the surface with chromium oxides. The dashed line shows the quantum yield of electron-hole pairs $\eta_{n,p}$

The effect of photolysis was also discovered later for other molecules adsorbed according to the donor-acceptor mechanism. For illustration let us consider photodehydrogenation of formic acid on a real silicon surface [9,80]:

$$\geqslant Si : HCOOH + h\nu \rightarrow H_2 + CO_2 \quad .$$

It was found that if adsorption of HCOOH takes place on a hydrated silicon surface having about 10^{13} coordination-bound water molecules per square centimeter which act as proton-donor centers [9.20], then pulsed illumination stimulates photodehydration:

$$\geqslant Si + HCOOH + h\nu \rightarrow H_2O + CO \quad .$$

This confirms the conclusion made in [9.77] regarding the role of free holes in acid-base catalytic reactions and agrees with the earlier discussed experiments on decomposition of HCOOH on the surface of p-n junctions (Sect.7.2)

In order to obtain further confirmation of the possibility of controlling the activity of proton-donor surface centers, GOLOVANOVA et al. [9.81] did experiments on photolysis of CCl_4 on real silicon surfaces. This reaction is known to proceed on proton-donor (Brönsted) centers. In [9.81] it was demonstrated that pulsed illumination with low-energy quanta ($h\nu \approx 2$ - 3 eV) leads to decomposition of CCl_4 with release of H_2, CO_2, HCl, and $COCl_2$. The dark reaction of decomposition of CCl_4 occurs at 480-700 K with activation energy $E_a^d = 0.6$ eV. The activation energy of photodesorption of the products of decomposition was found to be one-third of that value: $E_a^{ph} = 0.2$ eV. Photodissociation of CCl_4 was observed even at 77 K. It was demonstrated that the oxygen required for the production of CO_2 and $COCl_2$ comes from dissociation of coordination-bound H_2O molecules. Measurements of photoconductivity reveal that as in photolysis of water and HCOOH, decomposition of CCl_4 leads to the appearance of an additional chemical channel of surface recombination.

So far we have been considering photolysis in a vacuum when only firmly bound individual adsorptive complexes are present on the surface. It was pointed out in [9.67,68] that the presence of even a small concentration of weakly bound molecules nullifies the effect. According to Sect.8.2, the appearance of clusters allows some of the energy released in recombination to be dissipated into the molecular phase. The vibrational energy retained by the complex may be insufficient to start dissociation. On the other hand, because water in the adsorbed phase has a high dielectric constant, the dissociation constant for H_2O molecules must have a somewhat lower value. Apparently, the products of dissociation (H^+ and OH^-) interact with weakly bound water molecules, creating ions $H_5O_2^+$, $H_3O_2^+$, and the like which migrate over the surface altering the parameters of surface states and thereby affecting the electrophysical parameters of the surface (Sect.9.3).

Under certain conditions the suggested path of dissociation of adsorbed molecules (in particular, H_2O), which involves electrons and holes of the semiconductor can take place in the dark. BUTYAGIN and KOLBANEV [9.82] used this reaction to explain the mechanism of interaction between a newly created silicon surface and water during mechanical crushing. They detected the release of H_2 and H_2O_2 when the cleavage surface was wetted with water. The semiconductor's electrons and holes were assumed to be responsible for the generation of H_2O_2. Thermochemical estimates made in [9.82] indicate that in the presence of water film (owing to the high energy of hydration of H^+ ions) the dissociation of an H_2O molecule, participated in by electrons and holes of the semiconductor, becomes thermodynamically advantageous. This supports our conclusion regarding the creation of charged proton-containing ions in the adsorbed phase.

These phenomena are equally interesting for biophysics, and above all for the problem of photosynthesis, tissue respiratory processes, etc.) depend on combined electronic and ionic processes [9.83,84]. Experiments on photolysis of water on semiconductor surfaces furnish convenient models for studying the concatenation between electronic and molecular processes on the surface.

Illumination of chlorophyll and its man-made analogues, phthalocyanines, leads to positive charging of the surface (capture of holes) [9.56,59]. It is possible that photolysis of water further proceeds according to the donor-acceptor mechanism, as in the case of germanium. The presence of water or an electrolyte, owing to the high permittivity of the medium, will facilitate dissociation of protonized water molecules into H^+ and OH^-. The importance of holes in photolysis of molecules on semiconductor-electrolyte interfaces has been confirmed by experiments on decomposition of formic acid [9.85] and water. On the other hand, illumination of carotene films causes a buildup of negative charge [9.59]. The individual stages of the complicated processes which take place in contact areas between particles of chlorophyll and carotene (typical components of chloroplasts) can be modulated by simple processes in photodiode structures. Recall the experiments on decomposition of formic acid on germanium p-n junctions discussed in Sect.7.2.

Kinetic data on natural systems of photosynthesis indicate that at low temperatures (up to about 100 K) the rate of electron transfer does not depend on temperature [9.84]. DE VOULT and CHANCE [9.86], and later also other authors ascribed this to the tunnel mechanism of electron transitions. In [9.67], the signal from hydrogen was found to be independent of temperature in the range 77-300 K which was attributed to the tunnel mechanism of transfer of the charge carrier from the semiconductor to the center of photolysis of water.

The greatest activity in the decomposition of donor-type molecules (e.g., water) should be displayed by transition-metal atoms, which are found in many enzyme systems. BLUMENFELD [9.84] pointed out the role of Mn^{2+} ions in chloroplast-II photosystems in the genesis of O_2 molecules during photolysis of water. Atoms of Ge, even complexed with halogens, are less active than atoms of transition metals. Direct proof of strong deformation of water molecules on the surface of a typical complex-forming element, Ru, was obtained with the aid of ESDIAD (electron-stimulated desorption ion angular distribution) [9.87]. Estimates made in [9.88] indicate that the H-O-H angle in an H_2O molecule chemisorbed on Ru is increased by $3.1 \pm 0.5°$. This is consistent with our NMR data on deformation of H_2O molecules on oxide surfaces [9.20,28]. Active photolysis of water on a Ru-Co composite in the visible spectral range was observed in [9.89]. Experiments with semiconductors whose intrinsic oxide film is replaced by transition-metal oxides offer much promise as regards simulating systems of photosynthesis. This can be done with the molecular layering technique developed by ALESKOVSKY [9.90]. Investigations of the electrophysical characteristics of such surfaces have already been conducted [9.91]. Ions of transition metals can be introduced into the oxide layer of semiconductors (Ge, Si, etc.) by ion implantation. The photocatalytic activity of IS structures can be improved if ions in the insulator layer exchange charges with the allowed bands of the semiconductor.

As was pointed out in [9.83,92], IS, MIS, and EIS (electrolyte-insulator-semiconductor) structures are the most promising as regards simulating individual synthesis and receptor features of biological membranes. If the dielectric layer in sufficiently porous which is sometimes the case in real structures, then environmental molecules can diffuse through the pores and modify the parameters of electron states in the insulator and on the insulator-semiconductor interface (Sect.4.3). In particular, an immobilized enzyme system, sewn by chemical bonds to the semiconductor surface, can act as a dielectric layer. If a way can be found to control the charge status of active centers of the dielectric or interface by injecting charge carriers from the semiconductor or electrolyte, it will become possible to regulate dissociation and catalysis in such systems.

Of especial interest is double injection in MIS structures (simultaneous injection from the metal and the semiconductor). Here the volt-ampere characteristics contain nonlinear regions (S-shaped and N-shaped characteristics). Such systems can be employed as analogue oscillatory systems, generating the products of a catalytic reaction. Recall that synchronous oscillations of catalytic activity were discovered in real enzyme systems. Autooscillations were also observed in heterogeneous catalytic oxidation and have been extensively studied [9.93,94].

From a biological standpoint, the most controversial questions concern the mechanism of transport of charge carriers in protein enzyme systems and membranes. Electron, exciton, polaron, and proton mechanisms have been proposed [9.84,95], and are often contrasted with one another. As we pointed out in [9.83], this controversy arises mainly from a too "classical" understanding of semiconductor conduction as conduction over allowed bands. These concepts formulated in the 1930s and 1940s by JORDAN [9.96] and SZENT-GYORGUI [9.97], have remained unchanged despite the rapid development of the electron theory of semiconductors, and are used even now. Biological membranes contain highly ordered and liquidlike biopolymerous phases [9.98]. The theory of disordered systems, which assigns an important role to the hopping mechanism of charge movement over localized states, has not yet been applied to biological objects. However, such concepts are now being recognized in investigations of organic semiconductors [9.99]. The distinctive features of electron transport in biological membranes where the electric fields can be as strong as 10^5 V$\cdot$cm^{-1}, are discussed in [9.83]. To conclude this section, let us make one final remark. In stimulated photocatalysis and photolysis on semiconductor surfaces and in IS structures, the importance of electron processes can be paralleled by phonon excitations and the transfer of energy released in capture and recombination to vibrational

modes of adsorbed molecules (Sects.8.1,2). These processes, together with electron transitions can be of great importance in biological enzymic reactions,
in view of the extremely wide range of vibrational modes of bands in bioorganic
molecules.

9.3 The Effects of Protonic Processes on the Parameters of Surface States

As we demonstrated above, the H^+ and OH^- ions resulting from dissociation of
coordination-bound water molecules can greatly affect the parameters of surface
states. In order to varify this conclusion, systematic investigations were conducted in [9.100] on the effects of adsorption of polar donor-type molecules
on the rate of surface recombination, capture on fast states, and the surface
potential of real germanium surfaces. The adsorbates studied were water
($\mu = 1.8$ D), NH_3 ($\mu = 1.5$ D), pyridine Py ($\mu = 2.2$ D), acetone Ac ($\mu = 2.8$ D),
and acetylacetone Aca ($\mu = 2.8$ D), whose adsorption follows the donor-acceptor
mechanism [Ref.9.20,Chap.4].

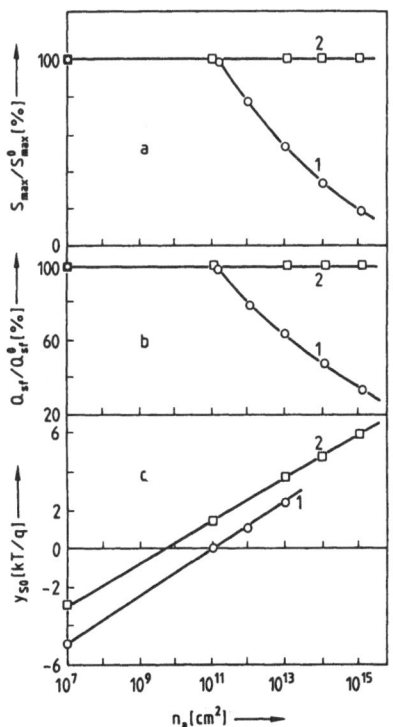

Fig.9.6. Surface recombination rate
S_{max}/S_{max}^0 (a), charge in fast states
Q_{sf}/Q_{sf}^0 (at $Y_s = 4$ kT/q) (b), and
surface potential Y_{s0} (c) versus
concentration n_a of H_2O (1) and
NH_3 (2) molecules adsorbed on germanium

235

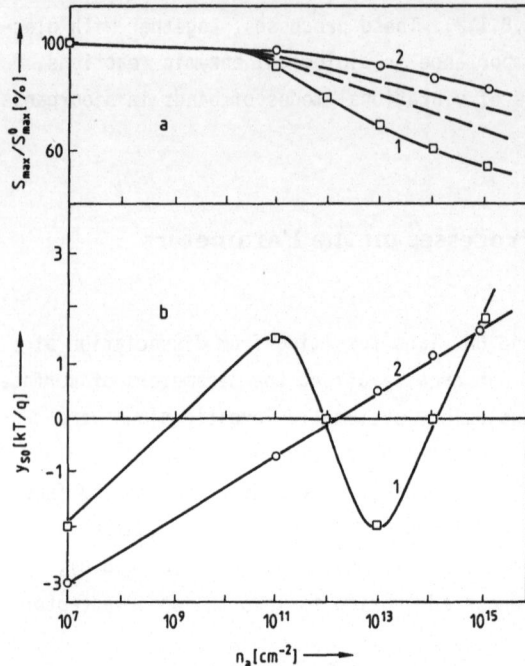

Fig.9.7a,b. Surface recombination rate S_{max}/S_{max}^0 (**a**) and surface potential Y_{s0} (**b**) versus concentration n_a of Aca (*1*) and Ac (*2*) molecules adsorbed on germanium

The relative values of the maximum rate of surface recombination S_{max}/S_{max}^0 and the charge in fast surface states Q_{sf}/Q_{sf}^0 (at $Y_s = 4$ kT/q) as functions of the concentration of adsorbed molecules n_a are plotted in Figs.9.6 - 8; the values S_{max}^0 and Q_{sf}^0 refer to the initial germanium specimen vacuumized at 500 K prior to adsorption. The same figures display curves $Y_s(n_a)$. Unlike the previously considered curves $S(n_a)$, $Q_{sf}(n_a)$, and $Y_s(n_a)$ (Sect.6.2, Fig.6.2), these curves were obtained with specimens in an ultrahigh vacuum ($\sim 10^{-8}$ torr), at a very low concentration of adsorbed molecules. The data in Figs.9.6 - 8 allow the following conclusions to be made:

1) Up to occupancies of about 10^{11} cm^{-2}, adsorption of all the molecules named has no effect on S_{max} or Q_{sf}. As we indicated in Sect.3.1, this implies the absence of direct interactions between adsorbed molecules and defects, which form the basis of centers of recombination and fast capture.

2) In contrast to Q_{sf} and S_{max}, surface potential Y_s starts to increase rapidly with the first admissions of vapors of the substances studied[4]. Since Q_{sf} remains constant, the growth of Y_{s0} is entirely determined by the increase

---

[4] In the case of Aca the N-shaped section on the $Y_{s0}(n_a)$ curve is apparently associated with the tautomeric effect. According to IR spectroscopy [9.101], the transition from adsorption of the enol form to adsorption of the keto form of Aca occurs at the same occupancy.

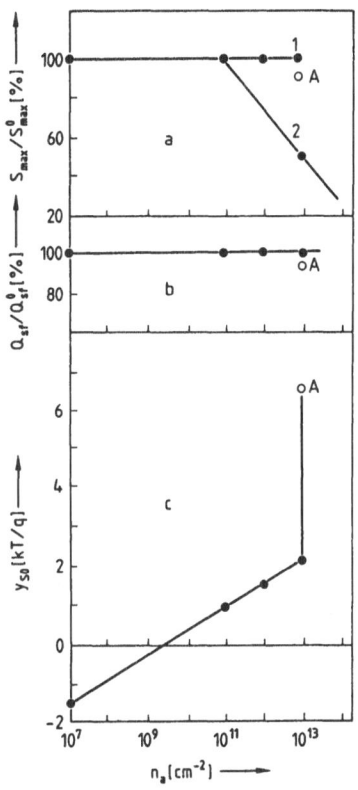

Fig.9.8a-c. Surface recombination rate S_{max}/S_{max}^0 (a), charge in fast surface states Q_{sf}/Q_{sf}^0 (at $Y_s = 4$ kT/q) (b), and surface potential Y_{s0} (c) versus concentration n_a of Py (1) and water molecules (2) adsorbed on germanium. Point A: adsorption of water on Ge surface with preadsorbed pyridine

in positive charge in slow adsorption states Q_{ss}. It is natural to assume that the most active primary centers of adsorption in the initial stage of occupation are those defects which form the basis of slow states.

3) Starting with $n_a = 10^{11}$ cm^{-2}, adsorption of H_2O and Aca results in a decrease in S_{max} and Q_{sf}. In the case of H_2O the changes are similar to those depicted in Figs.6.2,4, observed at higher occupancies. Adsorption of other molecules had either little effect (10 or 20 percent in the case of P_y and Ac) or no effect at all (e.g., NH_3) on the values of S_{max} and Q_{sf}.

The experimental data presented in Figs.9.6-8 provide another argument against the model that assumes a purely dipole mechanism of neutralization of centers of recombination and fast capture (Sects.3.1,6.2). Indeed, although H_2O and NH_3 have close values of dipole moment, their adsorption causes entirely different variations in S_{max} and Q_{sf} in each case (Fig.9.6). The same can be said of adsorption of other molecules.

The neutralization of centers of fast capture and recombination (drop in S_{max} and Q_{sf}) can be much more easily explained by means of protonic processes. According to Sects.9.1,2, the protons can be supplied by slow adsorptive states that have captured holes. It is significant that a noticeable drop in S_{max} and Q_{sf} during adsorption of H_2O (Fig.9.6) occurs at occupancies at which groups of adsorbed molecules start to fuse into polymolecular clusters. According to NMR data, intensive proton exchange takes place within such clusters. From the data presented in Sect.9.1 it follows that the dissociation constant for coordination-bound water is five orders of magnitude higher than that for ordinary water. Some of the protons that result from dissociation of water migrate over the surface to neutralize the centers of fast capture and recombination. By virtue of their small size, protons can creep through the imperfect lattice of silicon-oxygen (germanium-oxygen) tetrahedra into the region of heterojunction where the centers of recombination and fast capture are found (Fig.6.15).

Neutralization upon adsorption of Aca (Fig.9.7) is also connected with the presence of protons on the surface. These protons are generated during adsorption of the enol form of Aca, when Aca forms positively charged chelate complexes with hydrated surface germanium atoms [9.101].

In contrast to Aca, adsorption of acetone Ac (Fig.9.7) having the same value of dipole moment had little effect on S_{max}. Adsorption of Py and NH_3 whose dipole moment is smaller, did not change the values of S_{max} and Q_{sf} at all (Figs. 9.6,8). Dissociation of NH_3, Py, or Ac giving rise to H^+ is hardly probable. Recall that the dissociation constant for NH_3 molecules in liquid ammonia is 20 orders of magnitude smaller than that for water. Adsorption of ether did not affect S_{max} or Q_{sf}.

To substantiate the dipole mechanism of neutralization of centers of surface recombination, the authors of [9.102-104] provided data on adsorption of nitrobenzene and chlorobenzene polar molecules on germanium. In view of the high hygroscopicity of aromatic compounds, the observed effect of neutralization can be attributed to the presence of water molecules which are very hard to get rid of by simple vacuum distillation of the adsorbate (without zeolite traps) in installations containing vapors of organic grease. In fact, in experiments with acetone (Fig.9.7) the effect of neutralization became considerably lower judging from the value $(S_{max}^0 - S_{max})/S_{max}^0$ as the number of runs of adsorbate through a low-temperature zeolite trap increased (Fig.9.7, dashed curves; Curve 2 corresponds to six runs).

Direct proof of the protonic mechanism of neutralization was obtained in experiments on adsorption of Py. Molecules of Py are known to possess highly pronounced proton-acceptor properties. According to IR spectroscopy, PY interacts with coordination-bound molecules of water to produce PyH^+ ions [Ref.9.20, Chap.4]. As seen in Fig.9.8, adsorption of Py has almost no effect on S_{max} and

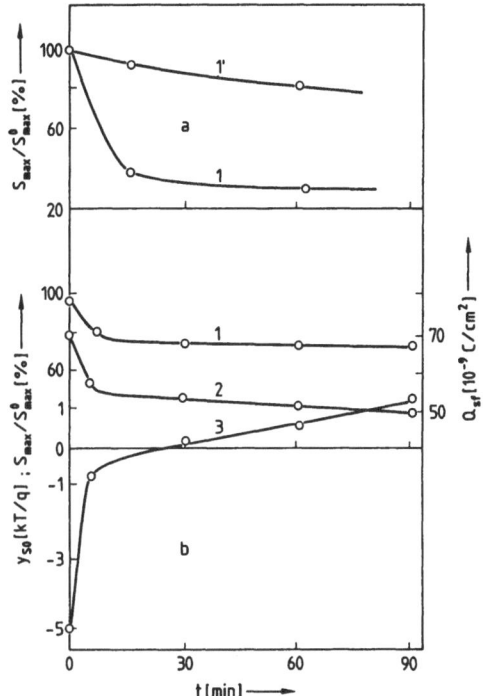

Fig.9.9. Kinetics of change in surface recombination rate S_{max}/S_{max}^0 $(1,1')$, change in fast states at $Y_s = 4$ kT/q (2), and surface potential (3) of germanium during adsorption of water; $(1')$ the same width preadsorbed pyridine. (a) $P_{H_2O} = 10^{-1}$ Torr, (b) 10^{-2} Torr

Q_{sf}. Admission of water vapor (at $p = 10^{-1}$ torr) to a surface with preadsorbed Py caused S_{max}/S_{max}^0 to decrease by only 10 to 20 % (Point A) while adsorption of the same quantity of water on the initial specimen reduced S_{max}/S_{max}^0 by 40 or 50 % (Fig.9.8, Curves 2). Similar passivative action of Py was also observed in measurements of Q_{sf}. When water was adsorbed on a surface with preadsorbed Py, the kinetics of S_{may}/S_{max}^0 as a function of time was much slower than when water was adsorbed on the initial specimen which must be attributed to decelerated diffusion of heavier PyH$^+$ ions (as compared with H$^+$ ions) towards centers of recombination (Fig.9.9).

Investigations of the kinetics of variation in $S_{max}(t)$, $Q_{sf}(t)$, and $Y_{s0}(t)$ during adsorption of H_2O in the region of low values of occupancy reveal that the rate of variation in these quantities increases in the order S_{max}, Q_{sf}, Y_{s0} (Fig.9.9). DORDA [9.105] associated the decelerated kinetics of Q_{sf} and Y_s with the diffusion of H_2O molecules through the oxide film. NOVOTOTSKY-VLASOV [9.104] considered the limiting stage of neutralization to be the process of fixing of dipoles near recombination centers.

In [9.100] it was demonstrated that the kinetic curves $S_{max}(t)$ for adsorption of both H_2O and Aca in the range of occupancies between 10^{12} and 10^{13} molecules $\cdot$ cm^{-2} are approximated sufficiently accurately by the expression

$$S_{max} = S_{max}^{\infty}[1 - \exp(-\alpha t^{0.5})] \quad , \tag{9.1}$$

where α is a coefficient that depends on the concentration of recombination centers ($N_r \leqslant 10^9$ cm^2) and the coefficient of surface diffusion. The same kind of equation was derived by BALAGUROV and VAKS [9.106] for random movement of particles on the surface with chaotic (Poisson) distribution of traps.

It was found that in the coordinates of this equation the kinetic curves $S_{max}(t)$ for H_2O and Aca coincide, although the masses of these molecules differ by a factor of five (Fig.9.10). Should the limiting stage be represented by diffusion of the molecules themselves, the bigger Aca molecules would be associated with slower kinetics (as represented by the calculated line 3 in the figure). The fact that the kinetic curves for H_2O and Aca coincide indicates that the same particles are responsible for neutralization in both cases. In light of the arguments developed above, these particles are most probably protons.

So far, in considering neutralization we have been exploring only the case of interaction between protons and discrete centers of recombination and capture. Many features of recombination and capture can be explained even more consistently (Sects.3.2,6.7) using the concept of a quasi-continuous energy spectrum which is typical of disordered systems. The protonic mechanism of neutralization is not bound to any particular type of spectrum. With a disordered surface, protons migrating over it will affect the amplitudes of ran-

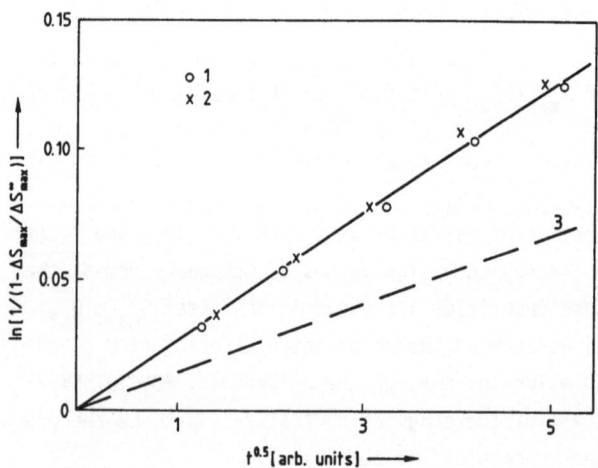

Fig.9.10. Kinetics of surface recombination rate in coordinates of (9.1) after adsorption of water (1) and Aca (2); calculated line (3)

dom fields and the probabilities of transitions in the quasi-continuous spectrum, thereby modifying the functions $S(Y_s)$ and $Q_{sf}(Y_s)$.

The ionic mechanism of neutralization evolved in [9.100] is sufficiently universal and can play an important role in MIS structures. Then neutralization can involve not only protons but also dopant atoms of alkali metals, found near the dielectric-semiconductor interface. FITZGERALD and GROVE [9.107] observed a considerable increase in the recombination rate when silicon-based MIS structures were treated thermoelectrically (negative potential was applied to the metal). According to the ionic mechanism of neutralization, this can be due to the removal of positive ions from the insulator-semiconductor interface.

It would be premature to speculate on the kinds of bonds developed by protons and other ions with defects since our knowledge of these defects is very limited (Sect.6.7). The parameters of defects can be modified both by the action of ions' Coulomb fields (Frenkel effect) and by the creation of chemical complexes. Much attention is currently being given to the idea that hydrogen atoms might passivate surface defects [9.108-111]. Treatment of silicon single crystals in hydrogen plasma improves the reverse bias leakage currents of p-n junctions [9.108], lowers the density of surface states generated by laser irradiation [9.109], and reduces the rate of surface recombination in polycrystalline silicon [9.110]. The authors of [9.111] attributed the reduction in the density of states on a silicon surface oxidized by the "chlorine" procedure (using HCl) to the interaction $Si^{3+} + H^{\cdot} \rightarrow SiH$. The authors of [9.111,112] reported that heat treatment of $Si-SiO_2$ structures results in a reduction in the density of surface states by an order of magnitude (to $5 \cdot 10^8$ cm^{-2}).

The influence of proton on the electrophysical parameters of real Si and Ge surfaces is not restricted to neutralization of fast and recombination states. BARDEEN and COOVERT [9.113] indicated long ago that prolonged application of an electric field to a semiconductor causes a buildup of slowly relaxing surface charge. When the field effect is studied using an ac or pulsed field, the positive charge accumulated when the field is on does not have time to relax during the break. This causes the field-effect curve to shift and deform. Charge accumulation was investigated by RZHANOV [9.103] and DORDA [9.114]. Rzhanov associated this effect with field-stimulated adsorption (electroadsorption, Sect.5.1) while Dorda assumed the accumulation to be connected with ejection of electrons from slow states into the semiconductor's conduction band. The accumulation effect is not limited to Ge and Si; it has also been observed on the surface of GaAs [9.115].

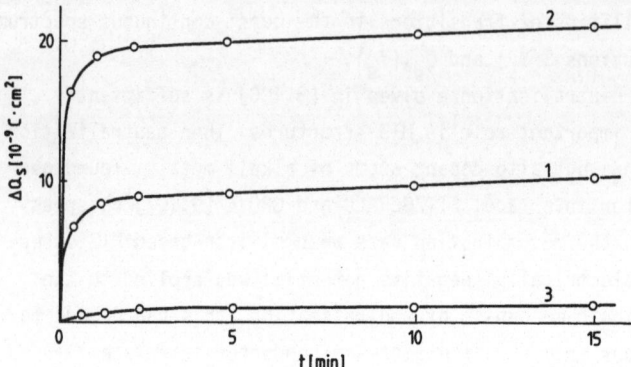

Fig.9.11. Kinetics of accumulation of surface charge Q_S on Ge upon application of 380 V sinusoidal transverse voltage (*1*), and with 60 V dc bias voltage (*2,3*). Field electrode negative (*2*) and field electroide positive (*3*). Pressure of water vapor p = 0.1 torr

Accumulation has been studied during adsorption of water, ammonia, ether, and pyridine on the surface of germanium [9.116,117]. As can be seen in Fig. 9.11, the application of a transverse electric ac field to the surface of a single germanium crystal resulted, in conformity with the cited works, in the accumulation of positive charge Q_S. When the ac field was biased with a permanent field (negative field electrode, which is equivalent to enriching the surface with holes), the value of Q_S increased (Curve 2). When the surface was enriched with electrons, the accumulation was quite low (Curve 3). The accumulation was zero when the experiment was carried out in an atmosphere of NH_3, $(C_2H_5)_2O$, and pyridine, up to high pressures of vapors.

As we demonstrated in Sect.7.2, mass-spectroscopic investigations carried out under conditions of nonzero accumulation effect failed to reveal electro-desorption upon application of electric fields of varying polarity to germanium specimens. The mechanism proposed by Dorda also does not explain the accumulation effect. As follows from Sect.6.4, adsorption of H_2O and NH_3 vapors creates slow adsorptive states having quite similar parameters (Table 6.1). The accumulation effect can be much more naturally explained from the standpoint of our donor-acceptor mechanism of adsorption and charging. When ac field effect is biased with a dc field (minus on the field electrode, when holes on the surface are more plentiful than electrons), coordination-bound water molecules tend to dissociate:

$$\equiv Ge : O\!\!\begin{array}{c} H \\ \diagup \\ \diagdown \\ H \end{array} + p \rightarrow \; \equiv Ge + H^+ + OH^- \quad .$$

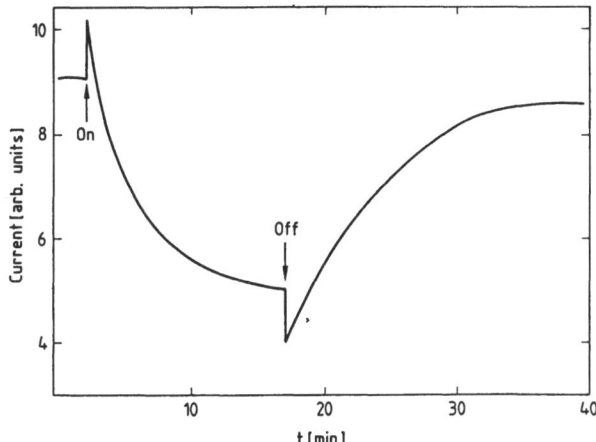

Fig.9.12. Kinetics of dark current and photoelectric current in anatase in an atmosphere of oxygen and water vapor at 23°C [9.18]. (Reproduced by permission of the authors)

Protons are retained by Coulomb forces in the heterojunction and do not leave the surface which is consistent with the data supplied by mass spectroscopy, while OH groups further oxidate the interface. The latter fact apparently explains the variations in the density of fast states as accumulation proceeds [9.103]. In contrast to adsorption of water, adsorption of NH_3, $(C_2H_5)_2O$, and Py does not lead to charge buildup, since the values of the dissociation constant for these molecules are many orders of magnitude lower than that for water. The accumulation effect can be termed a "monopolar" effect, as it depends on charge carriers of only one sign (holes). The photolysis effect depends on charges of either sign, so it is bipolar.

The products of heterolytic dissociation of H_2O molecules (and, generally speaking, of other molecules as well), H^+ and OH^- (as well as the products of their further hydration, H_3O^+ and $H_3O_2^-$), in migrating over the surface, can act as recombination centers; this is what we called the chemical channel of recombination in Sect.9.2. Owing to the existence of high potential barriers, the diffusion coefficients can be quite small, so diffusion will be the limit-stage of recombination. The slow photoconduction processes reportedly observed in a number of cases may be related to this effect. For instance, HAUFFE et al. [9.118] in their study of the electroconductivity of compressed anatase tablets in a humid oxygen atmosphere (740 torr O_2 and 20 torr H_2O), discovered peculiar variations in the current when UV light was switched on and off (Fig.9.12). The fast variations were justifiably attributed to electron photoconduction. The slow variations were explained by photodesorption of H_2O, which is reversed when the light is off. It is possible that heterogeneous photolysis of water molecules plays an important role in the slow kinetics. The resulting charged ions can act as additional centers of capture and recombination of charge carriers.

A subject that has recently been widely discussed is slow variations in the physical properties of certain semiconductor materials (primarily semiconductor films), caused by exposure to light. Reversible and irreversible changes in the structure and the optical and electric properties of amorphous films are usually associated with photostructural processes in which nonequilibrium charge carriers may play an active role [9.119-121]. To illustrate, let us discuss the processes that take place in amorphous silicon films obtained from decomposition of silanes a-Si:H.

In 1977 STABLER and WRONSKI [9.122,123] discovered a considerable reversible reduction in dark (σ_d) and photo (σ_{ph}) conductivity of a-Si:H films, caused by exposure to low-energy light quanta ($h\nu = 1.4 - 2.1$ eV). Heat treatment of the film at 450 K restored the initial value of σ_d. This effect is assumed to originate in the bulk, and is connected with the appearance of light-generated metastable charged states in the forbidden band of a-Si, which causes a shift in the Fermi level and consequently a change in σ_d [9.122-125]. The creation of metastable states is associated with reorientation of Si-H bonds [9.122, 123] or with self-capture of excitons [9.124]. The effects of prolonged exposure to light on the photoconductivity of a-Si were also reported in [9.126].

According to Sects.8.1, 9.2, at high levels of excitation the light-generated electrons and holes can stimulate defect formation and dissociation of adsorbed molecules. Indeed, the creation of new defects upon illumination of an a-Si film with an argon laser was proved in [9.127] using EPR and luminescence techniques. In [9.128] we investigated the influence of adsorption on the Stabler-Wronski effect. From Fig.9.13a it can be seen that a typical Stabler-Wronski effect was observed on an a-Si:H film at room atmosphere (temperature of film support : 500 K). When illumination was stopped, σ_{ph} decreased, and the linear section of plot $\sigma_{ph}(t)$ was the same as reported in [9.123,124]. After the light was switched off, the value of σ_{ph} was lower than σ_d^0. Vacuumization of the film at 300 K to 10^{-7} torr nullified the Stabler-Wronski effect (Fig.9.13b). Adsorption of water vapor at $p/p_s = 0.5$ was accompanied by slow variations in σ_d which indicates the creation of new surface states. The kinetics of variation of σ_d and σ_{ph} which is typical for the Stabler-Wronski effect, was studied after the system had come to equilibrium (Fig.9.13a). The kinetic curve turned out to be identical with the curve plotted for room atmosphere. Injection of saturated water vapor ($p / p_s = 1$) accelerated the kinetics of variation of

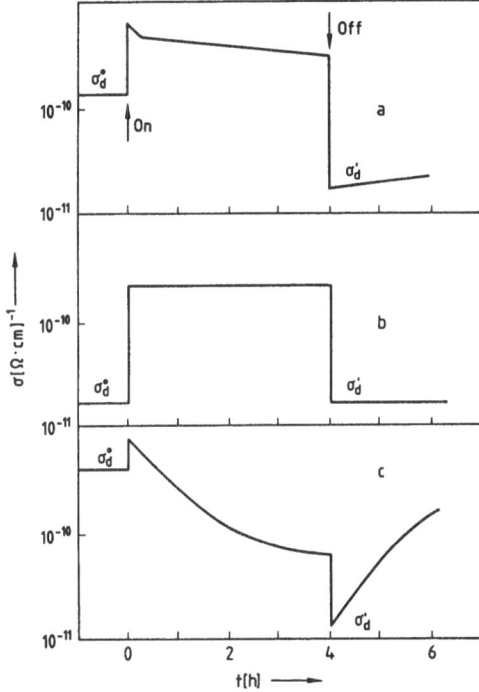

Fig.9.13a-c. Kinetics of variations in conductivity of amorphous silicon film a-Si during illumination, where σ_d is the dark conductivity. (**a**) Room atmosphere, (**b**) film vacuumized at $T = 300$ K (10^{-7} torr), (**c**) in saturated water vapor

σ_d. The photoconductivity slowly dropped to values lower than σ_d^0 (Fig.9.13c). Vaccumization at 300 K totally nullified the Stabler-Wronski effect. The kinetic curve in Fig.9.13c is similar to that obtained by Hauffe (Fig.9.12). The reversible variations in σ_{ph} during adsorption of H_2O in the case of a-Si:H films are most probably associated with the actuation of an additional channel of surface recombination, which is caused by the products of heterolytic dissociation of adsorbed molecules[5]. Note that adsorption of less readily dissociated molecules of NH_3 did not give rise to the Stabler-Wronski effect. Mass-spectroscopic investigations reveal that (as is the case with single crystals of silicon) illumination of a-Si:H films stimulates photolysis of adsorbed H_2O molecules and the release of H_2.

Experiments on adsorption will be necessary in order to assess the actual role played by bulk and surface processes in slow variations in the physical

---

[5] Note that the variations in σ_d and σ_{ph} observed during adsorption cannot entirely account for the Stabler-Wronski effect. When a-Si:H films obtained on supports having higher temperatures were used [9.122,123], adsorption of H_2O caused no changes in σ_d and σ_{ph}. The Stabler-Wronski effect was also observed after desorption of H_2O.

properties of amorphous semiconductors during illumination (in particular, the Stabler-Wronski effect). Such experiments have just started. KOLOMIETS and KOCHEMIROVSKY [9.129] were among the first to investigate the effects of adsorption on the electrophysical parameters of glasslike semiconductors (arsenic selenide). Adsorption of H_2O was found to raise the surface conductivity of glass by five or six orders of magnitude, and to affect the steady state photoconductivity and photoresponse. It was reported in [9.130] that adsorption of H_2O, NH_3, and alcohols has a strong effect on the photo emf of chalcogenide glasses. TANIELIAN and FRITZSCHE [9.125] discovered adsorption of H_2O and NH_3 to have a strong effect on the conductivity of a-Si:H. The mechanism by which adsorption influences the dark conductivity and photoconductivity of these substances is not yet completely understood. Owing to the high density of states in amorphous semiconductors, the Fermi level should be stabilized. The introduction of a relatively small number of new adsorptive states can hardly be expected to cause any considerable displacement of the Fermi level; more likely it is the influence of adsorbed molecules on the entire system or random fields which determine the energy spectrum of surface states. Apart from this, the electrophysical parameters of the surface are probably influenced by dissociative processes in the adsorbed phase that are associated with the capture and recombination of nonequilibrium charge carriers. Apparently, these processes are caused by slow adsorptive states, which have also been discovered on the surface of a-Si:H [9.131]. It is noteworthy that the kinetics of charge relaxation in these states during adsorption is similar to that for single crystals (Sect.6.5).

The close interrelations between ionic, adsorptive, and electron processes on semiconductor surfaces that have been elucidated in previous sections must also exist in MIS structures, especially at high levels of excitation of the semiconductor's electron subsystem. For instance, irradiation of silicon-based MIS structures with UV light with $h\nu \approx 5$ eV (which is considerably lower than the width of the forbidden band of SiO_2) is known to elevate the density of states on the Si-SiO_2 interface [9.132-135]. The electric properties of the oxide are also changed. As we mentioned in Sect.8.1, deterioration of MIS structures under the action of radiation and/or strong transverse fields is usually considered in terms of defect formation stimulated by nonequilibrium charge carriers in the semiconductor. However, in all these cases the possibility of proton processes must be taken into account. Our experiments, discussed in Sects.9.1,2, indicate that the most active suppliers of protons are coordination-bound water molecules. Protonic processes must also influence the radiative stability of MIS structures. The authors of [9.132,136,137] called attention to the role of humidity in the radiative stability of such structures, which they associated with the content of OH groups in the SiO_2 film. However, as we demonstrated in Sect.9.1, coordination-bound water molecules supply protons more readily than OH groups do. When the "chlorine" technique of oxidizing silicon is used, the activity of protonic processes is enhanced (see [9.20,Sect.4] which is reflected in the radiative stability. The resulting protons are probably also responsible for the buildup of positive charge in structures exposed to radiation; this conclusion is consistent with ideas expressed by WEINBERG et al. [9.138].

Drifting of dopant ions (e.g., alkali) and drifting of protons are usually looked upon as two independent processes in MIS structures. Mass-spectroscopic investigations [9.28] reveal, however, that the introduction of Na^+ ions onto a silicon surface entirely suppresses photolysis of H_2O and CCl_4 molecules. In catalysis this effect is known as poisoning of proton-donor centers. We see that alkali ions in MIS structures can suppress protonic processes.

References

Chapter 1

1.1 V.I. Lyashenko, V.G. Litovchenko, I.I. Stepko, V.I. Stricha, L.V. Lyashenko: *Elektronnye javlenia na poverchnosti poluprovodnikov* (Electronic Phenomena on Semiconductor Surfaces) (Naukova Dumka, Kiev (1968)

1.2 A.V. Rzhanov: *Elektronnye protsessy na poverchnosti poluprovodnikov* (Electronic Processes on Semiconductor Surface) (Nauka, Moscow 1971)

1.3 V.F. Kiselev: *Poverchnestnye javlenija v poluprovodnikach i dielektrikach* (Surface Phenomena in Semiconductor and Dielectric) (Nauka, Moscow 1970)

1.4 V.F. Kiselev, O.V. Krylov: *Adsorption Processes on Semiconductor and Dielectric Surfaces,* Springer Ser. Chem. Phys., Vol.32 (Springer, Berlin, Heidelberg 1985)

Chapter 2

2.1 I.E. Tamm: Sov. Phys. **1**, 753 (1932)

2.2 W. Shockley: Phys. Rev. **56**, 317 (1939)

2.3 ·V.F. Kiselev, O.V. Krylov: *Adsorption Processes on Semiconductor and Dielectric Surfaces I,* Springer Ser. Chem. Phys., Vol. 32 (Springer, Berlin, Heidelberg 1985)

2.4 V.V. Volkenstein: Zh. Fiz. Khim. **21**, 1317 (1947)

2.5 A. Many, Y. Goldstein, N.B. Grover: *Semiconductor Surfaces* (North Holland, Amsterdam 1965)

2.6 D.R. Frankl: Crit. Rev. Solid State Sci. **4**, 455 (1974)

2.7 V.I. Lyashenko, V.G. Litovchenko, I.I. Stepko, V.I. Stricha, L.V. Lyashenko: *Elektronnye Yavleniya na poverchnosti poluprovodnikov* (Electronic Phenomena on Semiconductor Surfaces) (Naukova Dumka, Kiev 1968)

2.8 A.V. Rzhanov: *Elektronnye protsessy na poverchnosti poluprovodnikov* (Electronic Processes on Semiconductor Surfaces) (Nauka, Moscow 1971)

2.9 M.J. Sparnaay: *The Electrical Double Layer* (Pergamon, Oxford 1972)

2.10 M. Schulz: Surf. Sci. **132**, 422 (1983)

2.11 C.G. Garret, W.H. Brattain: Phys. Rev. **99**, 376 (1955); J. Bell Syst. Tech. **35**, 104 (1956)

2.12 G.K. Pikus (ed.): *Fizika poverchnosti poluprovodnikov* (The Physics of Semiconductor Surfaces) (IL, Moscow 1959) supplement

2.13 J.R. Schrieffer: Phys. Rev. **97**, 641 (1955)

2.14 R.F. Green: Crit. Rev. Solid State Sci. **4**, 477 (1974)

2.15 V.N. Dobrovolski, V.G. Litovchenko: *Perenos electronov u dyrok u poverchnosti polyprovodnikov* (Transport of Electrons and Holes Below Semiconductors Surface) (Nauka Dumka, Kiev 1985)

2.16 R. Seiwatz, M. Green: J. Appl. Phys. **29**, 1034 (1958)
2.17 Yu.I. Gorkun: Fiz. Tverd. Tela **3**, 1061 (1961)
2.18 T. Ando, A. Fowler, F. Stern: Rev. Mod. Phys. **54**, 2 (1982)
2.19 W. Shockley, G.L. Pearson: Phys. Rev. **74**, 232 (1948)
2.20 H.C. Montgomery: Phys. Rev. **166**, 441 (1957)
2.21 A.E. Yunovich: Zh. Tekh. Fiz. **27**, 1707 (1957); Fiz. Tverd. Tela **1**, 908 (1959)
2.22 G. Rupprecht: Ann. N. Y. Acad. Sci. **101**, 960 (1963)
2.23 V.A. Zuev, A.V. Sachenko, K.B. Tolpygo: *Neravnovesnye pripoverkhnostnye protsessy v poluprovodnikakh i poluprovodnikovykh Priborakh* (Nonequilibrium Processes in the Subsurface Region of Semiconductors and Semiconductor Devices) (Sovetskoe Radio, Moscow 1977)
2.24 Yu.F. Novototsky-Vlasov: Trudy FIAN **48**, 3 (1969)
2.25 E.O. Johnson: Phys. Rev. **111**, 153 (1958)
2.26 H. Flietner, W. Füssel, K. Heilig: In *Physik der Halbleiteroberflächen*, Vol.16 (DDR Akademie der Wissenschaften, Berlin 1985) p.1
2.27 A. Goetzberger, E. Klausmann, M.J. Schulz: Crit. Rev. Solid State Sci. **6**, 2 (1976)
2.28 E.H. Nicollian, J.R. Brews: *MOS Physics and Technology* (Wiley, New York 1982)
2.29 S.M. Sze: *Physics of Semiconductor Devices* (Wiley, New York 1981)
2.30 M. Schulz, G. Pensl (eds.): *Insulating Films on Semiconductors*, Springer Ser. Electrophys., Vol.7 (Springer, Berlin, Heidelberg 1981)
2.31 K. Seeger: *Semiconductor Physics*, 3rd ed., Springer Ser. Solid-State Sci., Vol.40 (Springer, Berlin, Heidelberg 1985)
2.32 F.J. Himpsel: Appl. Phys. **A 38**, 205-212 (1985)

Chapter 3

3.1   A.V. Rzhanov: *Elektronnye protsessy na poverkhnosti poluprovodnikov*, (Electronic Processes on Semiconductor Surfaces) (Nauka, Moscow 1971)
3.2   Yu.F. Novototsky-Vlasov: Trudy FIAN **48**, 3 (1969)
3.3   Yu.A. Zarif'yants, V.F. Kiselev, S.N. Kozlov, Yu.F. Novototsky-Vlasov: Vestnik MGU, Fizika **16**, 84 (1975)
3.4   V.V. Murina, Yu.F. Novototsky-Vlasov: Sov. Phys.-Semiconductors **6**, 979 (1972)
3.5   E.I. Brilliantov, Yu.F. Novototsky-Vlasov: Fiz. Tekh. Polupr. **2**, 1393 (1968)
3.6   E.W. Kreutz, P. Schroll: Surf. Sci. **37**, 410 (1973)
3.7   N.L. Dmitruk, V.I. Lyashenko: Ukrain. Fiz. Zh. **16**, 104 (1971)
3.8   A.M. Murtasin, Yu.A. Zarif'yants: Sov. Phys.-Semiconductors **7**, 1364 (1974)
3.9   A. Many, Y. Goldenstein, N.B. Grover: *Semiconductor Surfaces* (North Holland, Amsterdam 1965)
3.10 V.I. Lyashenko, V.G. Lyashenko, V.G. Litovchenko, I.I. Stepko, V.I. Stricha, L.V. Lyashenko: *Elektronnye yavleniya na poverkhnosti poluprovodnikov* (Electronic Phenomena on Semiconductor Surfaces) (Naukova Dumka, Kiev 1968)
3.11 A.V. Rzhanov, V.P. Migal, N.N. Migal: Fiz. Tekh. Polupr. **3**, 231 (1969)
3.12 V.F. Kiselev, S.N. Kozlov, Yu.F. Novototsky-Vlasov, R.V. Prudnikov: Surface Sci. **11**, 111 (1968)
3.13 V.E. Primachenko, V.V. Antoshcuk, E.P. Mazas, V.V. Milenin: In *Problemi fiziki poverchnosti poluprovodnikov* (Problems of Physics of Semiconductor Surfaces), ed.by O.V. Snitko (Naukova Dumka, Kiev 1981) p.77
3.14 V.G. Litovchenko, V.S. Lysenko: Fiz. Tekh. Polupr. **4**, 72 (1970)
3.15 V.N. Drozdov, T.I. Kovalevskaya, A.V. Rzhanov, K.K. Svitashev: Mikroelektronika **2**, 154 (1973)

3.16 S.V. Pokrovskaya, V.N. Ovsyuk: Neorgan. Mater. **6**, 978 (1971)
3.17 I.G. Neizvestny, V.N. Ovsyuk: In *Tezisy 7-i konferentsii "Elektronnye prozessy na poverkhnosti poluprovodnikov* (7th Conference on Electronic Processes on Semiconductor Surfaces) ed. by A.V. Rzhanov (Nauka, Novosibirsk 1980) p.170
3.18 E.B. Gorokhov, E.L. Kamenkovich: In *Elektronnye protsessy na poverkhnosti poluprovodnikov* (Electronic Processes on Semiconductor Surfaces) (Nauka, Novosibirsk 1974) p.147
3.19 I.V. Korobov, I.I. Petruchuk: In *Elektronnye protsessy na poverkhnosti poluprovodnikov* (Electronic Processes on Semiconductor Surfaces) (Nauka, Novosibirsk 1974) p.119
3.20 T. Hattory: Jpn. J. Appl. Phys. **10**, 202 (1971)
3.21 M. Schulz: Surface Sci. **132**, 422 (1983)
3.22 E.H. Poindexter, G.J. Geraldi, M.E. Rueckel, P.J. Caplan, N.M. Johnson, D.K. Biegelsen: J. Appl. Phys. **56**, 2844 (1984)
3.23 Yu.M. Shirshov, V.A. Tyagai, O.B. Snitko: Fiz. Tekh. Polupr. **3**, 115 (1969)
3.24 B.I. Bedny, A.N. Kalinin, I.A. Karpovich: Sov. Phys.-Semiconductors **13**, 840 (1979)
3.25 V.G. Pankin, E.S. Rzhanova, A.V. Rzhanov: Sov. Phys.-Semiconductors **6**, 1302 (1972)
3.26 V.G. Prikhodko, I.B. Gulyaev: Fiz. Tekh. Polupr. **15**, 2400 (1981)
     V.G. Prikhodko, A.G. Zhdan: Surface Sci. **124**, 602 (1983)
3.27 V.F. Kiselev: *Poverkhnostnye yavleniya v poluprovodnikakh i dielektrikakh* (Surface Phenomena in Semiconductors and Dielectrics) (Nauka, Moscow 1970)
3.28 I.G. Neizvestny, S.W. Pokrovskaya, A.V. Rzhanov: Sov. Phys.-Semiconductors **6**, 281 (1972)
3.29 I.G. Neizvestny, V.N. Ovsyuk: Phys. Status Solidi **A 18**, 465 (1973)
3.30 V.F. Kiselev, O.V. Krylov: *Adsorption Processes on Semiconductor and Dielectric Surfaces I*, Springer Ser. Chem. Phys., Vol.32 (Springer, Berlin, Heidelberg 1985)
3.31 V.F. Kiselev, V.A. Matveev, R.V. Prudnikov: Phys. Status Solidi **A 50**, 739 (1978)
3.32 V.I. Grinev, V.F. Kiselev, V.A. Matveev: Sov. Phys.-Semiconductors **14**, 936 (1980)
3.33 V.I. Grinev, V.F. Kiselev: Phys. Status Solidi **A 66**, 493 (1981)
     V.I. Grinev, S.N. Karjagin, V.F. Kiselev, I.O. Ogloblin: Izv. Vyssh. Uchebn. Zaved. Fizika **9**, 94 (1982)
3.34 A.V. Sachenko, O.V. Snitko: *Photoeffekti v Pripoverchnostnich sloyach poluprovodnikov* (Photoeffects in the Subsurface Layers of Semiconductors) (Naukova Dumka, Kiev 1984)
3.35 C.G. Garret, W.H. Brattain: Phys. Rev. **99**, 376 (1955); J. Bell Syst. Tech. **35**, 104 (1956)
3.36 V.A. Zuev, A.V. Sachenko, K.B. Tolpygo: *Neravnovesnye pripoverkhnostnye protsessy v poluprovodnikakh i poluprovodnikovykh priborakh* (Nonequilibrium Processes in the Subsurface Region of Semiconductors and Semiconductor Devices) (Sovetskoe Radio, Moscow 1977)
3.37 D. Fitzgerald, A. Grove: Surface Sci. **9**, 347 (1968)
3.38 A.I. Gubanov: *Kvantovo-elektronnaya teoriya amorfnykh poluprovodnikov* (Quantum Electronic Theory of Amorphous Semiconductors) (AN SSSR, Leningrad 1963)
3.39 L. Banyai: *Physique de Semiconducteurs* (Masson, Paris 1969)
3.40 I.M. Lifshitz: Usp. Fiz. Nauk **83**, 617 (1964)
3.41 V.L. Bonch-Bruevich, I.P. Zwjagin, I.P. Kaiper, A.G. Mironov, P. Enderlain, B. Esser: *Elektronnaya teoriya neuporyadochennykh poluprovodnikov* (Electronic Theory of Disordered Semiconductors) (Nauka, Moscow 1981)
3.42 V.L. Bonch-Bruevich: In *Statisticheskaya fizika i kvantovaya teoriya polya* (Statistical Physics and the Quantum Theory of Fields) (Nauka, Moscow 1973) p.337

3.43 N.F. Mott, E.A. Davis: Electronic Processes in Non-Crystalline Materials (Clarendon, Oxford 1971)
3.44 V.L. Bonch-Bruevich: Zh. Eksp. Teor. Fiz. **61**, 1168 (1971)
3.45 W. Fahrner, A. Goetzberger: J. Appl. Phys. **44**, 725 (1973)
3.46 N.B. Goodman, H. Fritzsche: Phil. Mag. B **42**, 149 (1980)
3.47 A.V. Rzhanov, V.N. Ovsyuk: Fiz. Tekh. Polupr. 3, 294 (1969)
3.48 Yu.A. Zarif'yants, S.N. Karyagin, V.F. Kiselev: Vestnik MGU, Fizika **16**, 237 (1975)
3.49 V.L. Bonch-Bruevich: Probl. Kinet. Katal. **14**, 19 (1970)
3.50 G.G.E. Low: Proc. Phys. Soc. (London) B **68**, 10 (1955)
3.51 S.R. Morrison: In *Semiconductor Surface Physics*, ed. by R.H. Kingston (University of Pennsylvania Press, Philadelphia 1957) p.169
3.52 G.W. Pratt, H.H. Kolm: In *Semiconductor Surface Physics*, ed. by R.H. Kingston (University of Pennsylvania Press, Philadelphia 1957) p.297
3.53 O. Jäntsch: J. Phys. Chem. Solids **26**, 1233 (1965)
3.54 S. Koc: Appl. Phys. Lett. **4**, 151 (1964)
3.55 M.H. Pilkuhn: J. Appl. Phys. **34**, 3302 (1963)
3.56 S.N. Kozlov, V.F. Kiselev, Yu.F. Novototsky-Vlasov: Surface Sci. **28**, 395 (1971)
3.57 O.S. Frolov, O.V. Snitko, G.F. Romanova: Ukrain. Fiz. Zh. **14**, 585 (1969)
3.58 S. Koc: Czech. J. Phys. B 11, 193, 287 (1961)
3.59 G. Dorda: Czech. J. Phys. B 15, 581 (1965)
3.60 M. Lasser, C. Wysocki, B. Bernstein: Phys. Rev. **105**, 491 (1957)
3.61 O. Jäntsch: Z. Naturforsch. A 15, 141 (1960)
3.62 S.R. Hofstein, G. Warfield: Solid State Electr. **8**, 321 (1965)
3.63 P.K. Kashkarov, V.F. Kiselev, S.N. Kozlov, Yu.F. Novototsky-Vlasov: Sov. Phys.-Semiconductors **7**, 967 (1974)
3.64 S.N. Kozlov, N.L. Levshin: Vestnik MGU, Fizika **24**, 76 (1983)
3.65 H.J. van Hove: Surf. Sci. **22**, 76 (1970)
3.66 T.N. Povchan: Zh. Nauchn. Prikl. Fotogr. **19**, 221 (1974)
3.67 A. Deubner, F. Schulz: Ann. Phys. **5**, 113 (1960)
3.68 S.N. Kozlov, V.F. Kiselev: Fiz. Tekh. Polupr. **1**, 568 (1967)
3.69 L. Syrnev: Dokl. Bolg. Akad. Nauk 15, 719 (1962); **16**, 233 (1963)
3.70 A.P. Gorchakov, Yu.A. Zarif'yants: Sov. Phys.-Semiconductors **14**, 1203 (1980)
3.71 A.H. Boonstra: Philips Res. Rpt., Suppl. No.3, 1 (1968)
3.72 M.J. Sparnaay, A.H. Boonstra, J. van Ruller: Surface Sci. **2**, 56 (1964)
3.73 S.N. Kozlov, Yu.F. Novototsky-Vlasov, V.F. Kiselev: Sov. Phys.-Semiconductors **6**, 1788 (1973)
3.74 R.H. Kingston, A.L. McWhorter: Phys. Rev. **98**, 1191 (1955); **103**, 534 (1956)
3.75 Yu.F. Fedorovich, V.A. Vogel: Fiz. Tekh. Polupr. **3**, 840 (1969)
3.76 E.H. Niccolian, A.K. Sinha: In *Thin Films. Interdiffusion and Reactions*, ed. by J. Poate (Wiley, New York 1978) p.481
       E.H. Niccolian, J.R. Brews: *MOS Physics and Technology* (Wiley, New York 1982)
3.77 S.N. Kozlov, Yu.F. Novototsky-Vlasov: Phys. Status Solidi A **6**, 345 (1971)
3.78 S.N. Kozlov, N.L. Levshin: Phys. Status Solidi A **64**, 169 (1981)
3.79 V.F. Kiselev, Yu.A. Zarif'yants: Probl. Kinet. Katal. **16**, 221 (1975)
       V.F. Kiselev, S.N. Kozlov: Poverchnost. Fizika, Khimiya. Mechanika **1**, 13 (1982)
3.80 S.G. Davison, J.D. Levine: *Surface States* (Academic, New York 1970)
3.81 J.A. Appelbaum, D.R. Hamman: Crit. Rev. Solid State Sci. **6**, 357 (1976)
3.82 L.M. Falicov, F. Yndurain: J. Phys. C.: Solid State Phys. **8**, 147 (1975)
3.83 S.R. Morrison: *The Chemical Physics of Surfaces* (Plenum, New York 1977)
3.84 D.J. Chadi: Vacuum **33**, 613 (1983)
3.85 B.A. Nesterenko, O.V. Snitko: *Fizicheskie Svoistva atomarnochistich poverchnostei poluprovodnikov* (Physics Properties of Atomic-Clean Surfaces of Semiconductors) (Naukova Dumka, Kiev 1983)

3.86 M. Tomashek, J. Koutechy: In *Symposium on Electronic Phenomena in Chemisorption*, ed. by K. Hauffe (W. de Gruyter, Berlin 1969) p.41
3.87 B.A. Lippman: Ann. Physik **2**, 16 (1957)
3.88 S.G. Davison, K.P. Tan: Surface Sci. **27**, 297 (1971)
3.89 J. Koutecky: Kinet. Katal. **2**, 319 (1961)
3.90 M. Tomashek: Surf. Sci. **2**, 8 (1964); **4**, 485 (1966)
3.91 D.J. Chadi, M.L. Cohen: Phys. Rev. B **11**, 732 (1975)
3.92 J.D. Levine, P. Mark: Phys. Rev. **144**, 751 (1966)
3.93 W.E. Spicer, P.W. Chye, C.M. Garner, P. Pianetta: Surf. Sci. **86**, 763 (1979)
     K.C. Pandey: Phys. Rev. Lett. **49**, 223 (1982)
3.94 J.C. Phillips: Surface Sci. **63**, 1 (1977)
3.95 W. Mönch: Surface Sci. **63**, 79 (1977)
3.96 D.R. Frankl: *Electronical Properties of Semiconductors Surfaces* (Pergamon, Oxford 1967)
3.97 M.J. Sparnaay: *The Electrical Double Layer* (Pergamon, Oxford 1972)
3.98 D.R. Frankl: Crit. Rev. Solid State Sci. **4**, 455 (1974)
3.99 G.W. Gobeli, F.G. Allen: Surface Sci. **2**, 402 (1964)
3.100 H. Lüth, G. Heiland: Nuovo Cimento B **39**, 748 (1977)
3.101 J. von Wienskowski, H. Frotzheim: Phys. Status Solidi B **94**, 429 (1979)
3.102 B.A. Nesterenko, O.V. Snitko: Surface Sci. **5**, 380, 407 (1966)
3.103 F.G. Allen, G.W. Gobeli: Phys. Rev. **127**, 150 (1962)
3.104 H. Ibach (ed.): *Electron Spectroscopy for Surface Analysis*, Topics Current Phys. Vol.4 (Springer, Berlin, Heidelberg 1977)
3.105 A.W. Parke, A. McKinley, R.H. Williams: J. Phys. C: Solid State Phys. **11**, 993 (1978)
3.106 D.F. Eastman, W.D. Grobman, J. Precout, M. Erbudak: Phys. Rev. B **9**, 3473 (1974)
3.107 M. Henzler: Surface Sci. **9**, 31 (1968); **24**, 209 (1971)
3.108 M. Henzler: J. Appl. Phys. **40**, 3758 (1969)
3.109 M. Henzler: In *Advances in Solid State Physics*, ed. by O. Madelung (Pergamon, Oxford 1971) p.187
3.110 F. Baurle, W. Mönch, M. Henzler: J. Appl. Phys. **43**, 3917 (1972)
3.111 M. Erbudak, T.E. Fischer: Phys. Rev. Lett. **29**, 732 (1972)
3.112 M.F. Chung: J. Phys. Chem. Solids **32**, 475 (1971)
3.113 V.S. Kuznetsov, G.B. Demidovich, V.F. Kiselev: Vestnik MGU, Fizika **16**, 177 (1975)
3.114 G.B. Demidovich, V.F. Kiselev: Vestnik MGU, Fizika **14**, 158 (1973)
      G.B. Demidovich, S.N. Karyagin, V.F. Kiselev: Phys. Status Sol. (b) **114**, 705 (1982)
      S.N. Karyagin, P.K. Kashharov, V.F. Kiselev, A.V. Petrov: Surf. Sci. Lett. **146**, L-582 (1984)
3.115 J.E. Rowe, H. Ibach: Phys. Rev. Lett. **31**, 102 (1973); **32**, 421 (1974); Phys. Rev. B **10**, 710 (1974)
3.116 J. Assmann, W. Mönch: Surf. Sci. **99**, 34 (1980)
3.117 D. Haneman: Phys. Rev. **170**, 705 (1968)
3.118 S. Krueger, W. Mönch: Surf. Sci. **99**, 157 (1980)
      M. Henzler: Surf. Sci. **132**, 82 (1983); **152/153**, 963 (1985)
3.119 M.A. Van Hove, S.Y. Tong (eds.): *The Structure of Surfaces*, Springer Ser. Surface Sci., Vol.2 (Springer, Berlin, Heidelberg 1985)
3.120 D.E. Aspnes, A.A. Studna: Surf. Sci. **96**, 294 (1980)
      J.W.D. Martens, W.F. van den Bogert, A. van Silfhout: Surf. Sci. **105**, 275 (1981)
3.121 C. Sebenne, D. Bolmont, G. Guichar, M. Balkanski: Phys. Rev. B **12**, 3280 (1975)
3.122 B.P. Lemke, D. Haneman: Phys. Rev. B **17**, 1893 (1978)
3.123 D.E. Aspnes, P. Handler: Surf. Sci. **4**, 353 (1966)
3.124 D.E. Eastman, W.D. Grobman: Phys. Rev. Lett. **28**, 1378 (1972)

3.125 L.F. Wagner, W.E. Spicer: Phys. Rev. Lett. **28**, 1381 (1972)
3.126 C. Pardey, J.C. Phillips: Solid State Commun. **14**, 439 (1974)
3.127 Y. Yndurain, J. Rubbio: Phys. Rev. Lett. **29**, 732 (1972)
3.128 G.V. Gadyak, A.A. Karpushin, Yu. N. Morokov: In *Problemy fizicheskoi khimii poluprovodnikovykh poverkhnostei* (Problems in the Physical Chemistry of Semiconductor Surfaces), ed. by A.V. Rzhanov (Nauka, Novosibirsk 1978) p.72
3.129 M. Nishida: J. Phys. C: Solid State Phys. **11**, 1217 (1978)

Chapter 4

4.1 N.F. Mott, R.W. Gurney: *Electronic Processes in Ionic Crystals*, 2nd ed. (Oxford Univ. Press, Oxford 1948)
I. Bunget, M. Popescu: *Physics of Solid Dielectrics* (North-Holland, Amsterdam 1984)
4.2 W.A. Harrison: *Solid State Theory* (McGraw-Hill, New York 1970)
4.3 S.I. Pekar': *Issledovaniya elektronnoi teorii kristallov* (Investigations of the Electron Theory of Crystals) (AN SSSR, Moscow 1951)
4.4 V.F. Kiselev, O.V. Krylov: *Adsorption Processes on Semiconductor and Dielectric Surfaces I*, Springer Ser. Chem. Phys., Vol.32 (Springer, Berlin, Heidelberg 1985)
4.5 N.F. Mott, E.A. Davis: Electronic Processes in Non-Crystalline Materials (Clarendon, Oxford 1971)
4.6 S.T. Pantelides (ed.): *The Physics of SiO_2 and Its Interfaces* (Pergamon, Yorktown Heights 1978)
M. Schulz, G. Pensl (eds.): *Insulating Films on Semiconductors*, Springer Ser. Electrophys., Vol.7 (Springer, Berlin, Heidelberg 1981)
J.F. Verweij, D.R. Wolters (eds.): *Insulating Films on Semiconductors* (North-Holland, Amsterdam 1983)
4.7 A. Balzarotti, A. Bianconi: Phys. Status Solidi B **63**, 77 (1974)
4.8 R.Z. Bachrach, R.S. Bauer, J.C. McMenamin, A. Bianconi: In *Proc. Semiconductor Conference Edinburgh 1978* (Plenum. London) Preprint 76
R.S. Bauer, J.C. McMenamin, H. Petersen, A. Bianconi: In [4.6] p.170
A. Bianconi, L. Incoccia, S. Stipcich (eds.): *EXAFS and Near Edge Structure*, Springer Ser. Chem. Phys., Vol.27 (Springer, Berlin, Heidelberg 1983)
4.9 A. Szász, J. Kojnok: Appl. Surface Sci. **24**, 34 (1985)
4.10 R. Vanselov, R. Howe (eds.): *Chemistry and Physics of Solid Surfaces V*, Springer Ser. Chem. Phys., Vol.35 (Springer, Berlin, Heidelberg 1984)
4.11 M.G. Jani, L.E. Halliburton, E.E. Kohnkl: J. Appl. Phys. **54**, 6321 (1983)
4.12 N. Jakowski, H. Glaefeke, W. Wild: *Physik der Halbleiterfläche*, Bd.5 (Akad. Wiss. DDR, Berlin 1974) p.157
4.13 I.V. Krylova: In *Aktivnye poverchnosti tverdogo tela* (Active Solid State Surfaces) ed. by G.I. Distler, M.Ju. Butyagin (VINITI, Moscow 1976) p.22
4.14 V.A. Gritsenko: In *Elektronnye protsessy na poverchnosti poluprovodnikov* (Electronic Processes on Semiconductor Surfaces), ed. by A.V. Rzhanob (Nauka, Novosibirsk 1974) p.115
4.15 S.F. Timashev: Sov. Phys.-Solid State **14**, 136 (1972)
4.16 V.N. Abakumov, L.N. Kreshuk, U.N. Yasievich: Zh. Eksp. Teor. Fiz. **74**, 1019 (1978)
4.17 H.M. Pizowlocki: J. Appl. Phys. **57**, 5359 (1985)
4.18 E.W. Kreutz: Appl. Phys. **10**, 289 (1976)
4.19 G. Buchheim: Exp. Tech. Phys. **26**, 507 (1978)
4.20 M.A. Lampert, P. Mark: *Current Injection in Solids* (Academic, New York 1970)
E.H. Nicollian, J.R. Brews: *MOS Physics and Technology* (Wiley, New York 1984)

4.21 Yu.A. Zarif'yants, A.N. Nevsorov: React. Kinet. Catal. Lett. **4**, 215 (1976)
4.22 J. Gaznô: Phys. Stat. Sol. (a) **68**, 675 (1981)
4.23 N.B. Goodman, H. Fritzsche: Phil. Mag. B **42**, 149 (1980)
4.24 M.J. Sparnaay: *The Electrical Double Layer* (Pergamon, Oxford 1972)
4.25 K. Hauffe, R. Schmidt: Phys. Status Solidi A **3**, 173 (1970)
4.26 S.R. Morrison: Surf. Sci. **15**, 85 (1969)
4.27 R.P. Holmstrom, J. Lagowski, H.C. Gatos: Surf. Sci. **75**, 781 (1978)
4.28 G.H. Siegel, E.J. Friebele, R.J. Ginther, D.L. Griscom: IEEE Trans.
      NS-**21**, 56 (1974)
4.29 Y.H. Lee, J.W. Corbett: Phys. Rev. B **13**, 2653 (1976)
4.30 R.L. Pfeffer: J. Appl. Phys. **57**, 5176 (1985)
4.31 S.N. Karyagin, P.K. Kashkarov, V.F. Kiselev, S.N. Kozlov: Phys. Status
      Solidi A **37**, 17 (1976)
4.32 R.A.B. Devine, A. Golanski: J. Appl. Phys. **54**, 3833 (1983)
      A. Stesmans, J. Braet, J. Witters, R.F. Keermaecker: J. Appl. Phys. **55**,
      1551 (1984)
4.33 S.N. Karyagin, S.N. Kozlov, G.S. Plotnikov: Izv. Vyssh, Uchebn. Zaved.
      Fizika **2**, 88 (1981)
4.34 J.A. Reimer, R.W. Vaughan, J.C. Knights: Phys. Rev. Lett. **44**, 193 (1980)
      D.J. Leopold, J.B. Boyce, P.A. Fedders, R.E. Norberg: Phys. Rev. B **26**,
      6053 (1982)
4.35 S.V. Vintsent, V.F. Kiselev, G.S. Plotnikov: Phys. Stat. Sol. (a) **85**,
      273 (1984)
4.36 P.K. Kashkarov, V.F. Kiselev, S.N. Kozlov: Surface Sci. **75**, 231 (1978)
4.37 P.K. Kashkarov, S.N. Kozlov: Izv. Vyssh. Uchebn. Zaved. Fizika **2**, 79
      (1976); **5**, 36 (1982)
      P.K. Kashkarov, A.V. Petrov: Poverchnost. Fizka, Kchimia. Mechanika **3**,
      92 (1984)
4.38 E.W. Kreutz, P. Schroll: Surf. Sci. **37**, 410 (1973)
4.39 S.N. Kozlov, S.N. Kuznetsov: Sov. Phys. Semiconductors **12**, 995 (1978)
4.40 S.N. Kozlov, S.N. Kuznetsov: Izv. Vyssh. Uchebn. Zaved. Fizika **1**, 92 (1981)
4.41 V.F. Kiselev, G.S. Plotnikov, V.A. Bespalov, Ju. D. Fomin: Kinet. Katal.
      27, N⁰1 (1987)
4.42 R.C. Jaklevic, J. Lambe: Phys. Rev. Lett. **17**, 1139 (1966)
4.43 J. Lambe, R.C. Jaklevic: Phys. Rev. **165**, 821 (1968)
4.44 A. Bennett, C.B. Duke, S.D. Silverstuk: Phys. Rev. **166**, 969 (1968)
4.45 I.K. Janson: Zh. Eksp. Teor. Fiz. **60**, 1759 (1971)
4.46 B.F. Lewis, M. Mosesman, W.H. Weinberg: Surface Sci. **41**, 142 (1974)
4.47 J.D. Langan, P.K. Hansma: Surface Sci **52**, 211 (1975)
4.48 S. de Cheveigne, S. Gauthier, J. Klein, A. Leger: Surface Sci. **105**, 377
      (1981)
4.49 Sh. M. Kogan: Probl. Kinet. Katal. **10**, 52 (1960)
4.50 O. Peshev, G. Bliznakov: J. Catal. **7**, 18 (1967)
4.51 O. Peshev: Jpn. J. Appl. Phys. Suppl. **2**, 153 (1974)
4.52 O.M. Poltorak: Zh. Fiz. Khim. **29**, 1656 (1956); **32**, 534 (1958)
4.53 T.A. Goodwin, P. Mark: In *Progress in Surface Science* (Pergamon, New York
      1972) p.7
4.54 S.R. Morrison: Adv. Catal. **7**, 259 (1958)
4.55 N.B. Hannay (ed.): *Semiconductors* (Reinhold, New York 1960)
4.56 J.G. Slater: Phys. Rev. **103**, 1631 (1956)
4.57 R.D. Petriz: Phys. Rev. **104**, 1508 (1956)
4.58 S.R. Morrison: *The Chemical Physics of Surfaces* (Plenum, New York 1977)
4.59 V.G. Erofeichev, L.N. Kurbatov: Fiz. Tverd. Tela **3**, 595 (1961)
4.60 T.S. Moss, G.J. Burrel, B. Ellis: *Semiconductor Opto-Electronics* (Butter-
      worth, New York 1973)
4.61 L.I. Burbulyavichus, Yu. A. Zarif'yants, V.F. Kiselev, S.V. Khrustaleva:
      Kinet. Katal. **12**, 922 (1971)
4.62 V.N. Filimonov: Opt. Spektrosk. **13**, 709 (1958)

4.63 S.G. Kalashnikov, Ya.E. Pokrovsky: Zh. Tekh. Fiz. **22**, 883 (1952)
4.64 K.H. Seeger: *Semiconductor Physics* (Springer, Berlin, Heidelberg 1973)
4.65 A.S. Medeishis, Yu.K. Viehakas: Litovsk. Fiz. Sbornik **15**, 259 (1975)
4.66 J.W. Orton, B.J. Goldsmith: Appl. Phys. Lett. **37**, 557 (1980)
4.67 J.W. Orton: R. Progr. Phys. **43**, 1263 (1980)
4.68 M.M. Sayed, Ch.R. Westgate: Rev. Sci. Instr. **46**, 1074 (1975)
4.69 V.F. Kiselev: Dokl. Akad. Nauk SSSR **213**, 224 (1973)
4.70 J.N. Zemel: IEEE Trans. NS-**18**, 78 (1971)
4.71 G. Heiland, D. Kohl: *Proceed. Intern. Meeting on Chemical Sensors*,
     ed. by T. Seiyama (Elsevier, Amsterdam 1983)
4.72 I.A. Myasnikov, E.V. Bolshun, B.S. Agayan: Sov. Phys.-Doklady (DKPCAG)
     220, 136 (1975)
4.73 E.I. Goldman, A.G. Zhdan: Sov. Phys.-Semiconductors **12**, 491 (1978)
4.74 E.I. Goldman, I.B. Gulyaev: Sov. Phys.-Semiconductors **9**, 905 (1976)
4.75 E.I. Goldman, I.B. Gulyaev: Sov. Phys.-Semiconductors **11**, 300 (1977)

Chapter 5

5.1 A.F. Ioffe: *Informatsiya o nauchnykh issledovaniyakh v SSSR. Kataliz*
     (Information on Scientific Research in the USSR. Catalysis) (KTI,
     Moscow 1930) p.53
5.2 S.Z. Roginsky, E.I. Shulz: Ukrain. Khim. Zh. **3**, 177 (1928)
5.3 C. Wagner: Z. Phys. Chem. **B22**, 181 (1933)
5.4 C. Wagner, K. Hauffe: Z. Elektrochem. **44**, 172 (1938)
5.5 A.H. Wilson: Proc. Roy. Soc. **A133**, 458 (1931); **A134**, 277 (1932)
5.6 O. Peshev, V. Malakhov, Th. Volkenstein: in *Progress in Surface Science*,
     ed. by S.G. Davison (Pergamon, Oxford 1975) p.63
5.7 V.F. Kiselev, O.V. Krylov: *Adsorption Processes on Semiconductor and
     Dielectric Surfaces I*, Springer Ser. Chem. Phys., Vol.32 (Springer,
     Berlin, Heidelberg 1985)
5.8 V.F. Kiselev: *Poverkhnostnye yavleniya v poluprovodnikakh i dielektrikakh*
     (Surface Phenomena in Semiconductors and Dielectrics) (Nauka, Moscow
     1970)
5.9 W.H. Brattain, J. Bardeen: J. Bell Syst. Tech. **32**, 1 (1953)
5.10 R.H. Kingston: Phys. Rev. **98**, 1766 (1955)
5.11 G. Heiland: Z. Phys. **138**, 459 (1954); **142**, 415 (1955)
5.12 J.T. Low: J. Phys. Chem. **59**, 67 (1955)
5.13 S.R. Morrison: Adv. Catal. **7**, 259 (1955)
5.14 P. Mark: RCA Rev. **26**, 461 (1965)
5.15 S.M. Zenter: Zh. Eksp. Teor. Fiz. **3**, 682 (1938)
5.16 F.F. Volkenstein: Zh. Fiz. Khim. **21**, 1317 (1947)
5.17 V.L. Bonch-Bruevich: Zh. Fiz. Khim. **27**, 662 (1953)
5.18 J. Koutecky: Trans. Farad. Soc. **54**, 1038 (1958)
5.19 T.B. Grimley: J. Phys. Chem. Solids **14**, 227 (1960)
5.20 E.L. Nagaev: Zh. Fiz. Khim. **35**, 327, 2376 (1961)
5.21 F.F. Volkenstein: *The Electron Theory of Catalysis on Semiconductors*
     (McMillan, New York 1963); *Fizikokhimiya poverkhnosti poluprovodnikov*
     (The Physical Chemistry of Semiconductor Surfaces) (Nauka, Moscow 1973)
5.22 K. Hauffe: Adv. Catal. **7**, 213 (1955)
5.23 K. Hauffe: *Reaktionen in und an festen Stoffen* (Springer, Berlin, 1966)
5.24 A. Many: Crit. Rev. Solid State Sci. **4**, 515 (1974)
5.25 M. Tomashek, J. Koutecky: In *Symposium on Electronic Phenomena in Chemi-
     sorption and Catalysis on Semiconductors*, ed. by K. Hauffe, Th. Volken-
     stein (W. de Gruyter, Berlin 1969) p.41
5.26 C.G.B. Garrett: J. Chem. Phys. **33**, 966 (1960)
5.27 V.J. Lee: J. Catal. **17**, 178 (1970); J. Chem. Phys. **55**, 2905 (1971)

5.28 S.R. Morrison: J. Catal. **20**, 110 (1971)
5.29 Sh. M. Kogan, V.B. Sandomirsky: Zh. Fiz. Khim. **33**, 1129, 1709 (1959)
5.30 C.G.B. Garrett, W.H. Brattain: Phys. Rev. **99**, 376 (1955)
5.31 V.G. Baru, F.F. Volkenstein: Dokl. Akad. Nauk SSSR 167, 1314 (1966)
5.32 F.F. Volkenstein, I.V. Karpenko: J. Appl. Phys. **33**, 460 (1962)
5.33 F.F. Volkenstein, O. Peshev: J. Catal. **4**, 301 (1965)
5.34 A.E. Bazhanova, Yu. A. Zarif'yants, V.S. Kuznetsov: Zh. Fiz. Khim. **49**, 1771 (1975)
5.35 W.E. Garner: Adv. Catal. **9**, 169 (1952)
5.36 V.G. Litovchenko, V.I. Lyashenko, O.S. Frolov: In *Poverkhnostnye svoistva poluprovodnikov* (Surface Properties on Semiconductors) (AN SSSR, Moscow 1962) p.147
5.37 R.V. Prudnikov, Yu.F. Novototsky-Vlasov, V.F. Kiselev: Sov. Phys.- Solid State **8**, 1957 (1967)
5.38 V.F. Kiselev: Dokl. Akad. Nauk SSSR **176**, 124 (1967)
5.39 S.R. Morrison: In *Progress in Surface Science*, Vol.1, ed. by S.G. Davison (Pergamon, Oxford 1972) p.270
5.40 S.R. Morrison: *The Chemical Physics of Surfaces* (Plenum, New York 1977)
5.41 V.F. Kiselev, Yu.A. Zarif'yants: Probl. Kinet. Katal. 16, 221 (1975)
5.42 O.V. Krylov: *Catalysis by Nonmetals* (Academic, New York 1970)
5.43 V.L. Bonch-Bruevich: Zh. Fiz. Khim. **27**, 662 (1953)
5.44 R.P. Messmer: Surface Sci. **106**, 225 (1981)
5.45 A.J. Bennett, B. McCarrol, R.P. Messmer: Surface Sci. **24**, 191 (1971)
5.46 E.B. Chensky, Yu.Ya. Tkach: Sov. Phys.-JETP **79**, 1809 (1980)
5.47 F.I. Vilesov, A.N. Terenin: Dokl. Akad. Nauk SSSR **125**, 1053 (1959)
5.48 W. Göpel: J. Vac. Sci. Tech. **16**, 1229 (1979)
5.49 M. Green: Disc. Farad. Soc. **28**, 235 (1959)
5.50 S.R. Morrison: J. Phys. Chem. **57**, 860 (1953)
5.51 B.I. Boltaks: *Diffuziya i tochechnye defekty v poluprovodnikakh* (Diffusion and Point Defects in Semiconductors) (Nauka, Moscow 1973)
5.52 B. Henderson, A.E. Hughes: *Defects and Their Structures in Nonmetallic Solids* (Plenum, New York 1976)
      V. Paidar, L. Lejček (eds.): *The Structure and Properties of Crystall Defects* (Proc. of the Symp., Liblic. Czechoslovakia 1983) North-Holland Physics Publishing 1984
5.53 E.P. Matsas, L.L. Dyner, V.E. Primachenko, O.V. Snitko: Surface Sci. **19**, 109 (1970)
5.54 I.A. Karpovich, A.N. Kalinin, B.I. Bedny: Sov. Phys.-Semiconductors **10**, 832 (1976)
5.55 O.V. Romanov, M.A. Sokolov, P.R. Kokorov, A.D. Andreev: Mikroelektronika **6**, 263 (1977)
5.56 L.N. Abessonova, V.N. Dobrovolsky, Yu.S. Zharkikh: Sov.Phys.-Solid State **20**, 1431 (1978)
5.57 I.N. Levine: Photochem. Photobiol. **4**, 391 (1965)
5.58 V.G. Baru, F.F. Volkenstein: *Vliyanie radiatsii na poverkhnostnye svoistva poluprovodnikov* (The Influence of Radiation on the Surface Properties of Semiconductors) (Nauka, Moscow 1978)
5.59 T. Kwan: In *Symposium on Electronic Phenomena in Chemisorption and Catalysis on Semiconductors* (W. de Gruyter, Berlin 1969) p.184
5.60 E. Molinare: In *Symposium on Electronic Phenomena in Chemisorption and Catalysis on Semiconductors* (W. de Gruyter, Berlin 1969) p.167
5.61 L.L. Basov, G.N. Kuzmin, Yu.P. Solonitsyn: Usp. Fotoniki **6**, 82 (1977)
5.62 Y.H. Nym: Sov. Phys.-Semiconductors **8**, 1370 (1975)
5.63 F.F. Volkenstein, I.V. Karpenko, S.F. Timashev: Dokl. Akad. Nauk SSSR **206**, 13 (1972)
5.64 V.F. Kiselev: Z. Chem. **7**, 369 (1967)
5.65 V.F. Kiselev: In *Symposium on Electronic Phenomena in Chemisorption and Catalysis on Semiconductors* (W. de Gruyter, Berlin 1969) p.101

Chapter 6

6.1 V.F. Kiselev: Z. Chem. 7, 369 (1967)
6.2 V.F. Kiselev: In *Symposium on Electronic Phenomena in Chemisorption and Catalysis on Semiconductors* (W. de Gruyter, Berlin 1969) p.101
6.3 V.F. Kiselev, O.V. Krylov: *Adsorption Processes on Semiconductor and Dielectric Surfaces I*, Springer Ser. Chem. Phys., Vol.32 (Springer, Berlin, Heidelberg 1985)
6.4 E.N. Figurovskaya, V.F. Kiselev: Dokl. Akad. Nauk SSSR 181, 1365 (1968)
6.5 J.J. Chessick, A.C. Zettlemoyer, G.Y. Young: Canad. J. Chem. 33, 251 (1955)
6.6 J.A. Morrison, J.M. Los, L.E. Drain: Trans. Farad. Soc. 47, 1023 (1951)
6.7 V.V. Betsa, Yu.V. Popik: Sov. Phys.-Semiconductors 10, 1035 (1976)
6.8 B.V. Rosentuller, K.N. Spiridonov, O.V. Krylov: Dokl. Akad. Nauk SSSR 259, 859 (1981)
6.9 A. Rosenberg: J. Phys. Chem. 62, 1112 (1958)
6.10 A.B. Ephraim, Z. Calahorra, M. Folman: J. Chem. Soc. Farad. Trans. 73, 1668 (1977); 75, 158 (1979)
6.11 T.T. Bykova: In *Elektronnye protsessy na poverkhnosti i v monokristalli-cheskikh plenkakh* (Electronic Processes on Surfaces and in Monocrystalline Films) (Nauka, Novosibirsk 1967) p.181
6.12 A.N. Zhevno, V.V. Sidorik, V.D. Tkachev: Phys. Status Solidi B 102, 113 (1980)
6.13 M.J. Sparnaay: Surface Sci. 13, 99 (1969)
6.14 W. Göpel: J. Vac. Sci. Tech. 16, 1229 (1979)
6.15 V.F. Kiselev: *Poverkhnostnye yavleniya v poluprovodnikakh i dielektrikakh* (Surface Phenomena in Semiconductors and Dielectrics) (Nauka, Moscow 1970)
6.16 P. Mark: RCA Rev. 26, 461 (1965)
6.17 K. Hauffe: Adv. Catal. 7, 213 (1955)
6.18 K. Hauffe: *Reaktionen in und an festen Stoffen* (Springer, Berlin 1966)
6.19 S.R. Morrison: In *Progress in Surface Science*, Vol.1, ed. by S.G. Davison (Pergamon, Oxford 1972) p.270
6.20 V.F. Kiselev, Yu.A. Zarif'yants: Probl. Kinet. Katal. 16, 221 (1975)
6.21 V.I. Lyashenko, V.G. Litovchenko, I.I. Stepko, V.I. Stricha: *Elektronnye Yavleniya na poverkhnosti poluprovodnikov* (Electronic Phenomena on Semiconductor Surfaces) (Naukova Dumka, Kiev 1968)
6.22 F.F. Volkenstein: *The Electron Theory of Catalysis on Semiconductors* (McMillan, New York 1963); *Fizikokhimiya pverkhnosti poluprovodnikov* (The Physical Chemistry of Semiconductor Surfaces) (Nauka, Moscow 1973)
6.23 W.H. Brattain, J. Bardeen: J. Bell Syst. Tech. 32, 1 (1953)
6.24 R.H. Kingston: Phys. Rev. 98, 1766 (1955)
6.25 J.T. Low: J. Phys. Chem. 59, 67 (1955)
6.26 W.E. Garner: Adv. Catal. 9, 169 (1952)
6.27 V.G. Litovchenko, V.I. Lyashenko, O.S. Frolov: In *Poverkhnostnye svoistva poluprovodnikov* (Surface Properties on Semiconductors) (AN SSSR, Moscow 1962) p.147
6.28 Yu.F. Novototsky-Vlasov, M.P. Sinyukov: In *Poverkhnostnye svoistva poluprovodnikov* (Surface Properties of Semiconductors) (AN SSSR, Moscow 1962) p.69
6.29 V.A. Presnov, L.L. Lyuse: In *Poverkhnostnye svoistva poluprovodnikov* (Surface Properties of Semiconductors) (AN SSSR, Moscow 1962) p.217
6.30 V.V. Murina, Yu.F. Novototsky-Vlasov: In *Elektronnye yavleniya na poverkhnosti poluprovodnikov* (Electronic Phenomena on Semiconductor Surfaces) p.262 (Nauka, Novosibirsk 1974)
6.31 A.V. Rzhanov: *Elektronnye protsessy na poverkhnosti poluprovodnikov* (Electronic Processes on Semiconductor Surfaces) (Nauka, Moscow 1971)

6.32 O. Jantsch: J. Phys. Chem. Solids **26**, 1233 (1965)
6.33 G. Dorda: Czech. J. Phys. B **10**, 820 (1960); B **13**, 272 (1963)
6.34 G.M. Schwab: In *Semiconductor Surface Physics* (Univ. of Pennsylvania Press, Philadelphia 1957) p.283, ed.by R.H. Kingston
6.35 Yu.F. Novototsky-Vlasov: Trudy Fiz. Inst. AN SSSR **48**, 3 (1969)
6.36 R.V. Prudnikov, Yu.F. Novototsky-Vlasov, V.F. Kiselev: Sov. Phys.-Solid State **8**, 1957 (1967)
6.37 V.F. Kiselev, S.N. Kozlov, Yu.F. Novototsky-Vlasov, R.V. Prudnikov: Surface Sci. **11**, 111 (1968)
6.38 V.F. Kiselev: Kinet. Katal. **11**, 403 (1970)
6.39 S.N. Kozlov, V.F. Kiselev, Yu.F. Novototsky-Vlasov: Surface Sci. **28**, 395 (1971)
6.40 R.V. Prudnikov: Vestnik MGU, Fizika **11**, 126 (1970)
6.41 L.I. Burbulyavichus, Yu.A. Zarif'yants, S.N. Karyagin, V.F. Kiselev: Kinet. Katal. **14**, 1526 (1973)
6.42 V.N. Agarev, V.I. Stafeev: Sov. Phys.-Semiconductors **11**, 990 (1977)
6.43 J. Sochanski, H. Gatos: Surface Sci. **13**, 393 (1969)
6.44 V.A. Matveev, R.V. Prudnikov: Vestnik MGU, Fizika **33**, 56 (1978)
6.45 V.I. Grinev, V.F. Kiselev, V.A. Matveev: Sov. Phys.-Semiconductors **14**, 936 (1980)
6.46 M.J.D. Low, N. Madison, P. Ramamurthy: Surface Sci. **13**, 238 (1969)
6.47 V.F. Kiselev, V.A. Matveev, R.V. Prudnikov: Phys. Status Solidi A **50**, 739 (1978)
6.48 S.N. Kozlov, V.F. Kiselev: Kinet. Katal. **12**, 789 (1971)
6.49 V.I. Grinev, V.F. Kiselev: Phys. Status Solidi A **66**, 493 (1981)
6.50 M.J.D. Low, Kun-ichi-Matsushita: J. Phys. Chem. **73**, 908 (1969)
6.51 V.I. Grinev, V.F. Kiselev: Vestnik MGU, Fizika **22**, 79 (1981)
6.52 T.N. Kompaniets: Zh. Tekh. Fiz. **46**, 1361 (1976)
6.53 A. Metcalfe, S. Shankar: J. Chem. Soc. Farad. Trans. **75**, 962 (1979)
6.54 K.C. Panday: Phys. Rev. B **14**, 1557 (1976)
      S. Ciraci, R. Butz, E.M. Oellig, H. Wagner: Phys. Rev. B **30**, 711 (1984)
6.55 P.P. Konorov, Yu.A. Tarantov: Sov. Phys.-Semiconductors **9**, 648 (1975)
6.56 I.G. Neizvestny, S.V. Pokrovskaya, A.V. Rzhanov: Sov. Phys.-Semiconductors **6**, 281 (1972)
6.57 V.F. Kiselev: Dokl. Akad. Nauk SSSR **176**, 124 (1967)
6.58 V.I. Grinev, S.N. Karjagin, V.F. Kiselev, I.O. Ogloblin: Izv. Vyssh. Uchebn. Zaved. Fizika **9**, 94 (1982)
6.59 E.P. Matsas, L.L. Dyner, V.E. Primachenko, O.V. Snitko: Surface Sci. **19**, 109 (1970)
6.60 V.V. Milenin, V.E. Primachenko, O.V. Snitko: Sov. Phys.-Semiconductors **11**, 654 (1977)
6.61 I.A. Karpovich, B.I. Bedny, A.N. Kalinin: Sov. Phys.-Semiconductors **10**, 1107 (1976)
6.62 P.W. Schindler, B. Fürst, R. Dick, P.U. Wolf: J. Coll. Int. Sci. **53**, 478 (1976)
6.63 L.A. Ignatyeva: Dokl. Akad. Nauk SSSR **163**, 398 (1965)
6.64 G.F. Golovanova, V.F. Kiselev, A.S. Petrov, E.A. Silaev: Phys. Status Solidi A **64**, 177 (1981)
6.65 J. Puatu, K. Tu, J. Meier (eds.): *Thin Films. Interdiffusion and Reactions* (Wiley, New York 1978)
6.66 S.R. Morrison: *The Chemical Physics of Surfaces* (Plenum, New York 1977)
6.67 T.A. Goodwin, P. Mark: In *Progress in Surface Science*, Vol.1, ed. by S.G. Davison (Pergamon, Oxford 1973) p.3
6.68 D.J. Miller, D.H. Haneman: Surface Sci. **19**, 45 (1970)
6.69 D. Lichtman, Y. Shapiro: J. Nucl. Mater. **63**, 184 (1976)
6.70 M.N. Danchevskaya, E.N. Figurovskaya: Kinet. Katal. **10**, 930 (1969)
6.71 I.A. Myasnikov: Zh. VkhO im. Mendeleeva **20**, 19 (1975)

6.72 S.J. Teichner: Rev. Haut. Temp. **2**, 79 (1965)
6.73 Yu.P. Solonitsyn: Kinet. Katal. **6**, 79 (1965)
6.74 E.N. Figurovskaya, V.F. Kiselev: Dokl. Akad. Nauk SSSR **175**, 1336 (1967)
6.75 V.L. Rapoport: Dokl. Akad. Nauk SSSR **153**, 871 (1963)
6.76 G. Heiland, E. Mollwo, F. Stockman: Solid State Phys. **8**, 193 (1959)
      G. Heiland: Disc. Farad. Soc. **28**, 186 (1959)
      G. Heiland, P. Kunstmann: Surface Sci. **13**, 72 (1969)
6.77 D.A. Melnick: J. Chem. Phys. **26**, 1136 (1957)
6.78 D.B. Medved: J. Phys. Chem. Solids **20**, 255 (1961)
6.79 D. Eger, Y. Goldstein, A. Many: RCA Rev. **36**, 508 (1975)
6.80 B.M. Arghiropoulos, S.J. Teichner: J. Catal. **8**, 477 (1964)
6.81 S.K. Bhattacharyya, P. Mohanti: J. Catal. **20**, 10 (1971)
6.82 G.I. Golodets: *Heterogeneous Catalytic Reactions Involving Molecular Oxygen* (North-Holland, Amsterdam 1983)
6.83 D.A. King, D.P. Woodruff (eds.): *The Chemical Physics of Solid Surfaces and Heterogeneous Catalysis*, Chemisorption Systems (North-Holland, Amsterdam 1984)
6.84 A. Many: Crit. Rev. Solid State Sci. **4**, 515 (1974)
6.85 D. Eger, A. Many, Y. Goldstein: Surface Sci. **58**, 18 (1976)
6.86 G. Heiland, D. Kohl: Phys. Status Solid: **A 49**, 27 (1978)
6.87 O. Peshev: *Adsorptsiya i reaktsii na dispersnykh poluprovodnikakh* (Adsorption and Reactions on Disperse Semiconductors) (AN Bulgaria, Sofia 1980)
6.88 K. Hauffe, R. Schmidt: Phys. Status Solidi **A 3**, 173 (1970)
6.89 S.R. Morrison: Surface Sci. **15**, 85 (1969)
6.90 J.D. Levine: J. Vac. Sci. Tech. **6**, 549 (1969)
6.91 V.A. Tyagai, O.V. Snitko, Yu.N. Shirshov: Fiz. Tekh. Polupr. **2**, 1391 (1968)
6.92 L.J. Brilison: Surface Sci. **51**, 45 (1975)
6.93 A.M. Murtazin, Yu.A. Zarif'yants: Sov. Phys.-Semiconductors **7**, 1364 (1974)
6.94 P. Mark: J. Phys. Chem. Soc. **26**, 959, 1767 (1965)
6.95 S. Baidyaroy, P. Mark: Surface Sci. **30**, 53 (1972)
6.96 P. Mark: Surface Sci. **25**, 192 (1971)
6.97 A.E. Bazhanova, Yu.A. Zarif'yants: Vestnik MGU, Fizika **13**, 355 (1972)
6.98 E.F. Gross, R.I. Shekhmetev: Fiz. Tverd. Tela **5**, 502 (1963)
6.99 Yu.D. Pimenov: Vestnik LGU **10**, 69 (1968)
6.100 O.S. Sinez, G.P. Peka, Yu.I. Karchanin: Fiz. Tverd. Tela **6**, 3515 (1964)
6.101 Yu.D. Pimenov: Opt. Spektrosk. **32**, 92 (1972)
6.102 S.N. Kozlov, V.F. Kiselev, Yu.F. Novototsky-Vlasov: Sov. Phys.-Semiconductors **6**, 1788 (1973)
6.103 V.F. Kiselev: Probl. Kinet. Katal. **14**, 25 (1970)
6.104 Ya.I. Ryskin: Opt. Spektrosk. **4**, 532 (1958)
6.105 N.D. Sokolov: Usp. Fiz. Nauk **57**, 205 (1955)
6.106 H. Tsubomura: J. Chem. Phys. **24**, 977 (1956)
6.107 P.A. Thomas, C. Sebenne, M. Balkanski: Rev. Phys. Appl. **5**, 683 (1970)
6.108 E.H. Weber: Phys. Status Solidi **A 1**, 665 (1970)
6.109 E. Arijs, F. Cardon, W. Maenhout-Vandervorst: J. Solid State Chem. **6**, 310, 319 (1973)
6.110 F.F. Volkenstein, O. Peshev: J. Catal. **4**, 301 (1965)
6.111 Yu.P. Solonitsyn: Kinet. Katal. **6**, 433 (1965)
6.112 I.M. Podgorny, A.I. Gorbunov: Probl. Kinet. Katal. **14**, 148 (1970)
6.113 V.I. Zivenko, I.A. Myasnikov: Probl. Kinet. Katal. **14**, 198 (1970)
6.114 P.N. Savin, I.A. Myasnikov: Zh. Fiz. Khim. **48**, 173, 1262 (1974)
6.115 R. Glemsa, R. Kokes: J. Phys. Chem. **66**, 566 (1962)
6.116 J. Novotny: Bull. Acad. Polon., Ser. Sci. Chim. **17**, 173 (1969); **21**, 743, 751, 815, 821 (1973)
6.117 A. Bielansky, G. Deren, G. Gaber: Probl. Kinet. Katal. **10**, 37 (1960)
6.118 S.N. Kozlov, Yu.F. Novototsky-Vlasov, V.F. Kiselev: Fiz. Tekh. Polupr. **4**, 353 (1970)

6.119 A.E. Bazhanova, Yu.A. Zarif'yants, V.F. Kiselev: Sov. Phys.-Dokl. (Phys. Chem.) (DKPCAG) **217**, 763 (1974)
6.120 V.A. Sokolov, B.G. Govorunov: Izv. Vyssh. Uchebn. Zaved., Fizika **8**, 134 (1973)
N.N. Govorunov, V.A. Sokolov: Izv. Vyssh. Uchebn. Zaved., Fizika **8**, 93 (1974)
6.121 M.A. Magomedov, H.A. Magomedov: Izv. Vyssh. Uchebn, Zaved. Fizika **3**, 53 (1981)
6.122 T.I. Povchan: Zh. Prikl. Nauchn. Fotogr. **19**, 221 (1974)
6.123 Y.H. Nym: Sov. Phys.-Semiconductor **8**, 1370 (1975)
6.124 S.N. Kozlov: Izv. Vyssh. Uchebn. Zaved., Fizika **2**, 116 (1975)
6.125 A.E. Bazhanova, Yu.A. Zarif'yants: Vestnik MGU, Fizika **15**, 360 (1974)
6.126 V.N. Ageev, N.I. Ionov: In *Progress in Surface Science,* Vol.5, ed. by S.G. Davison (Pergamon, Oxford 1975) p.5
6.127 R.S. Mulliken, W.B. Person: *Molecular Complexes* (Interscience, New York 1969)
6.128 V.D. Kasarizky, S.N. Kozlov, Yu.F. Novototsky-Vlasov: Dokl. Akad. Nauk SSSR **195**, 115 (1970)
6.129 Yu.A. Zarif'yants, S.N. Karyagin, V.F. Kiselev: Dokl. Akad. Nauk SSSR **202**, 109 (1972)
6.130 E.I. Balabanov, B.R. Gakel, I.D. Mikhailov: Sov. Phys.-Dokl. **238**, 160 (1978)
6.131 L.I. Burbulyavichus, Yu.A. Zarif'yants, V.F. Kiselev: Vestnik MGU, Fizika **12**, 599 (1971)
6.132 V.A. Shwets, V.B. Kazansky: Kinet. Katal. **12**, 935 (1971)
6.133 J. Scheider, A. Racuber: Phys. Lett. **21**, 380 (1966)
6.134 Y. Kondratoff, C. Naccache, B. Imelik: J. Chem. Phys. **65**, 562 (1968)
6.135 V.E. Heinrich, H.J. Zeiger, G. Dresselhaus: Nat. Bur. Standards **455**, 133 (1975)
6.136 A.R. Gonzalez-Elipe, G. Manuero, J. Soria: Trans. Farad. Soc. **75**, 748 (1979)
6.137 A.A. Lisachenko, F.I. Vilesov: Dokl. Akad. Nauk SSSR **176**, 1103 (1967)
6.138 M. Watanabe: Jpn. J. Appl. Phys. **19**, 1853, 1863 (1980)
6.139 M. Green, K.H. Maxwell: J. Phys. Chem. Solids **11**, 915 (1960)
6.140 E.N. Guryanov, I.P. Goldstein, I.P. Romm: *Donorno-aktseptornaya svyaz'* (The Donor-Acceptor Bond) (Nauka, Moscow 1973)
6.141 E.L. Nagaev: Zh. Fiz. Khim. **35**, 327, 2376 (1961)
6.142 N.H. Fletcher: Phil. Mag. **16**, 159 (1967)
6.143 Yu.A. Zarif'yants, V.F. Kiselev: In *Problemy fiziki poluprovodnikov* (Problems in Semiconductor Physics) (University Press, Kaliningrad 1975) p.3
6.144 Yu.A. Zarif'yants, V.F. Kiselev, S.N. Kozlov, Yu.F. Novototsky-Vlasov: Vestnik MGU, Fizika **16**, 84 (1975)
6.145 R.S. Bauer, J.C. McMenamin, H. Petersen, A. Bianconi: In [4.6] p.170
A. Bianconi, L. Incoccia, S. Stipcich (eds.): *EXAFS and Near Edge Structure,* Springer Ser. Chem. Phys., Vol.27 (Springer, Berlin, Heidelberg 1983)
6.146 A. Bianconi, R.S. Bauer: Surface Sci. **99**, 76 (1980)
6.147 R.J. Grunthaner, P.J. Grunthaner, R.P. Vasquez: Phys. Rev. Lett. **43**, 1683 (1979)
R.P. Vasquez, F.J. Grunthaner: Surface Sci. **99**, 681 (1980)
6.148 A. Ishizaka, S. Iwata, Y. Kamigaki: Surface Sci. **84**, 355 (1979)
6.149 J. Finster, D. Schulz, F. Bechstedt, A. Meisel: Surface Sci. **152/153**, 1063 (1985)
G. Hollinger, F.J. Himpsel: Appl. Phys. Lett. **44**, 93 (1984)
6.150 J.S. Hohannessen, W.E. Spicer, Y.E. Strausser: J. Vac. Sci. Tech. **13**, 849 (1976)

6.151 C.C.Chang, D.M. Boulin: Surface Sci. **69**, 385 (1977)
6.152 T. Adachi, C.R. Helms: Appl. Phys. Lett. **35**, 199 (1979)
6.153 A. Benninghoven, J. Giber, J. Laszlo, N. Riedel, H.W. Werner (eds.):
      Secondary Ion Mass Spectroscopy (SIMS II), Springer Ser. Chem. Phys.,
      Vol.19 (Springer, Berlin, Heidelberg 1982)
      A. Benninghoven, J. Okano, R. Shimizu, H.W. Werner (eds.): *Secondary
      Ion Mass Spectroscopy (SIMS III)*, Springer Ser. Chem. Phys., Vol.36
      (Springer, Berlin, Heidelberg 1984)
6.154 W.L. Harrington, R.E. Honig, A.M. Goodman, R. Williams: Appl. Phys. Lett.
      27, 644 (1975)
6.155 A.E. Gershinskii, L.V. Mironova, E.I. Cherepov: Phys. Status Solidi A
      38, 369 (1976)
6.156 J.F. Wagner, C. Wilmsen: J. Appl. Phys. **50**, 874 (1979)
6.157 T. Sugano: Surface Sci. **98**, 145 (1980)
6.158 D.E. Aspnes, J.B. Theeten: Phys. Rev. Lett. **43**, 1046 (1979)
6.159 V.G. Litovchenko: Sov. Phys.-Semiconductors **6**, 696 (1972)
6.160 V.G. Litovchenko: *Osnovy fiziki poluprovodnikovykh sloistykh sistem*
      (Basis of Physics of Semiconductor Laminated Structures) (Naukova Dumka,
      Kiev 1980)
6.161 C. Fritzsche: Z. Angew. Phys. **24**, 48 (1967)
6.162 A.G. Revesz: Phys. Status Solidi A **57**, 657 (1980)
6.163 W. Kern: Solid State Tech. **17**, 35, 78 (1974)
6.164 V.G. Litovchenko, G.F. Romanova: Thin Solid Films **81**, 27 (1981)
6.165 M. Pijolat, G. Hollinger: Surface Sci. **105**, 114 (1981)
6.166 F. Herman, P. Batra, J.P. Kasovski: In *Proc. Int. Conf. on the Physics
      of SiO_2 and Its Interfaces*, ed. by S.T. Pantelides (Plenum, New York
      1978) p.333
6.167 S.T. Pantelides, M. Long: In *Proc. Int. Conf. on the Physics of SiO_2
      and Its Interfaces*, ed. by S.T. Pantelides (Plenum, New York 1978)
      p.339
6.168 K. Hübner: Phys. Status Solidi A **61**, 665 (1980)
6.169 N.B. Goodman, H. Fritzsche: Phil. Mag. **42**, 149 (1980)
6.170 R.B. Laughlin, J.D. Joannopoulos, D.G. Chadi: Phys. Rev. **B 21**, 5733 (1980)
6.171 D.J. Chadi: In *Proc. Int. Conf. on the Physics of SiO_2 and Its Interfaces*,
      ed. by S.T. Pantelides (Plenum, New York 1978) p.55
      R.N. Nucho, A. Madhukar: In *Proc. Int. Conf. on the Physics of SiO_2 and
      Its Interfaces*, ed. by S.T. Pantelides (Plenum, New York 1978) p.60
6.172 A. Stern, K. Hübner, Phys. Status Sol. **B 124**, K 171 (1984)
6.173 S.N. Kozlov, S.N. Kuznetsov: Sov. Phys.-Semiconductors **12**, 995 (1978)
6.174 N.F. Mott: In *Proc. Int. Conf. on the Physics of SiO_2 and Its Interfaces*,
      ed. by S.T. Pantelides (Plenum, New York 1978) p.1
6.175 K.L. Brover: Appl. Phys. Lett. **43**, 111 (1983)
6.176 A. Stesmans, J. Braet, J. Witters, R.F. Dekeermaecker: Surface Sci. 141,
      255 (1984)
6.177 H.R. Philipp: J. Phys. Chem. Solids **32**, 1935 (1971)
6.178 K. Hübner, R. Engelke: Phys. Status Solidi A 53, 79 (1979)
      W.Y. Ching: Phys. Rev. **B 26**, 6610 (1982)
6.179 T. Sugano, J.J. Chen, T. Hamano: Surface Sci. **98**, 154 (1980)
      M. Henzler, P.J. Marienhof: J. Vac. Sci. Techn. B **2**, 346 (1984)
6.180 J.D. Jorgensen: J. Appl. Phys. **49**, 5473 (1978)
6.181 D.J. Lepine: Phys. Rev. B **6**, 436 (1972)
6.182 S.N. Karyagin, V.V. Kurylev: In *Elektronnye protsessy na poverkhnosti
      poluprovodnikov* (Electronic Processes on Semiconductor Surfaces) ed. by
      A.V. Rzhanov (Nauka, Novosibirsk 1974) p.88
      V.V. Kurylev, S.N. Karyagin: Phys. Status Solidi A 21, 127 (1974)
      S.N. Karyagin: Phys. Status Solidi A **68**, 113 (1981)
6.183 E.H. Poindexter, P.J. Caplan, R.R. Razouk: J. Appl. Phys. **52**, 879 (1981)
6.184 P.M. Lenahan, P.V. Dressendorf: Appl. Phys. Lett. **44**, 96 (1984)

6.185 V.F. Kiselev, S.N. Kozlov, Yu.A. Zarif'yants: In *Problemy fizicheskoi khimii poluprovodnikovykh poverkhnostei* (Problems in the Physical Chemistry of Semiconductor Surfaces) (Nauka, Novosibirsk 1978) p.200
6.186 V.I. Stricha, S.S. Kilchitskaya: Izv. Vyssh. Uchebn. Zaved., Fizika **11**, 122 (1968)
6.187 A.I. Vovsi, L.P. Strakhov: Sov. Phys.-Solid State **14**, 2863 (1972)
6.188 J.D. Levine, S. Freeman: Phys. Rev. B **2**, 3255 (1970)
6.189 T.T. Bykova, E.F. Lasneva: Sov. Phys.-Semiconductors **6**, 1571 (1972)
6.190 V.A. Zuev, G.A. Sukach: Phys. Status Solidi A **18**, 57 (1973)
6.191 M. Hino, T. Sato: Bull. Chem. Soc. Jpn. **41**, 33 (1971)
6.192 Yu.A. Zarif'yants, V.F. Kiselev: In *Svyazannaya voda v dispersnykh sistemakh* (Bonded Water in Dispersed Systems), ed. by V.F. Kiselev (MGU, Moscow 1974) p.74
6.193 Yu.M. Tshekochikhin, A.A. Davidov, A.S. Kuznetsov: Zh. Fiz. Khim. **47**, 2499 (1973)
6.194 H. Ibach: Phys. Rev. Lett. **24**, 1416 (1970)
6.195 G. Heiland, H. Ibach: Surface Sci. **35**, 425 (1973)
6.196 V.V. Briksin, Yu.M. Gerbstein: Sov. Phys.-Solid State **14**, 453 (1972)
6.197 Yu.A. Pasechnik, O.V. Snitko: Pis'ma Zh. Eksp. Teor. Fiz. **17**, 583 (1973)
6.198 D.J. Evans, S. Ishioda, J.D. McMullen: Phys. Rev. **31**, 369 (1973)
6.199 V.A. Zuev, D.V. Korbutyak, V.G. Litovchenko: Surface Sci. **50**, 215 (1975)
6.200 V.G. Litovchenko, V.A. Zuev: Jpn. J. Appl. Phys. Suppl. **2**, 2 (1974)
6.201 P. Mark: Catal. Rev. **12**, 71 (1975)
6.202 P. Mark, S.S. Chang, W.F. Creighton, B.W. Lee: Crit. Rev. Solid State Sci. **5**, 189 (1975)
6.203 H. Wagner: In *Solid Surface Physics*, ed. by J. Hölzl, F.K. Shulte, H. Wagner, Springer Tracts in Modern Physics, Vol.85 (Springer, Berlin, Heidelberg 1979) p.123
6.204 V.T. Rajan, L.M. Falicov: J. Phys.C. Solid State Phys. **9**, 2533 (1976)
6.205 J.E. Rowe, S.B. Christman, H. Ibach: Phys. Rev. Lett. **34**, 874 (1975)
6.206 Yu.A. Zarif'yants, S.N. Karyagin, V.F. Kiselev: Vestnik MGU, Fizika **16**, 237 (1975)
6.207 G.A. Somorjai, M.A. van Hove: *Adsorbed Monolayers on Solid Surfaces, Structure and Bonding*, Vol. 38 (Springer, Berlin, Heidelberg 1979)
6.208 N.S. Lidorenko, S.L. Nagaev: Pis'ma Zh. Eksp. Teor. Fiz. **31**, 505 (1980)
6.209 M.F. Deigen, N.I. Kashirina, L.A. Suslov: Zh. Eksp. Teor. Fiz. **75**, 149 (1978)
6.210 H. Haken: *Synergetics*, 2nd ed. (Springer, Berlin, Heidelberg 1978)
6.211 F.J. Feigl, W.B. Fowler, K.L. Yip: Solid State Commun. **14**, 225 (1974)
6.212 R.A.B. Devine, A. Golanski: J. Appl. Phys. **54**, 3833 (1983)
6.213 A. Stesmans, J. Witters, J. Braet, R.F. Dekeermaecker: J. Appl. Phys. **55**, 1551 (1984)
6.214 S.N. Karyagin, P.K. Kashkarov: Phys. Status Solidi A **37**, 17 (1976)
6.215 S.N. Karyagin, S.N. Kozlov, G.S. Plotnikov: Izv. Vyssh. Uchebn. Zaved. Fizika **24**, 88 (1981)

Chapter 7

7.1 Yu.A. Zarif'yants, S.N. Karyagin, V.F. Kiselev: Dokl. Akad. Nauk SSSR **202**, 109 (1972)
7.2 H. Hartman, E. Eisenbraun, J. Heidberg: Z. Naturforsch. A **23**, 1689 (1968)
7.3 Li-Tun-Sin, V.L. Rappoport: Vestnik LGU **2**, 45 (1966)
7.4 A.I. Mashenko, V.B. Kazansky: Kinet. Katal. **8**, 853 (1967)
7.5 T. Kwan: In *Symposium on Electronic Phenomena in Chemisorption and Catalysis on Semiconductors* (W. de Gruyter, Berlin 1969) p.167
7.6 L.I. Burbulyavichus, Yu.A. Zarif'yants, V.F. Kiselev: Vestnik MGU, Fizika **12**, 599 (1971)

7.7   M. Setaka, K.M. Sancier, T. Kwan: J. Catal. **16**, 44 (1970)
7.8   M. Watanabe: Jpn. J. Appl. Phys. **19**, 1853, 1863 (1980)
7.9   V.D. Kasarizky, S.N. Kozlov, Yu.F. Novototsky-Vlasov: Dokl. Akad. Nauk
      SSSR **195**, 115 (1970)
7.10 S.N. Kozlov, Yu.F. Novototsky-Vlasov: Fiz. Tekh. Polupr. **4**, 356 (1970)
7.11 F.F. Volkenstein: *The Electron Theory of Catalysis on Semiconductors*
      (McMillan, New York 1963); *Fizikokhimiya poverkhnosti poluprovodnikov*
      (The Physical Chemistry of Semiconductor Surfaces) (Nauka, Moscow 1973)
7.12 Yu.D. Pimenov: Vestnik LGU **10**, 69 (1968)
7.13 L. Toko: Chem. Lett. **8**, 891 (1973)
7.14 V.F. Kiselev: Dokl. Akad. Nauk SSSR **176**, 124 (1967)
7.15 V.F. Kiselev: Z. Chem. **7**, 369 (1967)
7.16 V.F. Kiselev, O.V. Krylov: *Adsorption Processes on Semiconductor and
      Dielectric Surfaces I*, Springer Ser. Chem. Phys., Vol.32 (Springer,
      Berlin, Heidelberg 1985)
7.17 V.F. Kiselev: In *Symposium on Electronic Phenomena in Chemisorption and
      Catalysis on Semiconductors* (W. de Gruyter, Berlin 1969) p.101
7.18 V.B. Kazansky, N.D. Chuvylkin: Sov. Phys.-Dokl. (DKPCAG) **223**, 810 (1975)
      G.M. Zhidomirov, N.D. Micheikin: *Klasternoe priblizhenie v kwantovomecha-
      nithceskich issledovsniyach khemosorbcii i poverchnostnych struktur*
      (Cluster Approach in Quantum Chemistry Investigation of Chemisorption
      and Surface Structures), The Structure of Molecules and Chemical Bond,
      Vol.9 (VINITI, Moscow 1984)
7.19 J.P. Goldstein, L.F. Kucheruk, E.N. Guryanova: Sov. Phys.-Dokl. (DKPCAG)
      252, 430 (1975)
7.20 O.V. Krylov: *Catalysis by Nonmetals* (Academic, New York 1970)
7.21 V.F. Kiselev, Yu.A. Zarif'yants: Probl. Kinet. Katal. **16**, 221 (1975)
7.22 V.F. Kiselev: Kinet. Katal. **11**, 403 (1970)
7.23 O.V. Krylov: Probl. Kinet. Katal. **13**, 141 (1968)
7.24 E.P. Mikheeva, N.P. Keier: Kinet. Katal. **5**, 748 (1964)
7.25 O.S. Frolov, O.V. Snitko, G.F. Romanova: Ukrain. Fiz. Zh. **14**, 585 (1969)
7.26 E.P. Mikheeva, N.P. Keier: Sov. Phys.-Dokl. (DKPCAG) **219**, 1145 (1974)
7.27 N.P. Keier, E.P. Mikheeva: Kinet. Katal. **10**, 288 (1969)
7.28 S.A. Hoenig, J.R. Lane: Surface Sci. **11**, 163 (1968)
7.29 V.I. Lyashenko, A.A. Serba, I.I. Stepko: Kinet. Katal. **7**, 1095 (1966)
7.30 L.I. Ivankiv, M.V. Milyanchuk, A.K. Filatova: Probl. Kinet. Katal. **12**,
      327 (1968)
7.31 N.P. Keier: In *Proc. 4th Int. Congr. on Catalysis, Moscow 1968*, Vol.1
      (Akademiai Kiado, Budapest 1971) p.119
7.32 L.M. Usoltseva, E.P. Mikheeva, N.P. Keier: Kinet. Catal. **14**, 1197 (1973)
7.33 V.F. Kiselev: V.A. Matveev, A.S. Petrov: Kinet. Katal. **21**, 523 (1980)
7.34 A.G. Milnes: *Deep Impurities in Semiconductors* (Wiley, New York 1973)
7.35 G.G. Fedorov, R.V. Prudnikov, V.F. Kiselev: Phys. Status Solidi **A 14**,
      19 (1972)
7.36 V.F. Kiselev, R.V. Prudnikov, G.G. Fedorov: Dokl. Akad. Nauk SSSR **208**,
      387 (1973)
7.37 G.M. Schwab: In *Semiconductor Surface Physics*, ed. by R.H. Kingston
      (Univ. of Pennsylvania Press, Philadelphia 1957) p.283
7.38 V.I. Lyashenko, I.I. Stepko: Probl. Kinet. Katal. **10**, 111 (1960)
7.39 N.D. Chuvylkin, G.M. Zhidomirov, V.B. Kazansky: Kinet. Katal. **14**, 943,
      1579 (1973)
7.40 E. Molinare: In *Symposium on Electronic Phenomena in Chemisorption and
      Catalysis on Semiconductors* (W. de Gruyter, Berlin 1969) p.167
7.41 L.L. Basov, G.N. Kuzmin, Yu.P. Solonitsyn: Usp. Fotoniki **6**, 82 (1977)
7.42 V.S. Sacharenko, A.E. Cherkashin, N.P. Keier: Dokl. Akad. Nauk SSSR
      211, 628 (1973)
7.43 Y. Shapira, R.B. McQuistan, D. Lichtman: Phys. Rev. **B 15**, 2163 (1977)
      Nguen van Hieu, D. Lichtman: J. of Catal. **73**, 329 (1982)

7.44 V.G. Baru, F. F. Volkenstein: *Vliyanie radiatsii na poverkhnostnye svoistva poluprovodnikov* (The Influence of Radiation on the Surface Properties of Semiconductors) (Nauka, Moscow 1978)

7.45 S.A. Surin, V.G. Kazansky: In *Yavleniya fotosorbtsii i fotokataliza v geterogennykh sistemakh* (Phenomena of Photosorption and Photocatalysis in Heterogeneous Systems) Vol.4 (Nauka, Novosibirsk 1974) p.108, 120

7.46 A.A. Lisachenko, F.I. Vilesov: In *Yavleniya fotosorbtsii i fotokataliza v geterogennykh sistemakh* (Phenomena of Photosorption and Photocatalysis in Heterogeneous Systems) Vol.4, (Nauka, Novosibirsk 1974) p.3

7.47 L.L. Basov, Yu.P. Solonitsyn, G.N. Kuz'min: In *Yavleniya fotosorbtsii i fotokataliza v geterogennykh sistemakh* (Phenomena of Photosorption and Photocatalysis in Heterogeneous Systems) Vol.4 (Nauka, Novosibirsk 1974) p.9

7.48 A.R. Gonzalez-Elipe, G. Manuero, J. Soria: Trans. Farad. Soc. **75**, 748 (1979)

7.49 A.A. Lisachenko, F.I. Vilesov: Dokl. Akad. Nauk SSSR **176**, 1103 (1967)

7.50 V.N. Agarev, V.I. Stafeev: Sov. Phys.-Semiconductors **11**, 990 (1977)

7.51 J. Lagowski, H.C. Gatos: Surface Sci. **76**, 575 (1978)

7.52 T.T. Bykova, E.F. Lasneva: Sov. Phys.-Semiconductors **6**, 1193 (1973)

7.53 D. Eger, Y. Goldstein, A. Many: RCA Rev. **36**, 508 (1975)

7.54 V.A. Kotel'nikov, A.N. Terenin: Dokl. Akad. Nauk SSSR **174**, 1366 (1966)

7.55 V.A. Kotel'nikov, I.M. Prudnikov: Kinet. Katal. **10**, 1112 (1969)

7.56 V.F. Kiselev: *Poverkhnostnye yavleniya v poluprovodnikakh i dielektrikakh* (Surface Phenomena in Semiconductors and Dielectrics) (Nauka, Moscow 1970)

7.57 S.R. Morrison: In *Progress in Surface Science,* Vol.1, ed. by S.G. Davison (Pergamon, Oxford 1972) p.270

7.58 S.R. Morrison: *The Chemical Physics of Surfaces* (Plenum, New York 1977)

7.59 T. Freund, S.R. Morrison, Y.P. Gones: In *Proc. 4th Int. Congr. on Catalysis, Moscow 1968,* Vol.1 (Akademiai Kiado, Budapest 1971) p.72

7.60 A.M. Grizkov, V.A. Shvez, V.B. Kazansky: Kinet. Catal. **15**, 1114 (1974)

7.61 V.G. Baru, F.F. Volkenstein: Dokl. Akad. Nauk SSSR 167, 1314 (1966)

7.62 F.F. Volkenstein, I.V. Karpenko: J. Appl. Phys. **33**, 460 (1962)

7.63 V.F. Kiselev: Dokl. Akad. Nauk SSSR **213**, 224 (1973)

7.64 V.L. Bonch-Bruevich: Zh. Fiz. Khim. **27**, 662 (1953)

7.65 K. Tanabe: *Solid Acids and Bases* (Academic, New York 1970)

7.66 S.R. Morrison: Surface Sci. **50**, 329 (1975)

7.67 O.V. Krylov, S.A. Markova, I.P. Tretyakov: Kinet. Katal. **6**, 12 (1963)

7.68 A. Zecchina, M.G. Lofthouse, F.S. Stone: Trans. Farad. Soc. **71**, 1476 (1965)

7.69 D. Cordischi, V. Indovina: Trans. Farad. Soc. **72**, 2341 (1977)

7.70 G.M. Zhidomirov, A.A. Bagaturyants, I.A. Abronin: *Sovremennaya teoriya indeksov reaktsionnoi sposobnosti* (Modern Theory of Reactivity Indexes) (Khimiya, Moscow 1979)

Chapter 8

8.1 F.F. Volkenstein: *The Electron Theory of Catalysis on Semiconductors* (McMillan, New York 1963); *Fizikokhimiya pverkhnosti poluprovodnikov* (The Physical Chemistry of Semiconductor Surfaces) (Nauka, Moscow 1973)

8.2 K. Hauffe: Adv. Catal. **7**, 213 (1955)

8.3 C.G.B. Garrett: J. Chem. Phys. **33**, 966 (1960)

8.4 V.J. Lee: J. Catal. **17**, 178 (1970); J. Chem. Phys. **55**, 2905 (1971)

8.5 S.R. Morrison: In *Progress in Surface Science,* Vol.1, ed. by S.G. Davison (Pergamon, Oxford 1972) p.270

8.6 S.R. Morrison: *The Chemical Physics of Surfaces* (Plenum, New York 1977)

8.7   V.G. Baru, F.F. Volkenstein: Dokl. Akad. Nauk SSSR 167, 1314 (1966)
8.8   I.N. Levine: Photochem. Photobiol. 4, 391 (1965)
8.9   V.G. Baru, F.F. Volkenstein: *Vliyanie radiatsii na poverkhnostnye svoistva poluprovodnikov* (The Influence of Radiation on the Surface Properties of Semiconductors)(Nauka, Moscow 1978)
8.10 G. Heiland, E. Mollwo, F. Stockman: Solid State Phys. 8, 193 (1959)
      G. Heiland: Disc. Farad. Soc. 28, 186 (1959)
      G. Heiland, P. Kunstmann: Surface Sci. 13, 72 (1969)
8.11 Y. Shapira, R.B. McQuistan, D. Lichtman: Phys. Rev. B 15, 2163 (1977)
8.12 K. Hauffe, R. Stechemesser: In *Symposium on Electronic Phenomena in Chemisorption and Catalysis on Semiconductors in Moscow*, ed. by K. Hauffe, Th. Wolkenstein (W. de Gruyter, Berlin 1969) p.1
8.13 L.L. Basov, G.N. Kuzmin, Yu.P. Solonitsyn: Usp. Fotoniki 6, 82 (1977)
8.14 L.L. Basov, Yu.P. Solonitsyn, G.N. Kuz'min: In *Yavleniya fotosorbtsii i fotokataliza v heterogennykh sistemakh* (Phenomena of Photosorption and Photocatalysis in Heterogeneous Systems) Vol.4 (Nauka, Novosibirsk 1974) p.9
8.15 G.M. Zhabrova, V.I. Vladimirov: In *Symposium on Electronic Phenomena in Chemisorption and Catalysis on Semiconductors in Moscow*, ed. by K. Hauffe, Th. Wolkenstein (W. de Gruyter, Berlin 1969) p.231
8.16 O.V. Krylov: Probl. Kinet. Katal. 15, 85 (1973)
8.17 F.F. Volkenstein, I.V. Karpenko: J. Appl. Phys. 33, 460 (1962)
8.18 V.A. Zuev, V.G. Litovchenko, G.A. Sukach: Ukrain. Fiz. Zh. 20, 1147 (1975)
8.19 O.V. Krylov: *Catalysis by Nonmetals* (Academic, New York 1970)
8.20 V.F. Kiselev: Kinet. Catal. 19, 926 (1978)
8.21 V.F. Kiselev, S.N. Kozlov, N.L. Levshin: Sov. Phys.-Dokl. (DKPCAG) 256, 31 (1981)
8.22 V.F. Kiselev, S.N. Kozlov, N.L. Levshin: Phys. Status Solidi A, 66, 93 (1981)
8.23 I. Pankov: *Optical Processes in Semiconductors* (Prentice-Hall, Englewood Cliffs, NJ 1971)
8.24 W. Arnold, W.D. Compton: Phys. Rev. 116, 302 (1959)
8.25 C.W. Gwyn: J. Appl. Phys. 40, 4886 (1969)
8.26 J.D. Weeks, J.C. Tully, L.C. Kimerling: Phys. Rev. B 12, 3286 (1975)
8.27 E.H. Snow, A.S. Grove, D.J. Fitzgerald: Proc. IEEE 55, 1168 (1967)
8.28 H.W. Huang, C.W. Huang: Appl. Phys. Lett. 30, 533 (1977)
8.29 M. Nakagizi: Jpn. J. Appl. Phys. 13, 1610 (1974)
8.30 V.M. Maslovski, A.P. Nagin, V.V. Pospelov, V.M. Tulkin: Sov. Phys.-Tech. Phys. 24, 1044 (1979)
      A.P. Nagin, V.M. Tyulkin: Pisma v ZHTF 8, 1423 (1982)
8.31 V.Ya. Kiblik, V.G. Litovchenko, R.O. Litvinov: Ukrain. Fiz. Zh. 25, 1348 (1980)
8.32 C.T. Sah, J.Y. Sun, J.J. Tzou: J. Appl. Phys. 53, 8886 (1982)
8.33 M. Knoll, D. Bräunig, W.R. Fahrner: J. Appl. Phys. 53, 6946 (1982)
8.34 F.A. Kröger: *The Chemistry of Imperfect Crystals* (Wiley, NY 1964)
8.35 V.L. Vinezky, G.A. Cholodar: *Statisticheskoe vzaimodeistvie electronov i defektov v poluprovodnike* (Statistical Interaction of Electrons and Defects in Semiconductors) (Naukova Dumka, Kiev 1969)
8.36 G.F. Neumark: J. Appl. Phys. 51, 3383 (1980)
      Ya.I. Frenkel: *Sobranie nauchnykh trudov* (Selected works) Vol.2 (Nauka, Moscow 1975) p.217
8.37 L.C. Kimerling: In *Defects and Radiation Effects in Semiconductors*, Proc. of Inst. Phys. Conf. Ser. No. 46, London 1979, p. 56
8.38 P.J. Dean, W.J. Choyke: Adv. Phys. 26, 1 (1977)
8.39 G.E. Chaika, V.L. Vinetzkii: Phys. Status Solidi B 98, 727 (1980)
8.40 S. O'Hara, P.W. Hutchinson, P.S. Dobson: Appl. Phys. Lett. 30, 368 (1977)
8.41 P.W. Hutchinson, P.S. Dobson: Solid State Electronics 21, 1413 (1978)
8.42 J. Bardeen: Phys. Rev. 71, 717 (1947)

8.43 J.T. Gerstein, R. Janov, N. Tzoar: Phys. Rev. B **11**, 1267 (1975)
8.44 L.E. Brus, J. Comas: J. Chem. Phys. **54**, 2771 (1971)
8.45 V.F. Kiselev, A.S. Petrov, R.V. Prudnikov: Sov. Phys.-Dokl. (DKPCAG) **245**, 318 (1979)
8.46 V.F. Kiselev, S.N. Kozlov, A.S. Petrov, E.A. Silaev: Sov. Phys.-Dokl. (DKPCAG) **255**, 649 (1980)
8.47 Yu.A. Zarif'yants, S.N. Karyagin, V.F. Kiselev: Vestnik MGU, Fizika **16**, 237 (1975)
8.48 V.F. Kiselev, O.V. Krylov: *Adsorption Processes on Semiconductor and Dielectric Surfaces I*, Springer Ser. Chem. Phys., Vol.32 (Springer, Berlin, Heidelberg 1985)
8.49 J. Scheve, J.F. Schultz: In *Proc. 4th Int. Congr. on Catalysis, Moscow 1968*, Vol.2 (Akademiai Kiado, Budapest 1971) p.483
8.50 V.L. Bonch-Bruevich: In *Statisticheskaya fizika i kvantovaya teoriya polya* (Statistical Physics and the Quantum Theory of Fields) (Nauka, Moscow 1973) p.337
8.51 V.L. Bonch-Bruevich, V.D. Iskra: Sov. Phys.-Semiconductors **11**, 397 (1977)
8.52 Yu.D. Pimenov: Vestnik LGU **10**, 69 (1968)
8.53 Yu.D. Pimenov: Opt. Spectrosk. **32**, 92 (1972)
8.54 V.S. Vinogradov: Fiz. Tverd. Tela **13**, 3266 (1971)
8.55 S.F. Timashev: Sov. Phys.-Solid State **14**, 2267 (1972)
8.56 A.M. Stoneham: *Theory of Defects in Solids* (Clarendon, Oxford 1975)
8.57 V.L. Bonch-Bruevich, S.G. Kalashnikov: *Fizika poluprovodnikov* (Semiconductor Physics) (Nauka, Moscow 1977)
8.58 V.N. Abakumov, V.I. Perel, I.N. Jassievich: Sov. Phys.-Semiconductors **12**, 1 (1978)
8.59 S.F. Timashev: Sov. Phys.-Solid State **14**, 136 (1972)
8.60 M. Lax: Phys. Rev. **119**, 1502 (1960)
8.61 E.F. Smith, P.T. Landsberg: J. Phys. Chem. Solids **27**, 1727 (1966)
8.62 V.L. Bonch-Bruevich, V.G. Glasko: Fiz. Tverd. Tela **4**, 510 (1962)
8.63 A.V. Rzhanov: *Elektronnye protsessy na poverkhnosti poluprovodnikov* (Electronic Processes on Semiconductor Surfaces) (Nauka, Moscow 1971)
8.64 Yu.F. Novototsky-Vlasov: Trudy Fiz. Inst. AN SSSR **48**, 3 (1969)
8.65 Yu.A. Kursky: Fiz. Tverd. Tela **4**, 2620 (1962); **6**, 1485 (1964)
8.66 M.R. Belmont: Thin Solid Films **28**, 149 (1975)
8.67 I.G. Neizvestny, S.V. Pokrovskaya, A.V. Rzhanov: Sov. Phys.-Semiconcutors **6**, 281 (1972)
8.68 A.A. Karpushin, A.V. Chaplik: Fiz. Tekh. Polupr. **1**, 1043 (1967)
8.69 E.V. Vlasenko, R.A. Suris: Sov. Phys.-Semiconductors **12**, 888 (1978)
8.70 A.I. Gromov, P.P. Konorov, Yu.A. Tarantov: Vestnik LGU, Fizika **1**, 44 (1979)
8.71 V.I. Grinev, V.F. Kiselev, V.A. Matveev: Sov. Phys.-Semiconductors **14**, 936 (1980)
8.72 V.I. Grinev, V.F. Kiselev: Phys. Status Solidi A **66**, 493 (1981)
8.73 V.F. Kiselev, S.N. Kozlov: Vestnik MGU, Fizika **34**, 56 (1979)
8.74 V.F. Kiselev, S.N. Kozlov, N.L. Levshin: Sov. Phys.-Tech. Phys. Lett. **6**, 221 (1980)
      S.N. Kozlov, N.L. Levshin: Phys. Status Sol. (a) **83**, K149 (1984)
8.75 A. Maradudin: *Theoretical and Experimental Aspects of the Effects of Point Defects and Disorder on Vibrations of Crystals* (Academic, New York 1966)
8.76 K.K. Rebane: *Elementarnaya teoriya struktury kolebatel'nykh spektrov tsentrov kristallicheskikh defektov* (Nauka, Moscow 1968)
8.77 V.A. Zuev, A.V. Sachenko, K.B. Tolpygo: *Neravovesnye protsessy v poluprovodnikakh i poluprovodnikovykh priborakh* (Nonequilibrium Processes in Semiconductors and Semiconductor Devices) (Sovetskoe Radio, Moscow 1977)
8.78 H. Ueda: Phys. Status Solidi B **100**, 705 (1980)
8.79 F.F. Volkenstein, I.V. Karpenko, S.F. Timashev: Dokl. Akad. Nauk SSSR **206**, 13 (1972)

8.80 T.I. Povchan: Zh. Prikl. Nauchn. Fotogr. **19**, 221 (1974)
8.81 S.N. Kozlov, S.N. Kuznetsov: Sov. Phys.-Tech. Phys. Lett. **5**, 166 (1979)
8.82 A. Galwey: Adv. Catal. **26**, 247 (1977)
8.83 O.V. Krylov, M.U. Kislyuk: Kinet. Katal. **13**, 63, 588 (1972)
8.84 Probl. Kinet. Katal. **17** (1978)
8.85 F.F. Volkenstein, A.N. Gorban, V.A. Sokolov: *Radikalorekombinatsionnaya lyuminestsentsiya poluprovodnikov* (Radical Recombination Luminescence of Semiconductors) (Nauka, Moscow 1976)
8.86 Yu.N. Rufov: Probl. Kinet. Katal. **16**, 212 (1975)
8.87 I.V. Krylova: In *Aktivnaya poverkhnost' tverdykh tel*, ed. by G.I. Distler, P.Yu. Butyagin (VINITI, Moscow 1976) p.22
8.88 Yu.N. Rufov, A.A. Kadushin, S.Z. Roginsky: Dokl. Akad. Nauk SSSR **176**, 905 (1966)
8.89 V.P. Sakun, Yu.N. Rufov: Kinet. Catal. **20**, 360 (1979)
8.90 A.N. Ilyichev, Yu.N. Rufov, V.I. Vladimirova: Kinet. Catal. **20**, 356 (1979)
     V.P. Sakun, Yu.N. Rufov: Kchimicheskaja Fizika **1**, 435 (1982)
8.91 G.P. Peka, Yu.I. Karchanin: Dokl. Akad. Nauk SSSR **63**, 141 (1961)
8.92 O.S. Sinez, G.P. Peka, Yu.I. Karchanin: Fiz. Tverd. Tela **6**, 3515 (1964)
8.93 I.B. Ermolovich, G.P. Peka, M.K. Sheikman: Surface Sci. **24**, 229 (1971)
8.94 F.F. Volkenstein, G.P. Peka, V.V. Malakhov: J. Luminescence **5**, 252 (1972)
8.95 I.K. Vereshchagin: Opt. Spektrosk. **20**, 1066 (1966)
8.96 I.K. Vereshchagin, A.S. Kirichuk: Izv. Vyssh. Uchebn. Zaved., Fizika **4**, 160 (1966)
8.97 V.V. Mazhuga, N.D. Sokolov: Probl. Kinet. Katal. **14**, 73 (1970)
8.98 A. Maradudin, D.L. Mills, S.Y. Tong: In *Structure and Chemistry of Solid Surfaces*, ed. by G. Somorjai (Wiley, New York 1969) Chap.16
8.99 V.F. Kharlamov: Zh. Fiz. Khim. **50**, 2325 (1976)
8.100 V.A. Zuev, D.V. Korbutyak, V.G. Litovchenko: Surface Sci. **50**, 215 (1975)
8.101 V.G. Litovchenko, V.A. Zuev: Jpn. J. Appl. Phys. Suppl. **2**, 2 (1974)
8.102 H. Nagai, Y. Noguchi, Y. Mizushima: Jpn. J. Appl. Phys. Suppl. **18**, 377 (1979)
8.103 T. Suzuki, M. Ogawa: Appl. Phys. Lett. **34**, 447 (1979)
8.104 R.A. Street, R.H. Williams, R.S. Bauer: J. Vac. Sci. Tech. **17**, 1001 (1980)
8.105 C.A. Hoffman, H.J. Gerritsen, A.V. Nurmikko: J. Appl. Phys. **51**, 1603 (1980)
8.106 R.I. Mints, I.I. Millman, V.I, Kryuk: Usp. Fiz. Nauk **119**, 749 (1976)
8.107 I.V. Krylova: Phys. Status Solidi A **7**, 359 (1971)
8.108 E.I. Tolpygo, K.B. Tolpygo, M.K. Sheikman: Izv. Akad. Nauk SSSR, Fizika **30**, 1901 (1966)
8.109 L.P. Gursky, I.V. Rumak, A.I. Pokryshkin: Radiotekh. Elektronika **25**, 652 (1980)
8.110 L.V. Keldysh: Zh. Eksp. Teor. Fiz. **45**, 346 (1963)
8.111 I.A. Akimov, Yu.A. Cherkasov, M.I. Cherkashin: *Sensibilizirovannyi fotoeffekt* (Nauka, Moscow 1980)
8.112 R. Nelson, P. Yianoulis: Phot. Sci. Eng. **22**, 268 (1978)
8.113 B. Broich, G. Heiland: Surf. Sci. **92**, 247 (1980)
8.114 P.J. Regensburger, N.L. Petruzella: J. Non-Cryst. Solids **6**, 13 (1971)
8.115 B.W. Flynn, A.K. Owen, J. Malor: J. Phys. C. Solid State Phys. **10**, 4051 (1977)
8.116 L.N. Ionov, I.A. Akimov: Pis'ma Zh. Tekh. Fiz. **1**, 881 (1975)
8.117 N. Mott: J. Phot. **88**, 119 (1948)
8.118 V.F. Kiselev, S.N. Kozlov, G.S. Plotnikov: Pis'ma Zh. Tekh. Fiz. **7**, 937 (1981)
     S.V. Vintsents, S.N. Karjagin, G.S. Plotnikov: Vestnik MGU, Ser. Fizika **24**, 32 (1983)
     S.V. Vintsents, V.F. Kiselev, G.S. Plotnikov: Phys. Status Sol. (a) **85**, 273 (1984)
     V.A. Bespalov, V.F. Kiselev, G.S. Plotnikov, A.M. Saletski: Phys. Status Sol. (a) **85**, K 73 (1984)

V.A. Bespalov, P.K. Kashkarov, V.F. Kiselev, V.A. Matveev, G.S.
Plotnikov: Phys. Status Sol. (a) **92**, 315 (1985)

8.119 T.S. Moss, G.J. Burrell, B. Ellis: *Semiconductor Opto-Electronics*
(Butterworth, London 1973)

8.120 J.S. Blakemore: *Semiconductor Statistics* (Pergamon, London 1962)

8.121 V.A. Zuev, V.G. Litovchenko, G.A. Sukach: Sov. Phys.-Semiconductors **9**,
1083 (1976)

8.122 L.R. Weissberg: J. Appl. Phys. **39**, 6096 (1968)

8.123 T.T. Bykova, E.F. Lasneva, A.F. Tavasiev: *Tezisy 6-i konferentsii po
fizike poverkhnosti poluprovodnikov*, Vol.1 (6th Conference on the
Physics of Semiconductor Surfaces) (Naukova Dumka, Kiev 1977) p.51

8.124 V.D. Kasarizky, S.N. Kozlov, Yu.F. Novototsky-Vlasov: Dokl. Akad. Nauk
SSSR **195**, 115 (1970)

8.125 Yu.A. Zarif'yants, S.N. Karyagin, V.F. Kiselev: Dokl. Akad. Nauk SSSR
202, 109 (1972)

8.126 W. Lochmann: Phys. Status Solidi **A 40**, 285 (1977)

Chapter 9

9.1 W. Shockley, W.W. Kooper: Surface Sci. **2**, 277 (1964)

9.2 Yu.V. Fedorovich, L.K. Dumish: Mikroelektronika **2**, 159 (1973)

9.3 S.R. Hofstein: Appl. Phys. Lett. **10**, 95, 291 (1967)

9.4 E.H. Niccolian, A.V. Sinha: In *Thin Films. Interdiffusion and Reactions*,
ed. by J. Poate (Wiley, New York 1978) p.481

9.5 A.G. Revesz: In *Proc. Int. Conf. on the Physics of SiO_2 and Its Interfaces* (Plenum, New York 1978) p.222

9.6 T.W. Hickmot: J. Appl. Phys. **46**, 2583 (1975)

9.7 J.F. Verwey: In *Insulating Films on Semiconductors* (Inst. of Physics,
Bristol 1979) Ser. No. 50, p.114

9.8 J.P. Stagg, M.R. Boudry: In *Insulating Films on Semiconductors* (Inst.
of Physics, Bristol 1979) Ser. No. 50, p.75

9.9 S. Kurosaki, S. Saito, G. Sato: J. Chem. Phys. **23**, 1846 (1956)

9.10 A.Ya. Kuznetsov: Zh. Fiz. Khim. **27**, 657 (1953)

9.11 I.F. Abdrakhmanov, B.V. Deryagin: Dokl. Akad. Nauk SSSR **120**, 94 (1958)

9.12 M. Tschapek, R. Santamaria, I. Natal: Electrochim. Acta **14**, 889 (1969)

9.13 A.B. Kurzaev, S.N. Kozlov, V.F. Kiselev: Sov. Phys.-Dokl. (DKPCAG) **228**,
526 (1976)

9.14 A. Saffer, M. Folman: Trans. Farad. Soc. **62**, 3559 (1968)

9.15 J.H. Anderson, G.A. Parks: J. Phys. Chem. **72**, 3662 (1968)

9.16 N.B. Librovich, V.D. Mayorov, V.A. Savelyev: Sov. Phys.-Dokl. (DKPCAG)
225, 1383 (1975)

9.17 N.B. Librovich, V.P. Sakun, N.D. Sokolov: Zh. Eksp. Teor. Khim. **14**, 435
(1978)

9.18 L. Glasser: Chem. Rev. **75**, 21 (1975)

9.19 M. Tschapek, I. Natale, M.S. De Grossi: Anal. Assoc. Quim. Argentina **59**,
229 (1971)

9.20 V.F. Kiselev, O.V. Krylov: *Adsorption Processes on Semiconductor and
Dielectric Surfaces I*, Springer Ser. Chem. Phys., Vol.32 (Springer,
Berlin, Heidelberg 1985)

9.21 V.N. Pak, A.A. Malkov, N.G. Ventov: Elektrokhimiya **10**, 288 (1974)

9.22 Yu.V. Fedorovich, V.A. Vogel: Fiz. Tekh. Polupr. **3**, 840 (1969)

9.23 O.S. Frolov, O.V. Snitko, G.F. Romanova: Ukrain. Fiz. Zh. **14**, 585 (1969)

9.24 A.J. Bennett: Surface Sci. **50**, 77 (1975)

9.25 V.F. Kiselev, S.N. Kozlov, Yu.A. Zarif'yants: In *Problemy fizicheskoi
khimii poluprovodnikovykh poverkhnostei* (Problems in the Physical Chemistry of Semiconductor Surfaces) (Nauka, Novosibirsk 1978) p.200

9.26 J. Fripiat, J. Chaussidon, A. Jelli: *Chimie-Physique des Phenomenes de Surface* (Masson, Paris 1971) p.386
9.27 W.V. Moravia, R. Mills: Z. Phys. Chem. **79**, 2 (1972)
9.28 G.F. Golovanova, V.F. Kiselev, A.S. Petrov, E.A. Silaev: Phys. Status Solidi A **64**, 177 (1981)
9.29 P.P. Konorov, Yu.A. Tarantov, E.V. Kasyanenko: In *Problemy fizicheskoi khimii poverkhnosti poluprovodnikov* (Problems in the Physical Chemistry of Semiconductor Surfaces) (Nauka, Novosibirsk 1978) p.247
9.30 S.N. Kozlov, S.N. Kuznetsov: Sov. Phys.-Semiconductors **12**, 995 (1978)
9.31 V.V. Murina, Yu.F. Novototsky-Vlasov: In *Elektronnye yavleniya na poverkhnosti poluprovodnikov* (Electronic Phenomena on Semiconductor Surfaces) p.262
9.32 V.V. Murina, Yu.F. Novototsky-Vlasov: Sov. Phys.-Semiconductors **6**, 979 (1972)
9.33 K. Heilig, H. Flietner, J. Reineke: J. Phys. D. Appl. Phys. **12**, 927 (1979)
9.34 S.R. Ovchinski, A. Madan: Nature **276**, 482 (1978)
9.35 C.J. Hang, L. Ley, H.R. Shanks: Phys. Rev. B **23**, 6140 (1980)
9.36 R.J. Kriegler, Y.C. Cheng, D.R. Colton: J. Electrochem. Soc. 119, 388 (1972)
9.37 H. Frenzel, B.R. Singh, H. Haberie, P. Balk: Thin Solid Films **58**, 301 (1979)
9.38 G. Declerk, T. Hattory, G.A. May, I.D. Meidl: J. Electrochem. Soc. **122**, 436 (1975)
9.39 A. Rohatgi, S.R. Butler, F.J. Feigl: J. Electrochem. Soc. **126**, 143, 149 (1979)
9.40 M. Chen, J.W. Hile: J. Electrochem. Soc. **119**, 223 (1972)
9.41 V.A. Kotel'nikov, A.N. Terenin: Dokl. Akad. Nauk SSSR **174**, 1366 (1966)
9.42 E.H. Niccolian, C.N. Berglund, P.F. Schmidt, J.M. Andrews: J. Appl. Phys. **42**, 5654 (1971)
9.43 S.N. Karyagin, P.K. Kashkarov: Phys. Status Solidi A **37**, 17 (1976)
9.44 P.K. Kashkarov, V.F. Kiselev, S.N. Kozlov: Surface Sci. **75**, 231 (1978)
9.45 V.A. Kurashov, Yu.A. Plotnikov: In *Elektronnye protsessy na poverkhnosti poluprovodnikov* (Electronic Processes on Semiconductor Surfaces), ed. by A.V. Rzhanov (Nauka, Novosibirsk 1974) p.45
9.46 D.E. Carlson: IEEE Trans. **ED-24**, 449 (1977)
9.47 J. Shewchun, R. Singh, M.A. Green: J. Appl. Phys. **48**, 765 (1977)
9.48 W.E. Spear, P.G. Le Comber: Phil. Mag. **33**, 935 (1976)
9.49 D.E. Carlson, C.R. Wronski: RCA Rev. **38**, 211 (1977)
9.50 D.M. Mattox: J. Vac. Sci. Tech. **13**, 127 (1976)
9.51 K.I. Zamarev, V.N. Parmen: Uspechi Kchimii **52**, 1433 (1983)
9.52 T. Obnishi, Y. Nakato, H. Tsubomura: Ber. Bunsenges. Phys. Chem. **79**, 523 (1975)
9.53 M.S. Wrighton, D.L. Morse: J. Am. Chem. Soc. **98**, 44 (1976)
9.54 J.G. Mavroides, J.A. Kafalas, D.F. Kolesar: Appl. Phys. Lett. **28**, 241 (1976)
9.55 M. Fujihira, N. Onishi, T. Osa: Nature **268**, 226 (1977)
      J. Kivi: Chem. Phys. Lett. **83**, 594 (1981)
9.56 V.B. Evstigneev: In *Elementarnye fotoprotsessy v molekulakh* (Elementary Photoprocesses in Molecules) ed. by A. Krasnovsky (AN SSSR, Moscow 1976) p.243
      E.K. Putseiko: In *Molekulyarnaya fotonika* (Molecular Photonic) ed. by F.I. Vilesov (Nauka, Leningrad 1970) p.407
9.57 H. Tributsch, M. Calvin: Photochem. Photobiol. **14**, 95 (1971)
9.58 M. Calvin: Photochem. and Photobiol. **37**, 349 (1983)
9.59 G.G. Komissarov: In *Problemy biofotokhimii* (Nauka, Moscow 1973) p.116
9.60 J. Berkowitz: *Photoabsorption, Photoionization and Photoelectron Spectroscopy* (Academic, New York 1979)
9.61 C.R. Claydon, G.A. Segal, H.S. Taylor: J. Chem. Phys. **54**, 3799 (1971)

9.62 A.N. Terenin: Zh. Fiz. Khim. **6**, 189 (1935)
9.63 L.L. Basov, G.N. Kuzmin, Yu.P. Solonitsyn: Usp. Fotoniki **6**, 82 (1977)
9.64 L.L. Basov, V.A. Kotel'nikov, Yu.P. Solonitsyn: In *Spektroskopiya foto-prevrashchenii v molekulakh* (Spectroscopy of Photoconversion in Molecules) ed.by A.A. Krasnovsky (Nauka, Leningrad 1977) p.228
9.65 G.N. Lyalin, V.G. Sirota, V.E. Kholmogorov: Biofizika **20**, 535 (1975)
9.66 V.I. Korotkov, V.E. Kholmogorov: In *Yavleniya fotoadsorbtsii i fotokata-liza v geterogennykh sistemakh*, Vol.4 (Phenomena of Photoadsorption and Photocatalysis in Heterogeneous Systems) (Nauka, Novosibirsk 1974) p.96
9.67 V.F. Kiselev, A.S. Petrov, R.V. Prudnikov: Sov. Phys.-Dokl. (DKPCAG) **245**, 318 (1979)
9.68 V.F. Kiselev, S.N. Kozlov, A.S. Petrov, E.A. Silaev: Sov. Phys.-Dokl. (DKPCAG) **255**, 649 (1980)
9.69 V.F. Kiselev, A.S. Petrov, R.V. Prudnikov: Kinet. Catal. **19**, 1297 (1978)
9.70 V.S. Vavilov: *Effekty Radiatsii v poluprovodnikakh* (Effects of Radiation in Semiconductors) (Fizmatgiz, Moscow 1963)
9.71 F.F. Volkenstein: *The Electron Theory of Catalysis on Semiconductors* (McMillan, New York 1963); *Fizikokhimiya poverkhnosti poluprovodnikov* (The Physical Chemistry of Semiconductor Surfaces) (Nauka, Moscow 1973)
9.72 V.G. Baru, F.F. Volkenstein: *Vliyanie radiatsii na poverkhnostnye svoist-va poluprovodnikov* (The Influence of Radiation on the Surface Properties of Semiconductors) (Nauka, Moscow 1978)
9.73 S. Baidyarov, W.R. Bottoms, P. Mark: Surface Sci. **28**, 517 (1971)
9.74 Yu.P. Solonitsyn, G.N. Kuzmin: Kinet. Katal. **11**, 1267 (1970)
9.75 S.N. Kozlov, N.L. Levshin: Phys. Status Solidi A **64**, 169 (1981)
9.76 V.F. Kiselev, A.S. Petrov, R.V. Prudnikov: Sov. Phys.-Semiconductors **13**, 839 (1979)
9.77 V.F. Kiselev: Dokl. Akad. Nauk SSSR **176**, 124 (1967)
9.78 V.F. Kiselev, S.N. Kozlov: Poverkhnost'. Fiz. Khim. Tekh. **1**, 13 (1981)
9.79 G.F. Golovanova, V.F. Kiselev: Zh. Khim. Fiz. **1**, 1573 (1982)
9.80 G.F. Golovanova, V.F. Kiselev, E.A. Silaev, T.S. Stepanova: Kinetika i Kataliz **24**, 1173 (1983)
9.81 G.F. Golovanova, A.S. Petrov, E.A. Silaev: Kinet. Katal. **23**, 1275 (1982)
9.82 P.Yu. Butyagin, I.V. Kolbanev: Zh. Fiz. Khim. **55**, 1092 (1981)
9.83 V.F. Kiselev, V.V. Kurylev, S.F. Timashev: In *Aktivnaya poverkhnost' tver-dykh tel* (Active Solid-State Surfaces) (VINITI, Moscow 1976) p.56
9.84 L.A. Blumenfeld: *Progress in Biological Physics* (Springer, Berlin, Heidel-berg, New York 1981)
9.85 T. Freund, S.R. Morrison, Y.P. Gones: Proc. 4th Intern. Congress on Catalysis (Moscow 1968), Vol.1 (Akademie Kiado Press, Budapest 1971) p.72
9.86 D. De Voult, B. Chance: J. Biophys. **6**, 825 (1966)
9.87 T.E. Madey, J.T. Yates: Chem. Phys. Lett. **51**, 77 (1977)
9.88 F. Flores, I. Gabbag, N.H. March: Surface Sci. **107**, 127 (1981)
9.89 J.M. Lehn, J.P. Sauvage, R. Ziessel: Nouveau J. Chim. **3**, 423 (1979)
9.90 V.B. Aleskovsky: *Stekhiometryia i sintez tverdykh kombinatsii* (Stechio-metry and Synthesis of Solid Compositions) (Nauka, Leningrad 1976)
9.91 V.B. Aleskovski, V.E. Drozd: Sov. Phys.-Semiconductors **13**, 817 (1979)
9.92 V.F. Kiselev: Dokl. Akad. Nauk SSSR **213**, 224 (1973)
9.93 M.G. Slin'ko, M.M. Slin'ko: Catal. Rev. **17**, 119 (1978)
9.94 D.V. Altshuller, A.D. Berman: In *Konferentsiya po mekhanizmu kataliti-cheskikh reaktsii* (Conference on the Mechanism of Catalytic Reactions) (Inst. Khim. Fiz., Moscow 1974) pp.77
9.95 F. Gutman, L.E. Lyons: *Organic Semiconductors* (Wiley, New York 1967)
9.96 P. Jordan: Naturwiss. **26**, 693 (1938)
9.97 A. Szent-Gyorgui: Nature **148**, 157 (1941)
9.98 Y.K. Levine: In *Progress in Surface Science*, Vol.3, ed. by S.G. Davison (Pergamon, Oxford 1973) p.279

9.99 A. Bray, J.P. Farges: Phys. Status Solidi B **61**, 257 (1974)
9.100 V.F. Kiselev, V.A. Matveev, R.V. Prudnikov: Phys. Stat. Sol. (a) **50**, 739 (1978)
9.101 V.M. Klimovich, A.A. Tschuiko: In *Adsorbenty* (Adsorbents), ed. by A.A. Tschuiko, I.E. Neimark (Naukova Dumka, Kiev 1972) p.100
9.102 Yu.F. Novototsky-Vlasov, M.P. Sinyukov: In *Poverkhnostnye svoistva polu-provodnikov* (Surface Properties of Semiconductors)(AN SSSR, Moscow 1962) p.69
9.103 A.V. Rzhanov: *Elektronnye protsessy na poverkhnosti poluprovodnikov* (Electronic Processes on Semiconductor Surfaces)(Nauka, Moscow 1971)
9.104 Yu.F. Novototsky-Vlasov: Trudy Fiz. Inst. AN SSSR **48**, 3 (1969)
9.105 G. Dorda: Czech. J. Phys. B **10**, 820 (1960); B **13**, 272 (1963)
9.106 V.Ya. Balagurov, V.G. Vaks: Zh. Eksp.Teor. Fiz. **65**, 1939 (1973)
9.107 D.J. Fitzgerald, A.S. Growe: Surface Sci. **9**, 347 (1968)
9.108 J.I. Pankove, M.A. Lampert, H.I. Tarng: Appl. Phys. Lett. **32**, 439 (1978)
9.109 J.L. Benton, C.J. Doherty: Appl. Phys. Lett. **36**, 870 (1980)
9.110 C.H. Seager, D.S. Ginley: Appl. Phys. Lett. **34**, 337 (1979)
9.111 E.M. Lawson, S.J. Pearton: Phys. Status Sol. (a) **72**, K155 (1982)
9.112 M. Schulz: Surface Sci. **132**, 422 (1983)
9.113 J. Bardeen, R.B. Coovert: Phys. Rev. **104**, 47 (1956)
9.114 G. Dorda: Phys. Status Solidi (A) **3**, 1318 (1963); **5**, 157 (1964)
9.115 T.N. Sytenko: *Elektrofizicheskie svoistva poverkhnosti GaAs* (Electro-physical Properties of GaAs Surfaces) Politekh. Inst. Kiev, 1978) p.37
9.116 V.F. Kiselev, V.A. Matveev, A.S. Petrov: Kinet. Katal. **21**, 523 (1980)
9.117 G.F. Golovanova, E.A. Silaev: Poverkhnost'. Fiz. Khim. Tekh. **1**, 109(1982)
      V.A. Bespalov, V.F. Kiselev, P.K. Kashkarov, V.A. Matveev, G.S. Plotnikov: Phys. Status Sol. (a) **92**, 315 (1985)
9.118 K.Hauffe, H. Reveling, D. Rein: Naturwiss. **64**, 91 (1977)
9.119 T. Shimuzu, M. Kumeda, I. Watanabe, Y. Nakagaki: Solid State Commun. **27**, 223 (1978)
9.120 T. Takahashi, Y. Harada: J. Non-Cryst. Solids **35-36**, 1041 (1980)
9.121 K. Tanaka: J. Non-Cryst. Solids **35-36**, 1023 (1980)
9.122 D.L. Stabler, C.R. Wronski: Appl. Phys. Lett. **31**, 292 (1977)
9.123 D.L. Stabler, C.R. Wronski: J. Appl. Phys. **51**, 3262 (1980)
9.124 S.R. Eliott: Phil. Mag. **39**, 349 (1979)
9.125 M. Tanielian, H. Fritzsche: Appl. Phys. Lett. **33**, 353 (1978)
9.126 M.S. Said: Opt. Commun. **33**, 179 (1980)
      W.B. Jackson, R.A. Street, M.J. Tompson: Sol. State Commun. **47**, 435 (1983)
      W. Beyer, B. Hoheisel: Sol. State Commun. **47**, 573 (1983)
9.127 I. Hirabayashi, K. Marigaki, S. Nitta: Jpn. J. Appl. Phys. **19**, 357 (1980)
9.128 V.F. Kiselev, E.A. Silaev: Preprint No. 15 of Physics Dept., Moscow Univ. 1981
      A.G. Kazanskii, V.F. Kiselev, E.A. Silaev, V.S. Vavilov: Phys. Status Sol. (a) **76**, 337 (1983)
9.129 B.T. Kolomiets, A.S. Kochemirovsky: In·*Voprosy fiziki poluprovodnikov,* Vol.1 (Questions in Semiconductor Physics) (KGU, Kaliningrad 1975) p.97
9.130 H. Yhi-Wey: Phys. Rev. B **13**, 3495 (1976)
9.131 A.V. Bykov, Yu.A. Zarif'yants: Izv. Vyssh. Uchebn. Zaved., Fizika **4**, 117 (1981)
9.132 C.W. Gwyn: J. Appl. Phys. **40**, 4886 (1969)
9.133 J.D. Weeks, J.C. Tully, L.C. Kimerling: Phys. Rev. B **12**, 3286 (1975)
9.134 E.H. Snow, A.S. Grove, D.J. Fitzgerald: Proc. IEEE **55**, 1168 (1967)
9.135 V.Ya. Kiblik, V.G. Litovchenko, R.O. Litvinov: Ukrain. Fiz. Zh. **25**, 1348 (1980)
9.136 C.T. Sah, J.Y. Sun, J.J. Tzou: J. Appl. Phys. **53**, 8886 (1982)
9.137 J.S. van Wieringen, A. Kats: Philips Res. Dep. **12**, 432 (1957)
9.138 Z.A. Weinberg, D.H. Young, D.J. DiMaria, G.W. Rulloff: J. Appl. Phys. **50**, 5757 (1979)

Subject Index

Thermal stimulation
- of conductivity (TSC) 74,79
- of desorption (TSD) 74
- of exoelectron emission 74
Thermal velocity of free carriers
  175
Thermionic work function 10,11
Threshold
- of defect creation 187
- of photodissociation 228
Topological disorder 39,154
Transfer of energy 186
Tunneling of electrons
- in interface 156
- MIM structures 80

Urbach's rule 192
Ultraviolet photoelectron spectroscopy (UPS) 61

Vibrational modes of adsorbed
  molecules 198,202
Vibrationally excited molecules
  201-203
Vicinal surface 161
Voltage-current characteristic
  179-181

Wide-band semiconductors 67,180

X-ray spectroscopy 39,151